The Land Question in India

The Land Question in India

State, Dispossession, and Capitalist Transition

Edited by
Anthony P. D'Costa
and Achin Chakraborty

OXFORD
UNIVERSITY PRESS

OXFORD
UNIVERSITY PRESS

Great Clarendon Street, Oxford, OX2 6DP,
United Kingdom

Oxford University Press is a department of the University of Oxford.
It furthers the University's objective of excellence in research, scholarship, and education by publishing worldwide. Oxford is a registered trade mark of Oxford University Press in the UK and in certain other countries

First Edition published in 2017

Published in the United States of America by Oxford University Press
198 Madison Avenue, New York, NY 10016, United States of America

British Library Cataloguing in Publication Data
Data available

Library of Congress Control Number: 2016950870

ISBN 978–0–19–879244–4

Preface and Acknowledgments

The land question in India is a weighty subject and a single volume such as this is unlikely to do justice to the complexities of the topic. Nevertheless, there are some core dimensions around which land could be contextualized and the place of land in economic development re-conceptualized as new social forces impinge on the value and usage of land and the livelihoods that depend on them. India, like many other developing countries undergoing capitalist economic transition and transformation, offers a wide lens to view such change through the land question. More distinctively, India, with its particular form of democratic practices, provides a rich landscape in which the conflict between the social obligation of the state to protect its vulnerable small and marginal famers and the increasingly rapacious demand for land for nonagricultural purposes has become highly pronounced.

The replacement of the 1894 Land Acquisition Act with the 2013 Right to Fair Compensation and Transparency in Land Acquisition, Rehabilitation and Resettlement Act is reflective of this democratic sentiment. While the state has always exercised eminent domain and compensated those whose lands were appropriated for "public purpose," rehabilitation and resettlement have given new meanings to land acquisition. The 2013 act has certainly altered the terms and conditions by which land is to be transacted but has not eliminated the process of dispossession. However, it has meant long, drawn-out negotiations over "fair" prices and institutional demands for resettling and rehabilitating the dispossessed or "project-affected people." This has slowed the process of land transfer and thus presumably nonagricultural-based accumulation, while at the same time the Act has provided a legal apparatus by which the acquisition of land would inevitably be facilitated.

The significance of land in capitalist transition is well documented. Land, along with labor, is a vital factor of production. Historically, transforming the role of land from a source of subsistence livelihood to a source of capital accumulation has been fundamental to capitalist transition. Separating the direct producers from their land has been a structural imperative for primitive accumulation and agricultural sector dynamism and subsequent expansion and diversification of a capitalist industrial economy. Postcolonial societies such as

India, having inherited colonial structures, favored modern industry as a response to such structures; instead of the anticipated strong linkages between industry and agriculture, India has witnessed a lagging agricultural sector.

Where state-led land reforms aimed to induce rural dynamism, they instead led to supporting livelihoods (or coping mechanisms) for peasants and not rural accumulation to fund industrialization as experienced by other late industrializers such as South Korea and Taiwan. It appears transition in India has been stalled and land as a source of capital accumulation has been bypassed. India's capitalist transition, if it can be called that, is thus argued to be following a trajectory that does not entail rising agricultural productivity and labor-absorbing industrialization. Instead, income from the utilization of land through cultivation continues to be low, even though the market value of land in general has seen a meteoric rise due to nonagricultural usage such as industry, infrastructure, real estate (and thus speculation), in the context of growth-friendly economic reforms led by the state.

To tackle these different facets of the land question in India, including the changing role of the state in its pursuit of capitalist development and its tense relationship with democratic politics, this volume brings together multiple theoretical perspectives and a variety of empirical details to critically examine the place of land in the broader process of India's economic development under contemporary capitalism. The chapters are mostly drawn from a three-day conference held in March 2014 at the Institute of Development Studies Kolkata and co-organized by Anthony P. D'Costa of the Australia India Institute and the School of Social and Political Sciences, University of Melbourne; Achin Chakraborty of the Institute of Development Studies Kolkata; and Mritiunjoy Mohanty of the Indian Institute of Management, Calcutta.

This is the first major project on Contemporary India initiated by Anthony P. D'Costa since joining the University of Melbourne. The Faculty of Arts Committee, Dean Mark Considine, and Associate Dean (Research) Janet Fletcher all have been very supportive of this project by providing seed money to get the project off the drawing board. Amitabh Mattoo, then the director of the Australia India Institute, University of Melbourne, also contributed funds as part of the Institute's mission of working with Indian academic institutions. The Institute of Development Studies Kolkata and the Indian Institute of Management, Calcutta also contributed funds to meet all local expenses.

There have been a number of people who have helped with the logistical aspects of the conference. The operations staff at the Australia India Institute helped out with the paperwork for participating India-based as well as international scholars traveling to Kolkata and provided web-based publicity support. At IDSK, Sanchari Guha Samanta managed a number of organizational aspects of the conference ranging from keeping track of the abstracts and paper submissions to overseeing the arrangements at the venue.

We would also like to thank a number of people who helped with the call for proposals for the conference and intellectually contributed to the deliberations of the conference. Partha Chatterjee of Columbia University and a resident of Kolkata presented the opening paper, which has been subsequently reworked as the Prelude to the volume. Amiya Kumar Bagchi of IDSK actively engaged the participants. For agreeing to chair the sessions and discuss the papers we are grateful to Sushil Khanna, Hari S. Vasudevan, Anjan Chakrabarti, Manish Thakur, Manabi Majumdar, Priya Sangameswaran, Dwaipayan Bhattacharya, and Mritiunjoy Mohanty.

Anthony D'Costa would also like to thank Adam Swallow, the Economics Commissioning Editor at Oxford University Press, Oxford. He has thus far worked with Adam on four projects, the first one in 2004 when he was a Sabbatical Fellow at UNU WIDER in Helsinki and Adam was the publications manager. Last but not least, Janette Rawlings, as always, copyedited the entire manuscript in fine detail. To all these individuals and institutions we gratefully acknowledge their help and inspiration. The editors are responsible for all errors and omissions.

Anthony P. D'Costa
Development Studies Program, University of Melbourne

Achin Chakraborty
Institute of Development Studies Kolkata
June 2016

Contents

List of Figures

List of Tables

List of Abbreviations and Acronyms

ABD	accumulation by dispossession
AGL	Agricultural labor households
AHPL	Average Household Poverty Line
AIR	All India Review
APDR	Association for the Protection of Democratic Rights
AQ	agrarian question
AROGYSREE	Health Insurance Scheme of Government of Andhra Pradesh
ASSOCHAM	Associated Chambers of Commerce and Industry of India
BJP	Bharatiya Janata Party
BPO	Business Process Outsourcing
CEA	Central Electricity Authority
CHT	Chittagong Hill Tracts
CII	Confederation of Indian Industry
CMIE	Centre for Monitoring the Indian Economy
CPI (M)	Communist Party of India (Marxist)
CPI (ML)	Communist Party of India (Marxist–Leninist) (Maoist)
CTIL	Century Textiles and Industries Limited
DLF	Delhi Land and Finance
FAO	Food and Agriculture Organization
FDI	farm disposable income
FDI	foreign direct investment
FIAN	Food First Information and Action Network
FICCI	Federation of Indian Chamber of Commerce and Industry
FLI	farm labor income
FRA	Scheduled Tribes and Other Traditional Forest Dwellers (Recognition of Forest Rights) Act, 2006
GLR	Gauhati Law Review
GLT	Gauhati Law Tribunal
GNEA	Greater Noida Extension Area

GoM	group of ministers
GR	green revolution
GVO	gross value of output
HYV	high yielding variety
ICAR	Indian Council of Agricultural Research
ICDS	Integrated Child Development Services
IDFC	Infrastructure Development Finance Corporation
INR	Indian Rupee
IT	information technology
JPC	Joint Parliamentary Committee
JUSCO	Jamshedpur Utilities and Services Company
KARA	Kerala Agrarian Relations Act
KARB	Kerala Agrarian Relations Bill
KLRAA	Kerala Land Reforms Amended Act
LAA	Land Acquisition Act
LARR	Right to Fair Compensation and Transparency in Land Acquisition, Rehabilitation and Resettlement Act, 2013
LFG	Left Front Government
LWE	Left wing extremism
MASS	Manab Adhikar Sangram Sanity
MD	managing director
MLA	Member of the Legislative Assembly
MoEF	Ministry of Environment and Forest
MoU	Memorandum of Understanding
MP	Member of Parliament
MP	middle peasantry
MPCE	monthly per capita expenditure
MW	megawatts
MWC	Mahindra World City
MWCSEZ	Mahindra World City special economic zone
NAPM	National Alliance of People's Movements
NCEUS	National Commission for Enterprises in the Unorganised Sector
NCH	noncultivating landowning households
NDA	National Democratic Alliance
NEI	Northeast India
NER	Northeastern region

NFS	non-farm sector
NGO	non-governmental Organization
NOC	No Objection Certificate
NOIDA	New Okhla Industrial Development Authority
NREGA	(also MNREGA) Mahatma Gandhi Rural Employment Guarantee Act
NRRP	National Resettlement and Rehabilitation Policy
NSS	National Sample Survey
NSSO	National Sample Survey Organization
PA	primitive accumulation
PCP	petty commodity production/producers
PDS	public distribution system
PESA	Panchayats (Extension to Scheduled Areas) Act, 1996
POSCO	Pohang Iron and Steel Company
PP	poor peasantry
PPP	public-private partnership
PRI	Panchayati Raj institutions
RBI	Reserve Bank of India
RP	rich peasantry
RTI	Right to Information
SC	scheduled caste
SCR	Supreme Court Review
SEZ	special economic zone
SHG	self-help group
SOMALA	Social Organizations of Meghalaya Against Land Alienation
ST	scheduled tribe
THI	total household income
TISCO	Tata Iron and Steel Company
TL	Telengana
TMC	Trinamool Congress
UP	Uttar Pradesh
UPA	United Progressive Alliance
UPS	Usual Principal Status
USD	US dollar
WB	West Bengal
WBHDR	West Bengal Human Development Report
WBIDC	West Bengal Industrial Development Corporation

List of Contributors

Shapan Adnan obtained a PhD in Social and Political Sciences from the University of Cambridge. He is currently an independent scholar based in the UK. He has formerly taught at the National University of Singapore and Universities of Dhaka and Chittagong. He has been a visiting research fellow at the University of Oxford and is currently an Associate of its Contemporary South Asian Studies Programme (CSASP). He is a member of the international advisory board of the *Journal of Peasant Studies* and the Chittagong Hill Tracts Commission. His research and publications cover political economy, sociology, anthropology, and development.

Arindam Banerjee is Associate Professor in Economics at the School of Liberal Studies, Ambedkar University Delhi. He has completed his doctoral degree from the Centre for Economic Studies and Planning, JNU, New Delhi, on agrarian crisis and peasant accumulation. He has earlier worked at the Centre for Development Studies, Trivandrum, and Research and Information System, New Delhi. His primary areas of research interest are agrarian change, food security, political economy, and colonialism.

Rajesh Bhattacharya is Assistant Professor in the Public Policy and Management Group at the Indian Institute of Management Calcutta. He has also taught at South Asian University, University of Calcutta, and Presidency University. He obtained his PhD in Economics from University of Massachusetts, Amherst. His research areas include capitalism and dispossession, informal economy, urban political economy, and financialization of capital.

Snehashish Bhattacharya is Assistant Professor in the Faculty of Economics at South Asian University, New Delhi. He has also taught at Franklin and Marshall College, Pennsylvania, and The New School University, New York. His research interests include the political economy of development and the informal economy.

Achin Chakraborty is Professor of Economics and the Director of the Institute of Development Studies Kolkata (IDSK). Before joining IDSK in 2004 he had been an Associate Professor at the Centre for Development Studies, Kerala. Chakraborty received his PhD in Economics from the University of California at Riverside. He was a visiting professor at Sciences-Po, Paris. He has published widely in journals such as *Economic Theory, Journal of Environment and Development, Social Indicators Research*, and *Economic and Political Weekly* in the areas of development economics, welfare economics, and methodology of economics.

Gorky Chakraborty is Associate Professor of Economics at Institute of Development Studies Kolkata Kolkata. He works on issues related to development in Northeast India. His most recent book is *Look East Policy and Northeast India* (2014).

Partha Chatterjee is Professor of Anthropology and South Asian Studies at Columbia University, New York, and Honorary Professor, Centre for Studies in Social Sciences, Calcutta. Among his many books are *Bengal: The Land Question* (1984), *Nationalist Thought and the Colonial World* (1986), *The Nation and Its Fragments* (1993), *The Politics of the Governed* (2004), and *Lineages of Political Society* (2011).

Anthony P. D'Costa is Chair and Professor of Contemporary Indian Studies, Development Studies Program at the University of Melbourne. He has taught at the Copenhagen Business School and the University of Washington. He has written extensively on the political economy of steel, auto, and IT industries covering themes of capitalism and globalization, development, innovations, industrial restructuring, and international mobility of labor. Of his several books, his most recent ones are *After Development Dynamics: South Korea's Engagement with Contemporary Asia* (2015, edited) and *Changing Structures of Accumulation: Global Capitalism and the Mobility of IT Professionals from India to Japan* (2016). He is currently working on labor markets in high-tech cities.

Anirban Dasgupta is Associate Professor in the Faculty of Economics, South Asian University. He is a development economist with interests in agrarian studies and inclusive development strategies. He has published in *Conservation and Society, Development and Change, European Journal of Development Research*, and *Journal of Peasant Studies*. His current work is on the impact of global integration on Indian agriculture and the status of development theory in the contemporary world.

Kaveri Gill completed her Tripos, MPhil, and PhD, as well as a post-doctoral fellowship, at the University of Cambridge. She has worked with the Planning Commission; UNICEF; the Think Tank Initiative, IDRC; and Oxford Policy Management in New Delhi. Kaveri has published widely, including a monograph with Oxford University Press. Her research interests include the political economy of development and the informal sector; waste and sanitation; and public policy and its implementation in social sectors, especially health.

Abhijit Guha is Professor of Anthropology at Vidyasagar University, West Bengal, India. He has been studying land grabs by the government in West Bengal since 1994. Guha has uniquely combined anthropological fieldwork and archival data from departmental files and West Bengal Assembly Proceedings to construct ethnographies of land acquisition. His book *Land, Law and the Left: A Saga of Disempowerment of the Peasantry in the Era of Globalisation* (2007) is the first micro-level policy critique of the Left Government in West Bengal. He was selected as an expert in 2008 to give recommendations to the Parliamentary Standing Committee for the making of the new Land Acquisition Law of India.

Michael Levien is Assistant Professor of sociology at Johns Hopkins University. His articles on land dispossession in India have appeared in *Politics & Society, World*

Development, Economic and Political Weekly, Journal of Peasant Studies, Journal of Agrarian Change, Development and Change and *The Hindu*.

R. V. Ramana Murthy is Professor of Economics teaching at the School of Economics, University of Hyderabad. His current areas of research are agrarian transition and agrarian question in India, structural transformation, and political economy of global capitalism, and neoliberalism. He teaches research methodology, agricultural economics, and the Indian economy since independence.

Kenneth Bo Nielsen is a social anthropologist working at the Department of Sociology, University of Bergen, Norway. He also coordinates the Norwegian Network for Asian Studies hosted by the University of Oslo's Centre for Development and the Environment. His most recent book is a co-edited volume on *The Politics of Caste in West Bengal* (2016).

Alf Gunvald Nilsen is Associate Professor of Sociology at the University of Bergen, Norway, and a visiting senior researcher at the Society, Work and Development Institute at the University of the Witwatersrand, South Africa. His research focuses on social movements in the global south, with a particular concentration on India. He is the author of *Dispossession and Resistance in India: The River and the Rage* (2010) and co-author of *We Make Our Own History: Marxism and Social Movements in the Twilight of Neoliberalism* (2014).

Malabika Pal is Associate Professor in Economics at Miranda House College, University of Delhi. She obtained her PhD in Economics from Jawaharlal Nehru University, New Delhi. Her research interests are in the fields of Law and Economics and International Economics. She has worked on the economic analysis of tort law, the issues related to the acquisition of land and on financial crises. She is a recipient of the National Scholarship and the Ford Foundation Scholarship.

Mircea Raianu is a PhD Candidate in the History Department at Harvard University, completing a dissertation entitled "The Incorporation of India: The Tata Business Firm between Empire and Nation, *ca.* 1870–1960." He was the recipient of a Fulbright-Nehru Fellowship in 2013–14. His broader research interests include political economy and the legal and intellectual history of capitalism in modern South Asia.

Asok Kumar Ray was a visiting fellow at Omeo Kumar Das Institute of Social Change and Development, Guwahati. He works on political anthropology and institutions of the tribes in Northeast India and has written extensively on these issues.

R. Vijay is Professor of Economics at University of Hyderabad. He has completed his doctoral work on land markets with specific emphasis on peasant migration as a factor for an alternate land use pattern. His main research interests are the process of agrarian change in the rural economy and to analyze the impact of institutions on economic performance. He has published extensively on changing rural economic structure and nature of agrarian relations.

Prelude

Land and the Political Management of Primitive Accumulation

Partha Chatterjee

The People as a Political Construct

It is well known that the people first emerged in the history of the modern state sometime in the late eighteenth and early nineteenth centuries with the French, the American, and the Bolivárian revolutions in the Spanish colonies of South America. These were republican revolutions against monarchy and empire. They shifted the locus of legitimate sovereignty from the divine or dynastic rights of kings to that of the people. It is also well known that the idea of the people in this new concept of popular sovereignty was entirely abstract. In actuality, only a small part of the people was regarded as proper citizens with the right to be consulted in matters of law making or administration. But when new republican elites began to wield power, sometimes in thoroughly authoritarian ways, they always did so in the name of the people. And when the wave of democratic revolutions swept through Europe in the nineteenth century and anti-colonial revolutions became victorious in Asia and Africa in the twentieth, it was the idea of the people as the only legitimate foundation of sovereignty in the modern world that became universally recognized as a feature of modernity itself.

The idea of populism as we know it today arose in the twentieth century with the emergence of mass democracies, first in Europe and then in other parts of the world. In some ways, it is connected with a very old idea, going back to ancient Greece, which, while accepting the principle that regular public consultations among proper citizens made for good government, was at the same time deeply suspicious of the rule of the *demos*. The industrial societies of Western Europe and North America first established bourgeois norms and practices of representation and equality before the law within a limited section of the population and subsequently spread them to the rest of society through the extension of the suffrage to workers and women, the universalization of primary and secondary education, the growth of civil–social associations including national trade unions among the working classes, and the creation of national political parties with mass electoral following. For most of the twentieth century, liberal democracy in Europe and North America confidently asserted that the most efficient way to maintain the rule of capitalist property and merit-based reward was to do so within a system of competitive electoral democracy with universal adult franchise.

Populism has challenged this confidence, so that the orthodox view in political theory is to regard populism as a perversion of democracy. Populism points to the inequalities in wealth, income, and power that prevail in contemporary democracies and claims, on behalf of the authentic people, to attack those enemies of the people who rule by force, corruption, and falsehood. There have been several kinds of populism in the twentieth century, especially in Latin America but also more recently in Asia. Some have been right-wing, others left-wing; many defy classification in traditional ideological terms. In recent years, populism has become the subject of sophisticated theorization, especially following the writings of Ernesto Laclau (2005).

I bring up the phenomenon of contemporary populism in order to connect it to another recent development in the political economy of capitalist development in Asia and Africa. This is what Marx referred to in *Capital*, volume 1, as "the so-called primitive accumulation" (Marx 1990). Some recent scholars in India have revisited Marx on this topic and suggested that it has much to do with the recent spate of populist politics in postcolonial democracies in Asia and Africa. By connecting populism to primitive accumulation in the specific context of claims to the possession of land in contemporary India, I hope to frame the discussions in this volume within a broader political-economic field.

The Concept of Primitive Accumulation

The concept of primitive accumulation has recently reappeared in academic circles as a result of David Harvey's discussion of "accumulation by dispossession" in his book *The New Imperialism* (Harvey 2003). I will have more to say later about Harvey's argument. However, for our present discussion, an even more important recent intervention, in my view, was made in 2007 by the late Indian economist Kalyan Sanyal. But before I come to that subject, let me quickly review Marx's discussion of primitive accumulation.

In the last chapters of *Capital*, volume 1, Marx engaged in a historical discussion whose immediate object was to show the erroneous idea in Adam Smith of what Smith called "the previous accumulation" of capital as simply the accumulation of stock that precedes the division of labor. Marx showed that in order to emerge as industrial capital, there must first be a primitive or primary accumulation of capital in which three essential preconditions would have to be established. First, the direct producer, whether peasant or artisan, belonging to the precapitalist economy, must be dissociated from his or her means of production. Second, those means of production must come into the possession of the capitalist. Third, the dispossessed peasant or artisan must then have no option but to become a wage laborer and sell his or her labor-power to the capitalist. The emergence of the free wage laborer, freed from precapitalist bonds of servitude as well as from his own means of production, is an essential condition for industrial capitalism. Marx's historical chapters summarized the process in England from the end of serfdom to the expulsion of peasants from the land which created the mass of property-less workers that would form the labor force for the industrial revolution.

All of those who lost their means of production could not be immediately absorbed into the new factories. Marx did talk about the political management of primitive accumulation when he mentioned the laws against vagabondage, the workhouses for the poor, or packing off the destitute to Australia. But we will see why his discussion on this point is inadequate.

Since dispossessed workers no longer had any means of production of their own, they now had to buy all of their subsistence needs as commodities from the market. In other words, the pool of wage laborers also created a home market for the products of industrial capital. Historically, therefore, the destruction of handicrafts in general, and rural domestic industry in particular, was a necessary precondition for capitalist production.

In all of this, the role of the state was crucial. The destruction of feudal property required the force of law. In addition, Marx mentions the role of the public debt which had to be paid by increased taxation, which in turn led to the pauperization of small owners. But, as we will see, there are many other ways in which the state was instrumental in facilitating the primitive accumulation of capital.

In Western Europe, primitive accumulation was more or less accomplished by the middle of the nineteenth century. This was, we must remember, before the age of universal franchise and mass democracy. In the white settler colonies of the Americas and Australia, the indigenous populations were expropriated by force by the European colonists who seized their lands. In North America, the indigenous peoples were not included at all within the colonial economic formation, not even as an exploited workforce. The abundance of land and other means of production meant that most white settlers were small owners and few were wage laborers. Colonies in the US South and the Caribbean, which had large plantations, had labor consisting of African slaves. The rise of industrial capitalism in North America required the creation of a pool of free wage laborers. This was accomplished by the emancipation of slaves and the new influx of migrant laborers from Europe in the late nineteenth and early twentieth century. With the end of slavery in the Caribbean, plantation labor was supplied by free but indentured laborers, mostly from India. In all of this, once again, the role of the state was crucial.

To return to the story of primitive accumulation in Europe, not all those who were dispossessed of their means of production could be absorbed as wage laborers in industry. Marx underestimates the scale of this surplus population. Emigration to the Americas, Australia, and other colonies was a major part of the political management of primitive accumulation in Europe. Between 1815 and 1920, some 60 million Europeans migrated to the Americas. Besides, there were the millions of soldiers who died as cannon fodder in the endless European wars of the eighteenth, nineteenth, and twentieth centuries. A million people died in the Seven Years War in the middle of the eighteenth century, 5 million died in the Napoleonic wars in the early nineteenth century, 20 million died in World War I and at least 50 million died in Europe in World War II. Deaths by famine and epidemic were quite significant in European countries even in the nineteenth century, the most well known being the Irish famine of the 1840s, in which a million and a half people perished. The point is that in an age before mass democracy, such catastrophic deaths in wars, epidemics, or famines were politically far less damaging to the ruling powers than they would be now. That is a crucial point of difference in the conditions in which primitive accumulation is being managed in the countries of Asia and Africa today.

One should also mention here that in the so-called latecomer capitalist countries, the classic English way of the emergence of capital from below was not available. Germany and Japan industrialized in the late nineteenth and early twentieth century under a centralized authoritarian state extracting surplus from the agrarian sector to finance huge military expenditures to create demand for industrially produced consumption and capital goods. The dispossessed peasant population was—often coercively—mobilized into a labor force for state-financed infrastructure projects, state-sponsored industry, and, most significantly, the rapidly expanding armed forces. What is often called the second way of capitalism opened up the political possibility of the authoritarian management of primitive accumulation in the twentieth century. Indeed, the political management of primitive accumulation in the Soviet Union could be said to have been accomplished largely along the lines of the second way. This was theorized in the 1920s by the Soviet economist Yevgeny Preobrazhensky as "primitive socialist accumulation" (Preobrazhensky 1965). Primitive accumulation was carried out in the

1930s under Stalin's collectivization program using extremely coercive and violent methods.

Luxemburg's Elaboration

Rosa Luxemburg claimed in *The Accumulation of Capital* (1913) that a capitalist economy with expanded reproduction must necessarily face a crisis in realizing the surplus because the under-consumption by workers caused by their low wages would not provide sufficient markets for the additional products of expanded industrial manufacture. This crisis would have to be resolved by finding new markets in noncapitalist economies. Luxemburg's theoretical claim of a necessary tendency toward a realization crisis has been disputed. It has also been shown that if trade between the capitalist and noncapitalist sectors is balanced, it would not resolve the crisis by providing new markets.

Nevertheless, Luxemburg does provide a valid account of the actual historical process by which capital from advanced industrial countries dispossessed direct producers in the less advanced countries of Europe and in the European colonies of the East. She describes, first, the struggle against the natural economy through war and oppressive taxation in which the objects were to gain possession of land, forests, and minerals; to extract coerced labor; and to introduce a commodity economy. The examples were the British in India, the French in Algeria, and the Opium War in China. Second, she spoke of the attacks on the simple commodity economy of peasant societies. Third, she pointed out the international struggle for accumulation by export of capital to Russia, Turkey, Persia, India, Japan, China, and North Africa.

Primitive Accumulation in Large Agrarian Countries

Let me turn briefly to the history of primitive accumulation in the colonial countries of Asia, especially those with large agrarian populations. If we look at India, one of the most thoroughly administered European colonial possessions in the world, we find that primitive accumulation was first accomplished by colonial officials and European traders in the eighteenth and early nineteenth centuries who, by plundering the wealth of India, contributed, as Marx himself noted, to the formation of merchant capital in Britain. In the nineteenth century, Europeans acquired land in India for tea and coffee plantations as well as for mining, with the active support of the colonial state. British companies also built railways and other infrastructure in India with the legal and financial help of the state. Following the industrial revolution in Britain, imported manufactures, especially textiles, flooded the Indian market causing widespread destruction of the traditional crafts and forcing the dispossessed artisans into complete dependence on agriculture. This created a somewhat unique problem for the colonial government in India. The dispossessed peasantry had no alternative occupations. Hence, large-scale dispossession of peasants carried the constant threat of peasant revolt as well as famines. British rule in India is marked by a long series of local peasant uprisings as well as numerous famines, large and small. Consequently, while the colonial state promoted capitalist accumulation, it was also careful to protect the tenancy rights of peasants. As we will see, this set an enduring condition for postcolonial development.

After India gained independence in 1947, Indian capitalists initially looked to the new postcolonial state to mobilize the capital required for heavy industry and infrastructure. In the 1950s and 1960s, a large state sector in steel, heavy engineering, mining, oil and gas, roads, railways, airways, and even banking and insurance was created, much like in the second way of capitalist growth. The consumer goods sector was left to private capital. The interesting feature of the Indian case was the popular base of anti-colonial nationalism, organized through the Indian National Congress, a democratic political organization. This made it necessary for the postcolonial state to grant ownership rights to small farmers and protect the occupations of traditional craftspeople. Even when they were politically mobilized by the rural rich, small farmers exerted considerable pressure on government through the process of electoral democracy.

This regime of planned economy with a large state sector and strictly regulated private sector came to an end in 1991. The economy was opened up to the entry of foreign capital and consumer goods. Several state sectors such as transport, telecommunications, infrastructure, mining, banking, and insurance were opened up to private capital. Before the financial crisis of 2008, the annual growth rate of the Indian economy reached 8 or 9 percent. With this began a new phase of primitive accumulation, which has shaken the foundations of traditional small-peasant agriculture.

The key features of this new phase of primitive accumulation are the following. First, there is the growth of a vast informal economy. This is not the remnant of the traditional subsistence economy. Rather, it is the product of the new phase of capitalist accumulation. Official estimates show that more than 80 percent of enterprises in India are in the informal sector and most owners are self-employed, using family labor (National Commission for Enterprises in the Unorganised Sector (NCEUS) 2007). While many of these units are in the traditional sector such as subsistence agriculture, there is clear evidence that rural populations are being increasingly forced to abandon agriculture for informal nonagricultural activities.

Second, there is a migration of dispossessed people from rural areas, especially into medium- and small-sized towns. Recent censuses show that population growth in the large metropolitan cities of India has stabilized, while in smaller towns it is rising rapidly. Third, there is the spread of non-agricultural occupations in rural areas. In many regions, the majority of people living in villages no longer pursue agriculture as their main occupation. All of these are new features of primitive accumulation, posing completely new political challenges.

The Political Problem of Primitive Accumulation

Recent discussions on primitive accumulation in India follow from the theoretical arguments made by Kalyan Sanyal (2007). As we have said earlier, one of the historical conditions under which primitive accumulation is taking place today in postcolonial countries is the absence of some of the key methods of its political management that were available to European capitalism—mass emigration, conscription into the army, or deaths in epidemics and famines. As a result, postcolonial history has brought into the open a constitutive feature of capitalist development that was hidden in its European journey. The massive surplus population of the dispossessed that primitive accumulation in contemporary postcolonial countries is producing shows that in the process of its emergence, capital creates its own outside, which is not precapital but something entirely new. The new dispossessed population is not a reserve army of labor waiting to be absorbed into the industrial labor force: it is entirely redundant to the capitalist growth economy.

In other words, primitive accumulation largely destroys the old precapitalist economy, especially traditional small-peasant agriculture. But in its place, capitalist growth creates a huge sector of informal production and services that is not a vestige of precapital but is in fact a new outside of capital. Those in the informal sector compete in the same marketplace with formally incorporated firms. They manage to survive because their logic is not, as in the formal domain of capital, that of accumulation but sheer subsistence. A large number of informal units operate with family labor; where outside labor is employed, the owner is often himself or herself also a worker. Survival is frequently the result of self-exploitation.

How is the surplus population living under such precarious conditions in the zone outside of capital to be governed? Sanyal argues that even though capital is economically self-subsistent and does not need this outside to guarantee its conditions of reproduction, it is not politically or socially self-subsistent. That is to say, for reasons of social and political legitimacy, it cannot afford to ignore this outside and simply allow it to perish. Hence, Sanyal argues, it becomes incumbent upon postcolonial capitalism to reverse the effects of primitive accumulation.

How is this done? A part of the revenues that accrue to the state from taxes levied on the capitalist growth sector is used for social expenditures to support the livelihood of the dispossessed population. While primitive accumulation continues to destroy traditional occupations, its effects are sought to be reversed through the mediation of the state.

Several techniques have emerged for carrying out this process of reversal. First, there are direct transfers of money and commodities to the dispossessed groups through projects of poverty removal, for example subsidized food, guaranteed employment for specific periods, housing loans, and so on. Second, subsidized public services are provided for poor and disadvantaged groups in transport, health, schooling, and so on. Third, easy loans are provided for small businesses and self-employment. Fourth, production and service units in the informal sector are allowed to violate tax, labor, or pollution laws that apply to the corporate sector. Fifth, specific groups may be allowed to build unauthorized housing, vending stalls, production units, and so on. However, the crucial condition is that all of this must be done without jeopardizing the formal legal structure of property and civic norms. This condition is achieved in most cases by administrative decisions that treat these specific cases as exceptions to the law.

How are these governmental techniques of the reversal of the effects of primitive accumulation actually negotiated in the field of politics? It could be done through either authoritarian decisions or democratic compromise. Each has its own advantages and risks. The example of India may be useful in discussing these alternative strategies of political management of primitive accumulation.

Political Society as a Field of Negotiation

The state of emergency declared by Indira Gandhi in 1975–7 was the last sustained attempt in India to push through a developmental agenda by authoritarian bureaucratic methods of the kind that had been used, with varying degrees of effectiveness, in many other Third World countries. For India's governing classes, the failure of the Emergency drove home the lesson that biopolitical projects that have to do with the physical conditions of the life of the people could not be successfully pursued without passing them through the sieve of voluntary consent. Further, it also became clear that welfare measures could not be effectively administered except by opening their terms to negotiation with the affected population groups. The difficulty was that the ordinary conditions of citizenship and representation on which the constitutionally ordained structure of rights and duties was founded could not uniformly be made to apply to all population groups, especially not to the urban poor making a living in the informal economy. Most of the time, the poor in the city were not property owners, nor did they always observe the duties of proper urban citizens. What were the terms on which they might be recognized as parties to governmental negotiation?

The idea of political society (Chatterjee 2004, 2011) recognizes something *new* in the way governmental authorities began to negotiate with population groups. This no longer fitted the old form of patron–client relations between local notables and their protégés, nor was it quite the same as a political fixer getting things done at a government office on behalf of a local community. In the older models, the form of the local community was already given by the history of local social structures, and the function of the patron, too, was usually made possible by the prevailing hierarchies of class and status. Political society, on the other hand, often *creates* a community where none existed before, or else it *gives new form* to older community structures. That is to say, population groups by engaging with political society acquire the moral character of community. The phenomenon was particularly novel in the Indian city of the late twentieth century.

But the period following the end of the Emergency was also the time when postcolonial democracy gained wider and deeper foundations in rural India. The result was the emergence of political society around the new forms of negotiation between rural population groups and governmental agencies. Rapidly expanded governmental activities in rural India provided the grid for population groups to be mobilized into the moral form of communities with voice and identity. Consequently, the idea of political society marked a *new moment* in the democratization of Indian politics and society.

Take the familiar example of squatter settlements of the poor in numerous cities of India. These urban populations occupy land that does not belong to them and often use water, electricity, public transport, and other services without paying for them. But governmental authorities do not necessarily try to punish or put a stop to such illegalities, because of the political recognition that these populations serve certain necessary functions in the urban economy and that to forcibly remove them would involve huge political costs. On the other hand, they cannot also be treated as legitimate members of civil society who abide by the law. As a result, municipal authorities or the police deal with these people not as rights-bearing citizens but as urban populations who have specific characteristics and needs and who must be appropriately governed. On their side, these groups of urban poor negotiate with the authorities through political mobilization and alliances with other groups.

On the plane of governmentality, populations do not carry the ethical significance of citizenship. They are heterogeneous groups, each of which is defined and classified by its empirically observed characteristics and constituted as a rationally manipulable target population for governmental policies.

Consequently, if, despite their illegal occupation of land, they are given electricity connections or allowed to use municipal services, it is not because they have a right to them but because the authorities make a political calculation of costs and benefits and agree, for the time being, to give them those benefits. However, this can only be done in a way that does not jeopardize the legal order of property and the rights of proper citizens. The usual method is to construct a case such that the particular illegality associated with a specific population group may be treated as an exception that does not disturb the fundamental rule of law. Governmental decisions aimed at regulating the vast populations of the urban poor usually add up to a series of exceptions to the normal application of the law.

Populations respond to the regime of governmentality by seeking to constitute themselves as groups that deserve the attention of government. If as squatters they have violated the law, they do not necessarily deny that fact, nor do they claim that their illegal occupation of land is right. But they insist that they have a right to housing and livelihood in the city, and, if they are required to move elsewhere, they must be provided with rehabilitation. They form associations to negotiate with governmental authorities and seek public support for their cause. This becomes a major form of political participation for these groups, invoking their status as formal citizens but acting in ways that often contravene the approved practices of civic life. Their political mobilization involves an effort to turn an empirically formed population group into a virtuous community. The force of this moral appeal usually hinges on the generally recognized obligation of government to provide for the poor and the underprivileged.

Populism as the Political Form of Mass Democracy

This is where the politics of populism comes in. If we consider the example of elections in India, we will find that the overwhelming bulk of the political rhetoric expended in election campaigns concerns what governments have or have not done for which population groups. The function of rhetoric here is to turn the heterogeneous demands of populations into the morally coherent and emotionally persuasive form of popular demands. In this sense, populism is the only morally legitimate form of democratic politics under these conditions. It is important to emphasize that unlike traditional theories of modernization that would regard such populism as a perversion of modern democratic politics, our present approach would consider it with utter seriousness as a new and potentially richer development of democracy. It is also worth pointing out that one of the persistent findings of election studies in India is the relatively high electoral participation of voters belonging to the poor and underprivileged sections of the electorate (Stepan, Linz, and Yadav 2010).

If we follow Ernesto Laclau's argument on populist reason, the crucial condition is set by the sheer heterogeneity of demands made on government in any contemporary society. This heterogeneity is not unrelated to the fact that the massive social mobilizations that were effected in the twentieth century by giant trade unions and mass-based political parties have been largely dissolved in the last three or four decades by the administrative techniques of governmentality that no longer distribute universal welfare as an entitlement of citizenship but rather offer specific benefits to specific target population groups on the basis of specific empirically established needs. The heterogeneous population groups who are today the recipients of governmental benefits do not carry the moral significance of citizenship; they are merely population groups that need to be appropriately governed. The political strategy on the part of the ruling powers was clearly calculated to break up the large mobilizations that often challenged the stability of the capitalist order through paralyzing strikes by large trade unions and threatening electoral challenges by socialist or communist parties with mass electoral support. As Foucault (2010) explained in his lectures on biopolitics, the success of neoliberal governmentality was to put in place a different rationality of looking after populations—one that was flexible, attuned to individual needs and motivations, operating through a system of incentives and penalties that would achieve the desired social outcome by relying on the apparently free choice of individual beneficiaries.

But no system of governmentality can ever satisfy every need in society. If it did, government would be reduced to mere administration and there would be no place at all for any politics. In actual fact, demands are constantly made for governmental benefits, even though they are quite heterogeneous. It is in the domain of politics that they have to be turned into more comprehensive demands. Populism turns heterogeneous demands into the rhetorical fullness of popular demands. Laclau (2005) argues that chains of equivalence are established among these heterogeneous demands in order to claim that despite their substantive differences, they are all unfulfilled demands of the authentic people who are being denied those benefits by the ruling group who are none other than enemies of the people.

The field of electoral politics in postcolonial democracies like India is marked by competitive populism. Parties and leaders seek to give voice to a set of heterogeneous demands by different population groups. But they must also establish chains of equivalence among these demands in order to put together large voting blocs that would enable them to win seats and form governments. Populist rhetoric pitting the deprived people against their oppressive enemies is the usual way by which these electoral mobilizations are achieved.

From the point of view of those who survive in the informal sector and make claims through what I have called political society, the field of negotiation with governmental authorities is necessarily uncertain, laying down no firm principles, recognizing no definite rights, but leaving everything to the repeated and always temporary negotiation of claims. Groups in political society have to pick their way through this uncertain terrain by making a large array of connections outside the group—with other groups in similar situations, with more privileged and influential groups, with government functionaries, with political parties and leaders. They often make instrumental use of the fact that they can vote in elections. But the instrumental use of the vote is possible only within a field of strategic politics. This is the stuff of democratic politics as it takes place on the ground in India. It involves what appears to be a constantly shifting compromise between the normative values of constitutional propriety and the demonstrative assertion of popular demands.

Governmental authorities, on the other hand, when conceding a demand by making an exception to the law in a particular case, have to be careful that the interests of proper law-abiding citizens are not thereby threatened. Thus, squatters may be allowed water and electricity connections at specific negotiated rates as an exception to the usual structure of rates paid by regular customers, or vendors may be allowed to set up temporary stalls on the pavement without threatening the regular shops that have licenses and pay taxes, or small industries and services in the informal sector may be allowed to ignore labor laws and pollution regulations that apply to the formal sector. Declaring exceptions of this kind is always a balancing act and creates an unstable arrangement that may be disturbed either because the courts decide that the exception is unjustified or the political balance shifts against the population group concerned.

Given the huge number of demands that arise in an immensely heterogeneous society such as India where the vast majority of the population lives and works in the informal sector outside the properly regulated zones of civil society, the administrative response to these demands ends up in an array of temporary and often inconsistent exceptions. Population groups, too, do not seek to fundamentally change the existing structure of rules and regulations but claim that an exception be made in their case. As a result, the working of political society ends up in the piling up of exceptions.

The populist politics that tries to bring together the demands of disparate and localized groups does not constitute a revolutionary challenge to the structure of state authority. As a political imaginary, it seems to have little potential for radical democratic change. Many commentators think that the idea of political society empties the political actions of poor and exploited people of any concerted or sustained resistance to an oppressive and corrupt state machinery deeply involved in supporting the primitive accumulation of capital. Instead, it seems to focus exclusively on the negotiated transactions between government agencies and target population groups over the distribution of governmental benefits. It involves limited struggles for daily survival and prevents the emergence of a long-term strategy or vision of radical transformation (Gudavarthy 2012).

The charge is not entirely untrue. The form of politics spawned by political society has a horizon limited to the demands of particular groups and does not seek to generalize its claims to all citizens. But the techniques of struggle frequently go beyond the limits of the law and sometimes even use violence (or show of violence) in order to demonstrate extraordinary outrage or draw the attention of the wider public through the news media to the demands of the group. There is resistance in political society, sometimes even of a spectacular kind. But more often than not, it is resistance that tests rather than overtly violates the limits of conventional political practice. In so doing, it sometimes manages to induce responses from governmental agencies that change the familiar forms of the conventional.

Some of these changes happen cumulatively. Thus, repeated local struggles against eviction have led over the decades to a conventional view in most Indian cities that long-standing slums cannot be cleared. The recognition by the authorities of the claims of one group of pavement vendors becomes a precedent that can be used by other groups. Urban groups of this kind have managed to build national coalitions to coordinate their struggles, learn from one another and present to the authorities more coherent sets of demands. The long struggles against slum demolition have brought forth national policy statements from the government laying down a framework for the rehabilitation of evicted populations. But there are moments when apparently uncoordinated local struggles could, simply by their simultaneity, force the issue into the limelight and bring forth a policy response from the government. Thus, a spate of agitations in different parts of India against the acquisition by government of land for industry has led to new legislation offering better terms of compensation to those who lose their land. The present government in India, acting under pressure from corporate capitalist lobbies, has failed so far to push through new laws making it easier to acquire the land of small farmers for new industry and urban housing. Examples of such coordinated resistance in political society are clearly growing.

To Conclude

David Harvey has argued that the term "primitive" is an odd characterization for accumulation in today's context when capital is several centuries old and that "accumulation by dispossession" is a far more general description that could include older as well as recent instances of the history of accumulation

(Harvey 2003). But Harvey overlooks a central feature of Marx's definition. Harvey's focus is, of course, on the shift from manufacturing to finance capital in Western capitalist economies where much new accumulation is taking place through the acquisition of assets in real estate, housing, pension funds, and so on by speculative financial operations. The result is the dispossession of millions of people with small properties. This is undoubtedly a major new feature of contemporary capitalism. However, this dispossession does not involve the separation of the primary producer from his/her means of production. Those losing their homes in the United States because of the collapse of housing companies or financial corporations were not primary producers and had no means of production of their own. The theoretical as well as historical significance of this sort of dispossession is quite different from the dispossession of peasants and artisans in Asian and African countries.

The point is illustrated by a pervasive feature of contemporary postcolonial politics that would be hard to find in Europe or North America. There is a deep distrust in the popular mind in Asian and African countries of the procedures of justice established in the formal courts of law and administration. These are seen to be slow, opaque, and needlessly complicated that only the wealthy and the powerful can manipulate to their advantage. Instead of the procedural fairness promised by the courts, there is a deep yearning for an impartial judge whose authority would be accepted by all and who would have arbitrary powers to deliver justice. Much populist politics plays on this expectation widely shared among the people. As a result, once in power, populist leaders and parties will frequently violate procedure and use arbitrary power to deliver a supposedly good outcome. Needless to say, this tendency in populist politics only strengthens its authoritarian side. As we have said before, populism can come in many different ideological shapes.

Looking ahead, it seems certain that as industry, mining, commercial farming, exploitation of forest resources, and urban development proceed rapidly in Asian and African countries, primitive accumulation will also continue unabated. This will mean greater challenges for governments to supervise this process and minimize its political costs. The more authoritarian forms could deliver quicker growth but will involve high costs in terms of coercion, violence, and the stability of the state. The more democratic forms will be slow and messy but could afford greater flexibility and the possibility of course correction. Both strategies have their supporters. For those who value the norms of participation and consultation in the process of government, there is no question which strategy is to be preferred. However, political outcomes are not necessarily decided only by rational debate. They depend on movements, organization, and political struggle. The struggle over primitive accumulation, especially in relation to the possession of land, continues to be waged today in the countries of Asia.

References

Chatterjee, P. 2004. *The Politics of the Governed: Reflections on Popular Politics in Most of the World*. New York: Columbia University Press.

Chatterjee, P. 2011. *Lineages of Political Society: Studies in Postcolonial Democracy*. New York: Columbia University Press.

Foucault, M. 2010. *Birth of Biopolitics: Lectures at the Collège de France, 1978–1979*, translated by Graham Burchell. New York: Picador.

Gudavarthy, A. (ed.) 2012. *Re-framing Democracy and Agency in India: Interrogating Political Society*. London: Anthem Press.

Harvey, D. 2003. *The New Imperialism*. London: Oxford University Press.

Laclau, E. 2005. *On Populist Reason*. London: Verso.

Luxemburg, R. 1951 (1913). *The Accumulation of Capital*. London: Routledge and Kegan Paul.

Marx, K. 1990. *Capital*, vol. 1, part VIII, translated by Ben Fowkes. Harmondsworth: Penguin.

National Commission for Enterprises in the Unorganised Sector (NCEUS). 2007. "Report on Conditions of Work and Promotion of Livelihoods in the Unorganised Sector." Government of India, New Delhi.

Preobrazhensky, E. 1965. *The New Economics*. Oxford: Clarendon Press.

Sanyal, K. 2007. *Rethinking Capitalist Development: Primitive Accumulation, Governmentality and Post-colonial Capitalism*. New Delhi and London: Routledge.

Stepan, A., Linz, J. J., and Yadav, Y. 2010. *Crafting State-Nations: India and Other Multinational Democracies*. Baltimore, MD: Johns Hopkins University Press.

1

The Land Question in India

State, Dispossession, and Capitalist Transition

Anthony P. D'Costa and Achin Chakraborty

1.1 Introduction

The historical process of capitalist economic transformation, and occasionally transformation on socialist lines, has relied fundamentally on generating capital out of agriculture for nonagricultural purposes, mainly for industrialization. This simple routine of structural change has taken many forms over time and across space, sometimes organically but mostly through direct intervention. In either case, violent forms of change have accompanied capitalist transition from agriculture to industry. It continues to be that way today. More squarely, it has been the agrarian transition nested within the larger capitalist transformation process that lay behind the material and technological advancement of economies. The speed of change was dependent on the grip of past institutional arrangements governing land tenure systems, the rise of incipient capitalist classes in rural areas as well as in urban towns and mercantile cities, state intervention in creating capitalists by facilitating the enclosure movement, and the scale of benefits accruing to imperialist nations from colonialism. The rise of agrarian capitalists has historically spearheaded the accumulation process by transferring capital to industry in urban areas. The ensuing dispossession and displacement of peasants and small farmers from the land contributed to "free" labor, migration, and thus the rise of the urban industrial proletariat.

This in a nutshell is the transition *problématique*. Consequently, similar logic is applied to developing countries where the agriculture sector must undergo a capitalist revolution for the accumulation process to advance, leading to a diversified economy that includes strong intersectoral linkages between agriculture and industry. Rising agricultural productivity is key to pushing the economy forward, just as it pushes the rural workers away from it. Both Marxist and non-Marxist approaches have dwelt on the theoretical and practical aspects of agrarian transition in various ways deeming it a necessary condition for capitalist transformation. While the contemporary advanced capitalist countries may have undergone some version of this transition, with a scattering of similar experiences in a handful of East Asian late-industrializing countries, the position of most of the developing world (the former colonies and peripheries to metropolitan centers), does not approximate this trajectory. For some countries, such as India, the agrarian transition, or its absence, has been hotly debated and pursued piecemeal through state-mediated land reforms. Gunnar Myrdal in the 1960s had already proclaimed that India missed a transformational opportunity because of a poorly designed land reform program (Breman 2010: 322). However, today this classic form of transition has become moot due to the increasing insignificance of the agricultural sector in India in fueling overall capital accumulation. Some would go further and point out that the agricultural sector depends on public largesse for its viability. In fact, much of India's recent high growth rates have resulted largely from both rural and urban *nonagricultural* sectors, and curiously, from the growing services sector. The role of land as conventionally understood in he transition/transformation problem has taken on a different form and substance.

The objective of this volume is to take a fresh look at the land question in India. Rather than re-engage in the rich transition debate, which has reached a cul-de-sac, we go beyond it by critically examining both theoretically and empirically the role of land in contemporary India. We do not completely bypass the political economy discourse surrounding the classic question of capitalist transition in agriculture in India; in fact, our inquiry springs from it. It is not our intent to debate whether a socialist path exists from the transition trajectory (*à la* Brass 2011). However, our focus gravitates toward the development discourse that inevitably veers toward land and the role of the state in pushing a process of dispossession of peasants through direct expropriation for developmental purposes. By development we mean materialist progress on largely capitalist lines. Contemporary dispossession may look similar to the historical process of primitive accumulation whereby peasants were separated from their land to make room for capitalist agriculture and act as a source of economic surplus for expanded accumulation. But as we argue in this chapter, land in India is sought not only for accumulation proper but also for non-agricultural purposes. For example, farmers purchase land to reduce risk, real estate developers seek land for the well-off urban residents, and (local) states acquire land for infrastructure development and on behalf of business to support the latter's accumulation strategies such as special economic zones

(Banerjee-Guha 2013) while states anticipate legitimacy through economic progress. At another level, tribal communities (*adivasis*), who depend on land for their livelihoods and a moral economy that is independent of any price-driven markets, hold on to such property not as individual owners for profit but for collective security and to protect a way of life. Thus land, notwithstanding its historical and contemporary role in the accumulation process, has been and continues to be a turbulent arena in which classes, castes, and communities are in conflict with each other, with the state, and with capital, jockeying to determine the terms and conditions of land transactions or their prevention, through both market and non-market mechanisms.

The rest of this chapter is divided into three sections. In Section 1.2, we develop a pathway for analyzing the land question in contemporary India. We begin with a discussion of the broader transition question in the context of India's specific capitalist trajectory, position this volume in the recent literature that has dealt with the land question, and offer an alternative perspective as to why the land issue is no longer a transition question. In Section 1.3 we anticipate some of the political economic consequences of land as a commodity, such as dispossession and conflicts, and the role of the state in development today, especially when it comes to mediating land transactions. In this section we cover three overlapping areas that show the different ways the Indian state engages with the land question. The first is state intervention in land for development, the second is the intensification of conflicts over land today as land is converted into a commodity, and the third is the dual but contradictory role of the Indian state that gives, as part of its liberal-democratic responsibilities, and takes, as part of its aggressive pursuit of capitalist development on behalf of dominant classes. In Section 1.4 we present a brief sketch of the individual chapters, dividing them into three groups and integrating them into the wider context of the land question in India.

This volume goes beyond the perspective traditionally taken in the political economy of agrarian transition question, and deals with, inter alia, distributional conflicts arising from acquisition of land by the state for capital accumulation on the one hand and its commodification on the other. As a result, one would naturally expect inclusion of the issue of gender in the land question. We acknowledge that there is a need for "engendering the political economy of agrarian change" (Razavi 2009: 197), and a volume such as this using heterodox approaches should ideally include chapters focusing on gender issues when unpacking land institutions to make our understanding more comprehensive and inclusive. However, the chapters included in the volume contribute to a thematic unity in which the distributional conflicts have been viewed mainly through the lens of class and the transformational process or lack thereof through capital accumulation. Even when the identity of groups, such as the tribal population and violation of their rights to land, have been examined in this collection, they have been viewed primarily as belonging to the deprived and marginalized classes, without of course ignoring the tribal identity question. The issue of gender, in our opinion, would require a rather different analytical orientation to deal with questions such as how the prevailing laws and patriarchal social norms and practices form a self-reinforcing institutional complex that denies the right of women to own or control land. In fact, critics have rightly questioned the transformative and empowering potential of market-led agrarian reform policies. But the critiques sometimes embraced the "customary" or "traditional" property right institutions as an alternative. A gender lens would guard us against such a tendency, as it would help us see the disempowering implications of the return to the "customary" as exemplified by the experience of sub-Saharan Africa (Whitehead and Tsikata 2003). Inclusion of this dimension as part of the analytical framework would have made this volume unwieldy but intrinsically it very much remains worthy and a challenging subject for fuller exposition.

1.2 Land and Capitalist Transition Today

One starting point for discussing land is the agrarian transition question, which refers to a turning point when capitalist relations of production enter the countryside providing the motor force for further capital accumulation outside of agriculture. Land enters the transition discussion on account of primitive accumulation, a Marxist take on the process by which precapitalist peasants are separated from their land to make room for capitalist growth.[1] Agrarian transition refers to "those changes in the countryside, in the relevant economies, necessary to the overall development of capitalism and to its ultimate dominance in a particular national social formation" (Byres 2002: 55).

Primitive accumulation by definition is "non-market" driven and often equated with "extra-economic coercion" (Bernstein 2010: 27), which suggests violence accompanying dispossession. This is considered integral to the Marxist "origin" of capital, whose emergence creates free wage labor for capitalist industrialization as peasants become unhinged from their land (D'Costa 2014: 324). The process of primitive accumulation has varied temporally and spatially (see Allen 1999; Mezzadra 2011) but it is viewed as necessary for capitalist transformation, or alternatively, for socialist industrialization. Yevgeni Preobrazhensky's primitive "socialist" accumulation was pursued in the former Soviet Union by collectivizing agriculture and generating economic surplus for heavy industrialization (see Akram-Lodhi and Kay 2010: 19). Socialist or not, primitive accumulation in theory, experience, and practice entailed the destruction of precapitalist (ergo noncapitalist) social and economic systems to generate initial capital, which would then be reproduced and expanded on capitalist or socialist lines (see Saith 1985). In theory, the uprooted peasants become wageworkers or the industrial proletariat, while agricultural productivity rises as capitalist farmers introduce efficiency-enhancing production methods, accumulate, and contribute to economic diversification. Under this scenario, agrarian capitalists endogenously generate economic surplus, which then drives the nonagricultural sector suggesting the historical necessity of primitive accumulation in capitalist development.

Abstracting from this linear dynamic, Arthur Lewis (1954) suggested mechanisms, such as higher industrial wages to induce rural-to-urban migration, to absorb surplus labor from the countryside in urban industrial activities and thereby enhance agricultural productivity. Dynamic agriculture would then fund industry, which would in turn accommodate the migrants leaving the countryside. Efficient agriculture would keep prices of food low and thus hold down urban wages that are marginally higher than rural wages to induce migration. The ensuring rapid industrialization is expected to make a dent in the dual structure of the economy, a common feature of postcolonial societies. Furthermore, rising incomes in the countryside would become markets for industrial goods and the two sectors would positively reinforce each other as savings were directed toward capital formation.

At a general level the broader Marxist historical dynamic for some places and the Lewisian theoretical possibility of structural transformation of dualistic societies remain valid. However, both have three major shortcomings, theoretical and empirical, when we apply the role of land in the classical transition equation to contemporary postcolonial societies, and especially India. First, there is the issue of temporality, which means the genesis of capitalism from noncapitalist relations of production occurred at a particular historical moment. Today it is essentially complete. If the rise of capitalism, despite the precise dating problem, was in the long sixteenth century (Wallerstein 1979), is it possible to speak of primitive accumulation in postcolonial societies such as India in the twenty-first century, which has both capitalist and noncapitalist forms of production organization for surplus extraction? So, a basic question that requires addressing (other than the one attributed to it in the classical transition question) is, what role does land have in the accumulation process today? Is dispossession creating the space for dynamic capital accumulation in the countryside and thus for the economy as a whole in the future?

Second, the teleological narrative of capitalist logic in terms of the inevitability of structural transformation is wanting (D'Costa 2014; Sanyal 2007). For example, the persistence of a "subsistence economy" or the "need economy" in the milieu of a thriving and advanced capitalist sector belies the more mechanistic forms of structural transformation. This "incomplete transition" has raised doubts about the teleological underpinnings of the idea of transition itself. Contrary to the received Marxist interpretation, primitive accumulation for Sanyal (2007) is a permanent feature of capitalism in postcolonial economies, which requires continual dispossession of those who inhabit the spaces external to the circuits of capital proper (what he calls the "accumulation economy"). While the idea of being "outside of the circuits of capital" is contestable since the dispossessed are linked to formal production through petty commodity production (D'Costa 2016; Adnan, Chapter 3 this volume), the conditions for the hegemony of capital require diversion of some surplus from the accumulation economy toward the dispossessed in the "need economy" for their survival, or what approximates Chatterjee's "political society" (Chatterjee 2008). In other words, primitive accumulation in the postcolonial economy does not produce wage laborers to be employed by capital. It leads to confinement of the dispossessed in the informal sector. Therefore, instead of a capital–*pre*capital binary, Sanyal uses "capital–*non*capital" in order to free both capital and precapital from the historical or chronological ordering and renders the idea of transition in the classical sense irrelevant. To quote Sanyal:

> [T]he characterisation of the postcolonial economic as a complex of capital and non-capital, with the latter emerging in a space produced by the internal logic of the former, totally dispenses with the notion of transition. If there is a possible transition in this scenario, it is from pre-capitalism to the capital-non-capital complex... [Thus] *what we have is capitalism with an inherent heterogeneity.*
>
> (Sanyal 2007: 40; emphasis in original)

Whether we agree with this interpretation of Indian political economy (need economy and political society), what is empirically valid is India's capitalist trajectory that is a blended integration of primitive accumulation and capitalism proper. Here primitive accumulation is not precapital nor is the unfree labor associated with precapital temporary. Rather, it is part of mature capitalism (Brass 2011: 13). The dispossession arising from alienation of peasants from their land does not lead to the dissolution of the peasantry nor the creation of an industrial proletariat as experienced by early capitalist regions; it instead contributes to the coexistence of "surplus" labor, which is today approximated by the official understanding of the unorganized sector. In the words of Sanyal (2007: 249) the displaced represent a "dark space," whose reproduction, aside from contributing to absolute destitution, is absorbed through precarious forms of employment, such as self-employment and casual labor (National Commission for Enterprises in the Unorganised Sector (NCEUS) 2009).

Third, the empirical reality, at least for India, fundamentally challenges the received transition question simply because agriculture is not playing out its historic role in the transformation process. Cross-country econometric studies have demonstrated that initial differences in landholding patterns have long-lasting implications for growth and social development much later (Banerjee and Iyer 2005). That land is unequally distributed and becoming fragmented suggests differential rates of growth across classes and pronounced social differentiation in the countryside. More importantly, the peasantry has not completely disappeared nor has a large industrial proletariat appeared despite a fair degree of structural transformation in India through nonagricultural development. It appears that something internal to the agricultural system, although influenced by exogenous factors, could be contributing to the stalemate in India's capitalist transition.[2]

Byres (1993) had already identified the failure of capital accumulation in the Indian countryside thus signaling the absence of an endogenous dynamic. For one reason or another accumulation in agriculture was being blocked (see also Bharadwaj 1985). At least for a while there was some sort of a consensus that "semifeudal" relations persisted in the Indian countryside because of the particular way metropolitan capital during colonial rule reinforced rural class structures (Bagchi 2010). More recent developments in India, however, have questioned the existence of semifeudalism (Harriss 2013) and now there appears to be a consensus that capitalist relations characterize the Indian countryside even if vestiges of noncapitalist forms of production organization remain in many parts.

In a recent survey of farming conducted in the National Sample Survey (NSS) seventieth round (January–December 2013) the agrarian picture captured was quite dismal.[3] For example, during the agricultural year July 2012–June 2013, rural India had an estimated total of 90.2 million agricultural households.[4] These agricultural households were about 57.8 percent of the total estimated rural households of the country during the same period. In other words, about 42 percent of rural households earn almost no income from cultivation; their major source of income is either wage labor (farm and nonfarm) or nonfarm self-employment. The percentage of agricultural households in total rural households vary widely across the states of India—from a high of 78.4 percent in Rajasthan to a low of 27.3 percent in Kerala. Among the agricultural households (constituting 57.8 percent of all rural households), 63.5 percent of households reported that their principal source of income was cultivation (Table 1.1), which means only 36.7 percent of all rural households reported cultivation as their principal source of income. However, the importance of cultivation as the principal source of income sharply declines with the amount of land (size class) possessed. Households possessing less than 0.01 hectare principally depend on wage labor. Even among those who possess 0.01–0.40 hectares, only 42.1 percent of households reported cultivation as their principal source of income.

Table 1.1. Distribution of agricultural households by principal source of income during 365 days preceding the survey

Size class of land possessed (ha)	Cultivation	Livestock	Other agricultural activity	Non-agricultural enterprises	Wage/salaried employment	Others
<0.01	1.6	22.9	2.7	10.8	56.4	5.5
0.0–10.40	42.1	4.8	1.2	7.5	35.2	9.3
0.41–1.00	69.2	2.3	0.9	3.6	20.0	4.1
1.01–2.00	83.0	2.5	0.9	3.2	8.6	1.8
2.01–4.00	85.9	2.4	1.1	1.6	7.1	1.8
4.01–10.00	87.9	2.7	0.5	0.9	5.9	2.0
10.00+	89.4	5.5	1.5	1.8	1.7	0.1
All sizes	63.5	3.7	1.1	4.7	22.0	5.1

Source: NSS 70th Round

Table 1.2. Average monthly income from different sources, consumption expenditure and net investment in productive assets per agricultural household for each size class of land possessed (July 2012–June 2013)

Size class of land possessed (ha)	Income from wages/ salary (Rs.)	Net receipt from cultivation (Rs.)	Net receipt from farming of animals (Rs.)	Net receipt from non-farm business (Rs.)	Total income (Rs.)	Total consumption expenditure (Rs.)	Net investment in productive assets (Rs.)
<0.01	2902	30	1181	447	4561	5108	55
0.02–0.40	2386	687	621	459	4152	5401	251
0.41–1.00	2011	2145	629	462	5247	6021	540
1.01–2.00	1728	4209	818	593	7348	6457	422
2.01–4.00	1657	7359	1161	554	10,730	7786	746
4.01–10.00	2031	15,243	1501	861	19,637	10,104	1975
10.00+	1311	35,685	2622	1770	41,388	14,447	6987
All sizes	2071	3081	763	512	6426	6223	513

Source: NSS 70th Round

The fact that cultivation is not the principal source of income for an overwhelming majority of the rural households does not necessarily mean that they cannot earn a decent income from other sources. However, Table 1.2 shows a rather gloomy picture in this regard. Households belonging to the bottom three landsize classes earned on average Rs. 4,561, Rs. 4,152, and Rs. 5,247 respectively, per month in 2012–13. How good are these incomes? For 2011–12, the Planning Commission pegged the national poverty line at Rs. 816 per capita per month for rural areas, which means that a family of five would need at least need Rs. 4,080 to be considered not poor. In other words, the income that an agricultural household possessing less than one hectare of land would earn from all sources taken together would on average be slightly above the poverty line income. Furthermore, they are most likely to be all net dis-savers because their consumption expenditure would typically exceed income (Table 1.2). The precariousness of life in the rural areas is evident from above. What is striking is that even possession of land up to 0.4 hectare (one acre) does not make it better.

As is evident from the recent survey data, Indian agriculture is not dynamic in the capitalist sense. It does not generate a massive surplus for the rest of the economy. Furthermore, this rural stalemate is reproduced and reinforced by income from nonagricultural sectors. Several scholars have already posited (and we believe correctly, though others such as Banerjee, in Chapter 4 this volume, have a different perspective) that the "agrarian question has been bypassed" (Bernstein 1996, 2010; Lerche 2013; see also Dasgupta, Chapter 10 this volume).In fact, to a great degree, Indian agriculture as a whole is dependent on state subsidies rather than being a source of economic surplus. Not surprisingly, powerful rural social classes remain the principal beneficiaries. Others have argued that both the bypassing of agriculture as a source of economic surplus and the weakening of agriculture–industry linkages are due to India's insertion in the global circuits of capital since the 1980s (Harriss 2013: 359; Lerche 2013: 391). The paradox is that the weak links between industry and agriculture have been more than matched by stronger links between industry and the Indian services sector (Chandrasekhar 2007), thereby precluding the more classic variant of capitalist economic dynamics.[5] The net result of the lack of agricultural dynamism has been the freezing of social differentiation and polarization of peasants in the form of household subsistence farming. This has been reproduced through precarious forms of employment and reinforced by (remittance) income from nonagricultural employment (Harriss 2013: 358, 363). The inevitable consequence has been the decreasing significance of Indian agriculture (Figure 1.1). From 54 percent of GDP in 1950–1, the share of agriculture fell to 38 percent by 1980–1, which further declined to 16 percent in 2011–12.

This decline is consistent with the broader capitalist development in India outside of agriculture, namely relatively faster growth and expansion of non-agricultural sectors. It also reflects slow productivity growth in agriculture and thus the absence of the classic dynamism of rising productivity (and profitability) necessary for an agrarian transition. The continued reliance of a large fraction of the Indian population on agriculture is visible in Table 1.1. For example, soon after the economic reforms of 1991, agricultural employment in 1993–4 stood at a high of 65 percent of total employment. However, by 2011–12 the share of agricultural workers to total workers dropped to 50 percent (Table 1.3). This decline could be used to argue that agriculture *is* dynamic since fewer people are now engaged in agriculture, which presumably means higher productivity. While that may be the case in a few places and during certain years, the empirical reality suggests otherwise. Indian agricultural growth as a whole has fluctuated greatly, as evident by indices of

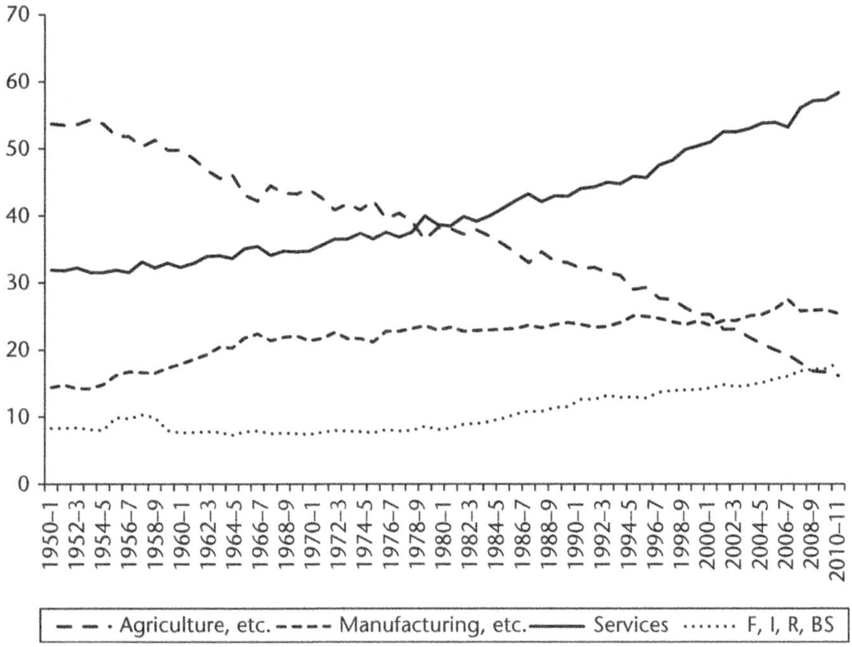

Figure 1.1. Decline of agriculture and the changing composition of the Indian economy (% of GDP)

Notes: "etc." includes other subsectors in each of the primary, secondary, and tertiary sectors. F = Finance, I = Insurance, R = Retail, and BS = Business Services.

Source: Government of India, Ministry of Finance (various issues), Economic Survey; available at: <http://indiabudget.nic.in/survey.asp> (accessed September 14, 2016).

Table 1.3. Changing distribution of the workforce across sectors in India

Sectors	Number of workers (in millions and %)				
	1993–4	1999–2000	2004–5	2009–10	2011–12
Agriculture	**241.5**	**246.6**	**268.6**	**244.9**	**231.9**
% of total	64.6	61.7	58.5	53.2	48.9
Manufacturing	38.9	42.8	53.9	50.7	59.8
% of total	10.4	10.7	11.7	11.0	12.6
Other industries	15.8	20.4	29.4	48.3	55.3
Services	77.7	89.8	107.3	116.3	127.3
% of total	20.8	22.5	23.4	25.3	26.8
Total	374.0	399.5	459.1	460.2	474.2

Notes: Until 2004–5, the absolute number of workers in agriculture in the country as a whole steadily increased, even though agriculture's share in total workers declined throughout the period from 1993–4 to 2011–12.

Source: From National Sample Survey (NSS) rounds 1993–4 to 2011–12

production and growth rates of agriculture at factor cost (Patnaik 2007: 115–50).[6]

Aside from the rural power structures that continue to direct state largesse to the dominant castes and classes, the grip of the dominant classes in the Indian countryside is reinforced by rental and interest income and speculative returns. The inadequate performance by a slow-growing agrarian economy owes its legacy to the continuing decline in public investments (Bardhan 2006). Furthermore, the sustained fragmentation of land due to coparcener inheritance practices has created uneconomically small landholdings thereby also contributing to subsistence, low-cash-income-based rural economies.[7] It is therefore no surprise that indebtedness of peasants and low productivity continue to characterize Indian agriculture today (Harriss 2013: 357).

Our take on the transition question is that there is little capitalist expansionary dynamic in Indian agriculture and hence it does not follow the teleological trajectory of the classical transition. Instead, a sharp but integrated dualism is generated between primitive accumulation and capitalism proper because of continuing dispossession on the one hand and the much higher growth in nonagricultural sectors in both rural and urban areas. This suggests, and Sanyal (2007) argues, that the historical process of primitive accumulation is also contemporary but because of its historical specificity is also quite Eurocentric (Marx (1958–71) in Mezzadra 2011: 308), meaning it is a European phenomenon rather than a universal one and certainly less applicable to former colonies. Thus, discussion of transition that is linked to primitive accumulation has generated different perspectives precisely because of the way primitive accumulation today is used. Our task, however, is not to settle this issue. We can simply use the term to denote the separation of peasants from their land through market and nonmarket mechanisms without imputing any implications for transition (Adnan 2013). In those instances where dispossession of *advasis* could clear the way for large-scale accumulation, such as mining and infrastructure projects, primitive accumulation could be applied (Banerjee-Guha 2013). However, by most accounts, primitive accumulation in the historical sense is complete.

To the extent the concept of primitive accumulation is used to characterize the so-called origin of capital, it sits uncomfortably with the contemporary rush to grab land by capital worldwide. One has to answer why capital requires forceful expropriation of land at a particular juncture to sustain accumulation (Levien 2012). Sustaining accumulation implies the nonprimitive variety. Some suggest that the concept of "accumulation by dispossession" put forward by David Harvey might be more appropriate (Banerjee-Guha 2013). One of the goals of contemporary scholarship in land dispossession is to explore how concepts such as primitive accumulation or accumulation by dispossession can be reassessed to illuminate the regionally diverse experiences of land expropriation. In this context we need to keep in mind India's empirical realities vis-à-vis the much-discussed phenomenon of global land grabbing in the contemporary world. Unlike the experiences of African and Latin American countries (Borras et al. 2012; Cotula 2012; Li 2011; Wily 2012: 768), large-scale expropriation of land in India is primarily done by domestic capital rather than foreign capital and the dispossession of peasants due to special economic zones in India stand out markedly from other experiences (Sampat 2010).[8]

It is also important to recognize that surplus labor in the countryside and in urban areas is a persistent feature of India and should not be viewed outside the orbit of contemporary capitalism; in fact, it is a product of capitalism (Adnan 2013; D'Costa 2014). Furthermore, the land question today is less about transition and more about the conflicts and the political economy of the differential benefits of circulation of land as a commodity. The land reality in India thus points to different sets of questions and demands a better understanding of the role of land today in capitalist development in postcolonial societies.

If agrarian transition, and by extension capitalist transition, is hobbled by low agricultural income, small plots of land, nonagrarian sources of income, and other factors such as declining water table, salinity, and soil erosion, then the place of land in the larger process of economic development is less significant in Indian agriculture. In this scenario rural dynamism is limited, unless offset by a massive expansion of off-farm activities in both rural and urban areas. In reality, off-farm rural economic activities are not deep, at least at this time, nor is contemporary industrialization sufficiently robust and widely spread to absorb rural labor en masse and alleviate rural misery. The persistence of a large unorganized sector in rural areas indicates the difficulties of reproducing capitalist production relations even if they are part of a capitalist India (Brass 2010: 27). Thus the transition question involving dynamic capitalism in Indian agriculture is currently moot. A different role must be attributed to land in India.

1.3 State, Development, and Dispossession

The centrality of land in contemporary Indian society and polity can hardly be exaggerated. As the land issue frequently overwhelms the popular discourse due to the development–dispossession dynamic, we must step back from the heat and tumble of the short run and take a hard look at the larger picture of how land fits into contemporary Indian political economy from alternative analytical perspectives. This volume, as the previous section indicated, goes beyond the political economy paradigm that centers on the structural aspects of the transition question. What we try to do is to read the evolution of the land question from multiple perspectives, which could broadly be called "institutional." This broader approach enables us to view public policies, and thus the role of the state, as the choices resulting from conflicts among competing interests that take place within political institutions. In this view, politics over land is seen as a distributional struggle (Grindle 2001) with the state and other institutions producing "audacious reforms" (Grindle 2000) and influencing policy formulation (Bates 1989). Both contestation and state intervention are political and are omnipresent in the land question in India today.

Today there is considerable pressure to transform land into a commodity to be bought and sold in the market for nonagricultural purposes. The motivation behind this dramatic turnaround in how land is used is driven by India's contemporary economic development concerns, such as industrialization, infrastructure, special economic zones (SEZs), and real estate expansion. This is not inconsistent with capitalist development thus far and the possibility of even greater capital accumulation over the longer term. What is different fundamentally is that land no longer acts as the source of economic surplus nor is the motivation of the state rooted in making agriculture dynamic in the capitalist sense. On the contrary, acquisition of land today is directed at nonagricultural development. For example, witness the well-known Singur case where the local state government acquired land on behalf of private industry to set up an automobile factory (Nielsen 2010; Guha, Chapter 12 this volume). Furthermore, the conversion of agricultural land or dispossession of *adivasis* for various industrial and infrastructural projects is also not new. What has changed is the role of the state in securing land by way of "eminent domain" or using "public purpose" arguments, in some instances coercively, as in West Bengal (see Guha, Chapter 12 this volume), less for itself and more on behalf of business (see also Bhattacharya, Bhattacharya, and Gill, Chapter 7 this volume). For example, there is evidence, even if scattered, that the state (and its representative politicians such as Narendra Modi, now India's prime minister) aided local businesses in the state of Gujarat such as Adani and others by acquiring land at well-below-market prices (Jaffrelot 2015). Additionally, the state has also introduced measures that try to shield the dispossessed from the worst effects of losing land and livelihoods but in effect has created the structural and legal basis to facilitate land transactions more transparently and more definitively. In other words, an incipient land market is increasingly becoming a realized one.

The operative phrase here is "capitalist context," meaning, in the absence of land as a dynamic source of accumulation, it is being transformed into a commodity, to be bought and sold in the market. This is akin to Karl Polanyi's description of the "great transformation" (2001) where land gradually becomes like any other product. However, far from the invisible workings of the market through the price system, there is also institutional intervention, namely, by the state, which acts as an owner, intermediary, buyer, and distributor all rolled into one in the emerging land markets in contemporary India. It is not the market supplanting the state but the state becoming more business friendly (Sud 2014). The analytical as well as the actual economic, political, and legal implications of this transformation are manifold. For the purposes of this volume, we identify three overlapping areas in which the role of the state is revealed in its many forms when engaging with the land question. The first is state intervention in land for development, the second is the intensification of conflicts over land today, and the third is the dual contradictory role of the Indian state that gives as part of its liberal-democratic responsibilities and takes as part of its aggressive pursuit of development on behalf of dominant classes.

First and foremost, India, as a late entrant to industrialization, has had an interventionist state. Since independence it has tried transforming the economy through planning, public investment, and detailed economic management and social development. While heavy industry under infant industry protection received substantial state attention, agriculture with some exceptions was largely neglected. A few local states such as West Bengal and Kerala introduced land reforms aimed at both economic development and social justice but the reforms failed to induce the classical transition dynamic (Dasgupta, Chapter 10 this volume). Some states such as Punjab and Haryana experienced a green revolution that contributed to an agrarian transition of sorts. Overall, Indian agriculture remained structurally underdeveloped. The Indian state, however, remained developmentalist, at least in intent if not in outcomes. With economic reforms that started in the 1980s and accelerated since 1991, the state has been withdrawing from the market and facilitating more intense working of the market (see Walker 2008). It is in this context that land takes on a different role, namely, it becomes a commodity and the role of the state is substantial because it wants to facilitate development by making land available for a variety of nonagricultural purposes.

Economic development remains an aspiration in India and as Marx, with reference to India opined in a different context, advanced capitalist countries are a mirror image of India's future (in Larrain 1989: 47). The future, as imagined by the state and expected by the society, is the transformation of the material basis of the country. Development promises jobs, fewer everyday struggles from rising incomes, better health, and all the modern trappings of consumption. Capitalism as an economic system has excelled in its ability to produce but on the consumption side has been less successful. In fact, the idea of development is so ingrained institutionally and in the psyche of people that it is impossible to separate development from the Indian state, despite the actual functioning of and results achieved by the state. The idea of development is further reinforced by multilateral and national organizations such as the World Bank and the former Planning Commission. In the absence of agrarian transition, development has been increasingly funneled through nonagricultural activities such as urban development, industrialization, and modern infrastructure. It is here that land comes to the fore and the state is directly involved in ensuring that land can be transacted in the market, secured, and made available for profit-based economic activities by private capital. These uses include SEZs, infrastructure, real estate, mining, and manufacturing. Such private accumulation is equated with development, especially when India's economic growth rate has been halting and the jobs outlook remains bleak for the millions of its youth. One might add that land transactions are never a one-way process; rather, multiple actors impinge on the process, including the owners of the land who could very well be willing sellers provided the price is right.

Second, by addressing why conflicts around land have intensified the role of the state is revealed. After all, both in pre-independent and post-independent India, both the private sector and the Indian state did usurp land for public and private purposes, for dams, roads, steel mills, and other infrastructural projects (see Raianu, Chapter 11 this volume). There was also popular resistance in the past where people were dispossessed and displaced but then the state using its power of eminent domain could get away with it easily. Today, however, it is a different matter; "public purpose" of the state has taken on "private benefit" and is rationalized with economic development arguments (Pal, Chapter 6 this volume). Land disputes have arisen today principally because of increasing commodification of land that has been responsible for the recent surge in land-related conflicts. The uncertainty associated with the sale of land, and in most cases loss of livelihoods, and the asymmetry of price information between eager buyers and reluctant sellers contribute, often violently, to the contestation among the different players. This occurs when transactions are not transparent, titling is unclear, laws and regulations governing land acquisitions are poorly designed or do not adequately recognize community rights of common property, evictions are coerced, land is undervalued, and when livelihoods are seriously at stake. Consequently, many large-scale industrial and infrastructural projects in recent years have been held up and in some instances cancelled outright.

The consequences of converting land to a commodity are many as dispossession and displacement are integral to this process. However, the use of the term "commodity" to characterize land at the present juncture is still contentious. A commodity is typically bought and sold in the market at a price determined by the interaction of supply and demand. But the heavy presence of the Indian state in acquiring land from the peasants and landowners in rural areas, based on ad hoc administered prices, without adequate compensation, and handing it over to capitalists almost free of cost does not conform to the standard notion of a commodity. However, at the same time, land is now being bought and sold in rural India much more frequently than before, both as an asset and a factor of production. Consequently, there are many concerns surrounding the land acquisition process. These include but are not limited to: interclass and intra-society conflicts arising from the increasing commodification of land; distributive implications after the transactions; the real and latent coercive evictions of tribal communities whose attachments to land and forests go beyond subsistence livelihood dependence; the legal and ethical implications of acquisition under both market and extra-market forces, especially under a democratic political system where certain rights are recognized and exercised; and the role of the state vacillating between recognition and denial of rights in its quest for development.

Land acquisition and dispossession as practiced and experienced typically proceed by diluting or denying both individual and communal property rights to land under the rubric of "public interest." Here the primary role of the state has not been to promote efficient working of the price mechanism in the land market; rather it has been to suppress the price mechanism through acquisition with unilaterally determined "compensation." Payment of compensation is typically bureaucratically governed and poor farmers (sellers) are always at the receiving end. The resulting conflict of interest between the farmers and the capitalists may be reminiscent of the well-known Ricardian conflict between capitalists and landlords but the Indian experience shows a different kind of conflict. In the case of state-mediated acquisition and dispossession, the losers in the game are the marginal and small farmers, poor tenants, and agricultural workers, who are most likely to lose their livelihood options along with their land. The conflict thus revolves around the state and owners of land. Furthermore, unlike the Ricardian dynamics of falling land prices over the long haul due to nonagricultural sources of income and wealth (see Piketty 2014: 6), in India as elsewhere there is an upward pressure on land prices. The irony is that this price increase is for the same reason, namely, wealth and income accrued from nonagricultural activities, which is increasingly invested in land to cater to urban–industrial demand.

In post-independent India, the need for modernization and industrialization within the model of a mixed economy with the state sector to attain "the commanding heights" in the economy was articulated by both big business and the government. This approach also drove the state's attitude toward land as private property. Land was needed for rapid industrialization and urbanization, and the state engaged in land acquisition for "public purposes." The colonial Land Acquisition Act of 1894 has since been invoked to acquire land by the national and subnational governments in India. The history of development in postcolonial India cannot be discussed without reference to the large-scale displacement of people, a disproportionate number of whom happen to be tribal people. Over the past six decades in independent India a significant amount of land has been acquired by the Indian state for infrastructure and industrial development. Large dams and multipurpose projects, such as Bhakra Nangal, Hirakud, Damodar Valley, and Narmada Valley led to large-scale dispossession and displacement. Big public sector industrial projects and the townships around them in Bhilai, Rourkela, Bokaro, Durgapur, and so on, also entailed the acquisition of land.[9] What distinguishes the more recent rush for land acquisition from earlier ones is that the state has increasingly stretched the definition of public purpose or "eminent domain" to acquire land for private capital. For example, the Singur land acquisition was justified as public purpose as the resulting industrial jobs would meet that social goal. Consequently, agricultural land since the 1991 economic reforms has succumbed to the exponentially increasing demand by private capital for industry, shopping malls, office complexes, and residential enclaves for the burgeoning middle and upper-middle classes. By extension, speculation in land has risen, which involves not only brokers, agents, and real estate companies but also landowners. The latter resist parting with land now due to unattractive prices but may be willing sellers in the future when prices are higher. Thus the friction surrounding market non-clearing prices and coerced clearing prices lie at the root of these conflicts.

In view of this uncertainty caused by lack of clarity in property rights and the inefficiency of the land administration system, the World Bank's policy recommendations concerning rural land in India make sense (World Bank 2007). The Bank's recommendations focus on the need to improve the land administration system and land policy to reduce the high costs of transferring land in India. The inefficiency of the land policy and administration system is also responsible for restricting the development of a rental market for land. A rental market can be beneficial for the rural landless poor and thus has equity consequences as well. Therefore, little objection can be raised against the efficiency-enhancing administrative reforms that the World Bank has recommended. However, the neoclassical economic argument for allocative efficiency that underlies these recommendations is not consistent with the Indian state's forceful acquisition of land. It is hard to argue that acquisition of large tracts of land for the benefit of private capital improves allocative efficiency for the simple reason that market signals are completely ignored by the state and sometimes extra-economic influences exercised, not to mention the reluctance of the dominant private interests to accept "true" market forces.

The recent timing of intensified land disputes is also explained by the rapid increase in land prices throughout India, with some locations and regions experiencing multifold increases within the space of a few years because of uneven capitalist development and differential links with the world economy. The greater Bangalore area is a good example of globally influenced real estate prices, driven by the export-oriented information technology (IT) industry. The intensity of conflicts over land acquisition is more severe today partly because of the frequency and scale of dispossession but also because of better knowledge of land prices due to the spread of information (Chakravorty 2013). This makes the conflicts double sided, in that not all farmers are unwilling sellers since prices are moving up and they prefer to sell at even higher rates. Consequently, it becomes difficult to settle at a reasonable market-clearing price acceptable to both transacting partners. As one study of special economic zones shows, people likely to be displaced were willing to sell land, provided the price was "right" (Cross 2014). This also suggests that land transactions are not just a matter of unwilling sellers hijacking development projects but rather the inability to arrive at mutually agreeable prices and the non-availability of alternative livelihood options.

The third area is the contradictory role of the Indian state that both gives to the people and takes away from them. Thus far we have presented the intermediary role of the state in acquiring land for development purposes. The Indian state, however, is also subject to India's democratic practices and the electoral significance of people that often bear the brunt of land disputes. Today the rights of those who might be dispossessed and displaced are recognized and they are protected through the Right to Fair Compensation and Transparency in Land Acquisition, Rehabilitation and Resettlement (LARR) Act, 2013. The Act was introduced by the United Progressive Alliance of the Congress Party, which pursued various "rights-based" legislations. This "rights-based" act, consistent with others such as Forest Rights, NREGA, and Education (Jenkins 2013: 608) has replaced the colonial Land Acquisition Act of 1894, which effectively allowed the government to acquire any privately owned land for "public purpose" in exchange for a fixed compensation. Interestingly, the LARR was up for revision under the new pro-business government elected in 2014, which was determined to make land acquisition less cumbersome. However, lacking a majority in the upper house, the ruling Bharatiya Janata Party (BJP) could not muster sufficient support to alter the 2013 LARR Act, indicating a strong sentiment against land acquisition without fair rehabilitation and resettlement.

The conflicts are as much about land prices as they are about compensation, which in a dynamic environment can fluctuate based on structural and electoral power. The National Democratic Alliance today under the BJP is exploring ways to alter the act to facilitate business investments and induce higher economic growth rates. However, the resistance offered by farmer and peasant groups has been strong enough for the government to take a softer "pro-farmer" position. The question is how much the state will give if the act is diluted against farmers and what might be the legal and livelihood implications for the dispossessed.

The vacillation on the part of the state arises for both political expediency as well as civil rights action based on democratic aspirations and practices. Acquiring land for development without politically negotiating "price" and "rehabilitation" of the displaced is fraught with conflicts, which the Indian state has learned the hard way. It is also evident that capitalist accumulation proper can be slowed down or deferred as resistance to land acquisition intensifies, a small space afforded to India's "political society" (Chatterjee 2008). The agrarian crisis and farmer suicides that plague India cannot be ignored by any government that is elected but if dispossession and displacement without adequate rehabilitation and resettlement goes unaddressed, land acquisition will be stymied. As it is now evident, investments in large nonagricultural projects necessarily require land and the acquisition of land has become a critical bottleneck to seeing these projects through. In addition to sources of conflict mentioned above, the difficulty in aggregating fragmented landholdings into larger holdings contributes to the complexity of the acquisition process even if land has increasingly become a commodity to be bought and sold in the market where mutually agreed upon prices have a significant bearing in clearing markets. However, as it was also indicated, these are imperfect markets because of the various institutions governing and impeding land transactions. Resistance to ill-functioning markets is a political process as well as a result of institutional responses to the inherent tension between the goals of capitalist expansion outside of agriculture and a democratic polity. The broader question for future research is not just the trajectory of Indian capitalism, where land is either taken away or protected on a large scale, but how land disputes are to be settled while not giving up the quest for capitalist economic development.

1.4 Brief Chapter Descriptions and Integration

There are fourteen chapters in this book, sandwiched between a Prelude and a Postscript to the land question. In all the chapters, the land question is viewed with reference to the property rights regime, its connection with the structure of accumulation, and the role of the Indian state in shaping development through the land acquisition process. These chapters reveal the conflict-ridden dynamics among chief stakeholders such as the owners and buyers of land, the various private intermediaries, and the state. Aside from theoretically analyzing the role of land today in India's capitalist development and social transformation, these chapters also address how conflicts arising from land acquisition are negotiated, resolved, or left unresolved. Furthermore, these chapters theoretically and empirically discuss the role of legal and policy interventions on acquisitions and thus the differential impact on dominant and marginalized groups, the stand of the state on the issues of land and livelihood rights for those displaced through acquisition, and how they conflict with alternative interpretations of rights and practices.

The chapters are divided into three groups. The first two groups comprise mostly theoretical chapters and provide a grand narrative of the land question in contemporary India. They cover primitive and contemporary accumulation and legal–institutional dimensions of "regimes of dispossession." The third group of six chapters is informed by different theoretical and historical understanding and provides a variety of regional empirical perspectives on the contemporary land question in India. Similar to the theoretical positions taken in several chapters, the empirical chapters go beyond the transition question to include new forms of socioeconomic dynamics and political responses to the pressures of accumulation, dispossession, and loss of livelihoods.

1.4.1 *Primitive and Contemporary Accumulation*

As indicated earlier, there seems to be an agreement that the classical agrarian question pertaining to the mobilization of agricultural surplus for industrialization is less relevant for India. Therefore, we go beyond this discourse. The existing theoretical perspectives on dispossession of land, such as different variants of Marx's "primitive accumulation" or David Harvey's "accumulation by dispossession," provide insights into the contemporary processes of dispossession in India. But how adequate are they in explaining the relationship between land dispossession and capitalism? While acknowledging that Harvey's accumulation by dispossession marks a significant advancement on primitive accumulation as it attempts to capture diverse examples of contemporary dispossession, Levien (Chapter 2) observes that Harvey underemphasizes the fact that land dispossession is a deeply political process in which owners of the means of coercion redistribute assets from one group to another and in which the role of the state remains central. He examines this question and advances the concept of "regimes of dispossession" as an "alternative way of understanding how dispossession is politically organized in different phases of capitalism." A regime of dispossession is an institutionalized way of expropriating land from the current owners or users. Levien observes that India has passed from a regime that dispossessed land for state-led industrial and infrastructural expansion to one that dispossesses land for private capital. He argues that beginning in the early 1990s, India shifted from a regime that dispossessed land for state-led projects of material expansion to one that dispossesses land for private and decreasingly productive investments. This new regime of dispossession, in which states have become mere land brokers for private capital, is arguably less "developmental" than its Nehruvian predecessor. The upshot is that India's "land wars" are unlikely to dissipate any time soon, and the "land question" may be the largest contradiction for Indian capitalism for the foreseeable future.

Shapan Adnan (Chapter 3), takes a different theoretical approach, in which the mechanisms of primitive accumulation or accumulation by dispossession (ABD) are not restricted to the use of force, but also include land transfer by agreement, as well as indirect mechanisms that are concerned with very different objectives. Analyzing evidence on land alienation, resistance, and workforce trends, Adnan shows that the outcomes of primitive accumulation in India under neoliberal globalization do not accord fully with the classic Marxian schema of transition to capitalism. Critical assessment of trends in employment status over 1999–2012 indicates that the self-employed accounted for at least half of the Indian workforce, whereas workers in regular wage employment remained restricted to 20 percent or less. These trends suggest that primitive accumulation in neoliberal India has not been substantially followed by the absorption of the dispossessed in regular capitalist employment. Adnan puts forward three hypotheses to explain these trends based upon: (i) inability of the formal and the informal capitalist sector to generate adequate incremental employment; (ii) limited options of survival other than self-employment for dispossessed groups, particularly given the predominance of unproductive capital and precapitalist rent extraction over productive capital; and (iii) the ability of such groups to shift to forms of "distress employment" under worsening economic conditions. Moreover, the parallel trend of Indian migrant workers finding wage employment in the international labor market suggests that the prospect of the transition to capitalism in India remains an open question.

If the historical role of the peasantry is reduced to simply abdicating claims on land and other natural resources for the purpose of industrialization, then "should one categorize peasant resistance to land acquisitions as a Luddite, anti-progressive, anti-capitalist position?" This is the central question raised by Arindam Banerjee (Chapter 4). This is important since the official stance against farmers' resistance to land acquisition seems to be an affirmative answer to this question, which has been forcefully asserted by the mainstream media in India. Banerjee critically interrogates the historical trajectories of capitalism and the resolution of the agrarian question therein. The paths of early capitalist development are not possible to replicate in developing countries, more so under contemporary globalization. However, far from the agrarian question being "dead" for these countries, it is all the more relevant within a complex of neoliberal exploitation of developing world peasantry. The chapter studies the patterns of accumulation under neoliberalism in Indian agriculture based on primary and secondary data. The accumulation process under agrarian crisis in India illustrates the new delineations within the peasantry and possible alliances against neoliberalism. This chapter links the agrarian question with that of land and draws insights on the different streams of growing resistance to land acquisition in recent times.

1.4.2 *Legal–Institutional Dimensions of "Regimes of Dispossession"*

Kenneth Bo Nielsen and Alf Gunvald Nilsen (Chapter 5) take the idea of "regimes of dispossession" further by looking at one of the means through which such a regime is institutionalized: passing laws either to facilitate land acquisition or to contain the political voices against the means of acquisition the state often resorts to. For Nielsen and Nilsen, lawmaking serves as a modality through which a "compromise equilibrium" can be established between dominant and subaltern groups, which facilitates the neoliberal restructuring in the longer term. They argue that the institutions, discourses, and technologies of rule that attach to the state are not just sites in which hegemonic power is exercised, but also nodes of contention where subaltern resistance can be articulated and pursued. With the concrete case of the land acquisition attempt in Singur in the state of West Bengal, and drawing on Gramsci's concept of hegemony, they show that the modes of resistance against land acquisition form a complex process that leads to "unstable equilibria" of compromise "in which the interest of the dominant social groups prevail." These processes seem to culminate in a "compromise equilibrium" that would reconcile the demands of industry with the demands of social movements for greater protection of the rights of the individual landowner within a legal framework. Nielsen and Nilsen pursue this dialectic that gave rise to LARR Act of 2013, which was passed by the Lok Sabha in August 2013 and came into effect on January 1, 2014, and which was later subject to a set of ordinance-driven amendments by the Modi government in early 2015.While the authors acknowledge the progressive measures contained in the act, they also suggest that it may nonetheless, in the long run, facilitate the process of neoliberal social and economic restructuring in India.

Malabika Pal (Chapter 6) examines the fairness claim of the LARR Act of 2013. She uses the utilitarian fairness standard proposed by one of the most influential American constitutional scholars of the twentieth and twenty-first centuries, Frank Michelman. His study of judicial decisions from an ethical perspective introducing the concept of "demoralization costs" has shaped the interpretational debate on takings law in the United States. His analysis is particularly relevant for the land question in India today since there is widespread feeling that millions of people have been unfairly deprived of their land and livelihoods. Pal looks at the role of the Indian judiciary in interpreting the land acquisition legislation since landmark judgments affect the morale of society. She concludes that using Michelman's standard would help in bringing about greater "fairness" than what the new legislation has achieved.

A series of legislative actions like the Panchayats (Extension to Scheduled Areas) Act, 1996 (PESA) and the Scheduled Tribes and Other Traditional Forest Dwellers (Recognition of Forest Rights) Act, 2006 (FRA) have been passed to institutionalize *adivasi* (tribal communities) autonomy and recognize their customary and community rights over natural resources. However, these have not protected the *adivasis* from the risks of dispossession as the state can still take away land and natural resources for private capital. This contradiction between the "giving" state and the "grabbing" state in the particular social setting of *adivasis'* customary rights over the forest lands has been dealt with by Rajesh Bhattacharya, Snehashish Bhattacharya, and Kaveri Gill (Chapter 7). They argue that a series of legislations enacted by the state granting autonomy to tribal communities over land have facilitated access to natural resources by capital. These acts are often celebrated as a victory for such struggles, leading to a circumscription of the ability of capital to access natural resources at will. In Chapter 7, Bhattacharya, Bhattacharya, and Gill, however, argue that such politico-juridical interventions may also point to the emergence of a more protean neoliberal governance structure. Given the ambiguous character of land as a *commodity*, particularly *adivasi* land that refuses to be encompassed by the logic of the market, the more protean neoliberal order may seek to instrumentally use these legislative acts to clearly define property rights over resources, which can then form the basis of negotiations with the *adivasi* communities over land for the benefit of capital.

1.4.3 *Regional Perspectives*

It is a common observation that different regions in India follow different trajectories as far as development dynamics are concerned. Even in the old agrarian transition debate in India of the 1970s, accumulation of agrarian surplus and signs of capitalist transition were never uniform across the country. Punjab and parts of western Uttar Pradesh manifested accumulation processes that were quite different from what was observed in the mainly rice-growing eastern states. It is therefore obvious that for more illuminating discussions on the land question one has to be sensitive to the diversities observed in different regions of India. Although the phenomenon of the erstwhile "peasants" being constantly transformed into part-time cultivators, landless workers, or self-employed in petty production and services is common throughout India, the processes are not identical everywhere and therefore it is worth looking at how certain factors external to the immediate surroundings of a peasant production-based economy, such as increasing penetration of the market and changing role of the state, influence this process. The role that the states of the Indian federation play is particularly important in this context since agriculture as well as land are constitutionally (local) "state subjects."

The increase in noncultivating rural households owning land may be a common phenomenon all over rural India, based on primary data collected from nine villages in Andhra Pradesh. Vijay analyzes this phenomenon (Chapter 8). These households continue to exert influence on the land and land-lease markets and shape the production structure of agriculture. These households are noncultivators as they themselves do not cultivate the land but are peasants to the extent that they are moving away from cultivating practices to nonfarm activities without completely breaking from the peasant origin by holding on to the land. On the other hand, in the prosperous coastal areas of Andhra Pradesh, the evolving rural economic structure is witnessing the emergence of a new class of landlords. They own land but do not cultivate the land themselves and derive a ground rent by leasing land out. Unlike the earlier landlords who were from "landlord families" and owned large tracts of land, these new landlords could be erstwhile cultivators who have diversified out of agriculture but continue their interest in land. As Vijay shows in Chapter 8, these households do not sell their land either because of increasing land prices or rental income earned from such land. Thus a revival of absentee landlords of a different kind and a revival of tenancy are apparent in some of these areas, implying a departure from the classic transition dynamic.

These village-level changes need to be juxtaposed with the widespread phenomenon of petty commodity production that gives rise to the paradoxical phenomenon of weak livelihood sustainability of the households in the rural areas. Based on field data, Ramana Murthy's analysis (Chapter 9) points to self-exploitation by small and marginal farmers, suggesting that the resolution to the agrarian question no longer seems necessary for capitalism. Similar to Vijay's findings, Ramana Murthy shows that there are indications that the owners of land are leaving the production space to petty commodity producers through tenancy. However, unlike the findings in the previous chapter Ramana Murthy shows that owners also sell their land to become suppliers of various agricultural and nonagricultural inputs and earn interest, rents, and quasi-rents. This chapter shows that this petty commodity producing class and the area operated by them is increasing, in spite of the fact that their agriculture is essentially unviable. Farmer suicides reflect the distress faced by this class. However, this chapter also shows that this class survives through various strategies, including considerable diversification away from their dependence on farming. As a result, land becomes no longer central to their survival, while the postcolonial state manages their desperate condition through welfare transfers. This chapter provides evidence for the changing profile of rural farm households, landholding structure, farm and non-farm incomes of rural farm households, and welfare transfers. The changing portfolio of rural incomes brings out the shifts in dependence on agriculture for the rural households and their implications for the land question.

A comparative perspective sometimes may be more illuminating if the entities to be compared are chosen based on strong similarities. The two Indian states of Kerala and West Bengal have apparent commonalities as both have tried to radically change their agrarian institutions, although there are important differences in the specific methods of intervention and their implementation. Anirban Dasgupta (Chapter 10) revisits the experience of land reforms in Kerala and West Bengal to provide a comparative analysis of the impact of left reformism on the nature of capital accumulation in these two states. The chapter builds on a conceptual framework combining a contemporary Marxist reading of the agrarian question and the theoretical justification of land reforms from a developmentalist perspective. The analysis in the chapter shows that land reforms were not able to generate a process of inclusive industrial development in either state. While the reform programs of leftist governments in these two states were not driven by a clear development agenda, fundamental changes in agrarian relations brought about by successful implementation of land reforms had potential to kick-start a process of inclusive industrial development. The analysis in this chapter demonstrates that this did not happen in either case. In Kerala, land reforms did not revitalize agricultural production primarily because of a powerful trade union movement leading to overpricing of labor and resistance to technological upgrading. With the absence of surplus generation in agriculture, the intersectoral linkages for industrial growth could not be operationalized. In contrast, West Bengal witnessed a significant increase in agricultural productivity at the initial stages of land reform implementation but a variety of institutional factors prevented the transmission of this success to the overall economy including the manufacturing sector. Dasgupta interprets the experience of the two states as a disarticulation of the accumulation problematic of the agrarian question in the sense that agriculture was not able to contribute toward capital accumulation and therefore long-term industrial development in spite of major interventions in land tenure relations.

To return to the regional perspectives on regimes of dispossession the two case studies across time and space show interesting contrasts. Mircea Raianu (Chapter 11) goes back to the colonial period, and examines patterns of land acquisition in the establishment of two enterprises that would obtain the status of proto-national industries before independence in 1947: the Tata hydroelectric power companies in the Western Ghats, and the Tata Iron and Steel Company at Jamshedpur. Adopting a comparative regional perspective, he shows how the legal instruments of dispossession varied according to the distribution of power. He argues that the entry of Indian capital in the industrial sector in the early twentieth century made possible two seemingly contradictory but mutually constitutive trends: the legal designation of private capital as capable of fulfilling a "public purpose," and the increasingly direct involvement of the state in resource capture and management for the purpose of industrial development. The chapter uncovers the origins of key aspects of the "land question" in India, including the predominance of domestic over foreign capital, the enabling role of the state, and the persistence of surplus labor, dimensions that are also present today in postcolonial India.

The ethnographic account of land acquisition in Medinipur district of West Bengal presented by Abhijit Guha (Chapter 12) shows how the administrative wing of the (local) state rationalizes acquisition of land for private interests and how the state has dealt with the feeble voices of protest. The irony is that the West Bengal government has been by and large pro-peasant and yet sided with industrial capitalists to foster economic development and manufacturing employment in the state. The chapter is written in a reflexive personal narrative form in which the author travels in time and space juxtaposing his direct fieldwork experiences and encounters around land acquisition with archival data collected from the land acquisition files in the department and the assembly proceedings. The chapter reveals a clear mismatch in the behavioral pattern between the actual cultivators of land and the policymakers and executers of land acquisition operating under market forces. The transition from a land-based rural economy towards an industrial regime was not smooth. It was characterized by protest, resistance, and bargaining by the peasants. There was also government and corporate failure, since the official claims of industrial development and employment were not met during the transition. These episodes finally led to a stalemate in the transition from agriculture to industry in particular rural areas of West Bengal.

The hill areas of Northeast India provide a substantially different context for the land question. The sixth regional case, presented by Gorky Chakraborty and Asok Kumar Ray (Chapter 13), discusses the state's role in changing property right institutions in the hills of the Northeastern states. The authors examine the appropriation of community land and subsequent dispossession of the tribes embedded in the social economy of the hills of Northeast India. They point out that the historico-epistemological hiatus between the customary law-abiding tribes vis-à-vis positive law imposed by the state is fundamental to understand the difference between the two contrasting interpretations of property rights enacted in the highlands of the region. Traditionally, shifting cultivation was the main system of agricultural production in this region, which was not for surplus generation, unlike the mode prevalent in the wet-field-dominated plains. Shifting cultivation was discouraged by both the colonial and postcolonial state. Thus the policy shift toward settled cultivation led to large-scale dispossession among the tribal groups by the settlers from neighboring regions. The chapter highlights the implications of trivialization of community land and dispossession of the tribal masses in Manipur, Meghalaya, and Tripura and suggests that the constitutional protections have fallen short in protecting the community resources of the tribes in the region. To make their point, the authors argue that the emerging issue of hydropower in Arunachal Pradesh and the associated threats of dispossession of the communities (through wet rice and commercial crop cultivation) under the hegemony of state-business collusion are the manifestation of the continuing process of appropriation of community resources in the hill areas of Northeast India. This is another instance of dispossession where the state played an indirect role and was not directly involved in acquiring land for private interests.

Notes

1. In earlier debates, transition also referred to the possibility of socialist development. The term "transition" today is deployed to refer to former Soviet Union, East European countries, and China, which are presumably in various stages of moving toward capitalist market economies.
2. Exogenous here does not mean international influences as was the case in Latin America's agrarian outcomes, where both history and exogenous international factors have contributed to a "disarticulated accumulation" model (de Janvry 1990: 32–45). Rather it is the Indian nonagricultural sector that is influencing rural outcomes.
3. The *Situation Assessment Survey of Agricultural Households*.
4. For this survey, NSSO defined "agricultural household" as a household receiving a value of produce more than Rs. 3,000 from agricultural activities (e.g., cultivation of field crops, horticultural crops, fodder crops, plantation, animal husbandry, poultry, fishery, piggery, bee-keeping, vermiculture, sericulture, etc.) and having at least one member self-employed in agriculture either in the principal status or in subsidiary status during the 365 days preceding the survey.
5. This could be due to the fact that almost all manufacturing requires dedicated services (see Cohen and Zysman 1988).
6. Government of India, Ministry of Finance, 2015. Economic Survey, 2014–15, available at: <http://indiabudget.nic.in/survey.asp> (accessed May 19, 2015).
7. Coparcener refers to the inheritance custom of property being divided equally among (generally) male siblings of a household, thereby generating smaller plots of land over time for those inheriting land.
8. In the former cases, land grabbing is driven by food insecurity, access to raw materials, or presumably fostering more efficient (large-scale) agriculture.
9. One estimate puts the number of people displaced by the state under the aegis of eminent domain (a hangover of the colonial government's Land Acquisition Act of 1894) to at least 50 million (Sampat 2010: 170–1).

References

Adnan, S. 2013. "Land Grabs and Primitive Accumulation in Deltaic Bangladesh: Interactions Between Neoliberal Globalization, State Interventions, Power Relations and Peasant Resistance," *Journal of Peasant Studies*, 40(1): 87–128.

Akram-Lodhi, A. H. and Kay, C. 2010. "The Agrarian Question: Peasants and Rural Change." In *Peasants and Globalization: Political Economy, Rural Transformation and the Agrarian Question*, edited by A. H. Akram-Lodhi and C. Kay. Abingdon: Routledge, pp. 3–34.

Allen, R. C. 1999. "Tracking the Agricultural Revolution in England," *Economic History Review*, 52(2): 209–35.

Bagchi, A. K. 2010. "Nineteenth Century Imperialism and Structural Transformation in Colonized Countries." In *Peasants and Globalization: Political Economy, Rural Transformation and the Agrarian Question*, edited by A. H. Akram-Lodhi and C. Kay. Abingdon: Routledge, pp. 83–110.

Banerjee, A. V. and Iyer, L. 2005. "History, Institutions, and Economic Performance: The Legacy of Colonial Land Tenure Systems in India," *The American Economic Review*, 95(4)(September): 1190–213.

Banerjee-Guha, S. 2013. "Accumulation and Dispossession: Contradictions of Growth and Development in Contemporary India," *South Asia*, 36(2): 165–79.

Bardhan, P. 2006. *The Political Economy of Development in India*. Delhi: Oxford University Press.

Bates, R. H. 1989. *Beyond the Miracle of the Market: The Political Economy of Agrarian Development in Kenya*. Cambridge: Cambridge University Press.

Bernstein, H. 1996. "Agrarian Questions Then and Now," *Journal of Peasant Studies*, 24(1–2): 22–59.

Bernstein, H. 2010. *Class Dynamics of Agrarian Change*. Halifax: Fernwood Publishing.

Bharadwaj, K. 1985. "A View on Commercialisation in Indian Agriculture and the Development of Capitalism," *Journal of Peasant Studies*, 12(4): 7–25.

Borras, S. M., Franco, J. C., Gómez, S., Kay, C., and Spoor, M. 2012. "Land Grabbing in Latin America and the Caribbean," *Journal of Peasant Studies*, 39(3–4): 845–72.

Brass, T. 2010. "Unfree Labour as *Primitive* Accumulation?" *Capital and Class*, 35(1): 23–38.

Brass, T. 2011. "Primitive Accumulation, Capitalist Development and Socialist Transition: Still Waiting for Godot," *Journal of Contemporary Asia*, 41(1): 1–24.

Breman, J. 2010. "The Political Economy of Agrarian Change in India." In *Routledge Handbook of South Asian Politics: India, Pakistan, Bangladesh, Sri Lanka, and Nepal*, edited by P. R. Brass. Abingdon: Routledge, pp. 321–36.

Byres, T. J. 1993. "State, Class and Development Planning in India." In *The State and Development Planning in India*, edited by T. J. Byres. Delhi: Oxford University Press, pp. 5–50.

Byres, T. J. 2002. "Paths of Capitalist Agrarian Transition in the Past and in the Contemporary World." In *Agrarian Studies: Essays on Agrarian Relations in Less-Developed Countries*, edited by V. K. Ramachandran and M. Swaminathan. New Delhi: Tulika Books, pp. 54–83.

Chakravorty, S. 2013. *The Price of Land: Acquisition, Conflict, Consequence*. New Delhi: Oxford University Press.

Chandrasekhar, C. P. 2007. "The Progress of 'Reform' and the Retrogression of Agriculture," available at: <https://www.academia.edu/3338396/The_Progress_of_Reform_and_the_Retrogression_of_Agriculture> (accessed March 18, 2015).

Chatterjee, P. 2008. "Democracy and Economic Transformation in India," *Economic and Political Weekly*, 43(16): 53–62.

Cohen, S. S. and Zysman, J. 1988. *Manufacturing Matters: The Myth of the Post-Industrial Economy*. New York: Basic Books.

Cotula, L. 2012. "The International Political Economy of the Global Land Rush: A Critical Appraisal of Trends, Scale, Geography and Drivers," *Journal of Peasant Studies*, 39(3–4): 649–80.

Cross, J. 2014. *Dream Zones: Anticipating Capitalism and Development in India*. London: Pluto Press.

D'Costa, A. P. 2014. "Compressed Capitalism and Development: Primitive Accumulation, Petty Commodity Production, and Capitalist Maturity in India and China," *Critical Asian Studies*, 6(2): 317–44.

D'Costa, A. P. 2016. "Jobless Growth Even in Labor-Abundant India: Capitalist Transition and the Challenges of Employment." Paper presented for the Panel on "Global Capitalism, National Experiences, and Changing Labor Markets in Asia," at the *Association of Asian Studies Annual Conference*, Washington State Convention Center, Seattle, March 31–April 3, 2016.

de Janvry, A. 1990. *The Agrarian Question and Reformism in Latin America*. Baltimore, MD: Johns Hopkins University Press.

Grindle, M. S. 2000. *Audacious Reforms: Institutional Innovation and Democracy in Latin America*. Baltimore, MD: Johns Hopkins University Press.

Grindle, M. S. 2001. "In Quest of the Political: The Political Economy of Development Policymaking." In *Frontiers of Development Economics*, edited by Gerald M. Meier and Joseph E. Stiglitz. Washington, DC, and New York: The World Bank and Oxford University Press, pp. 345–80.

Harriss, J. 2013. "Does 'Landlordism' Still Matter? Reflections on Agrarian Change in India," *Journal of Agrarian Change*, 13(3): 351–64.

Jaffrelot, C. 2015. "Business-Friendly Gujarat under Narendra Modi: Towards a New Political Economy?" Paper presented at the South Asia Studies Center, University of Washington, Seattle, November 6, 2015.

Jenkins, R. 2013. "Land, Rights and Reform in India," *Pacific Affairs*, 86(3): 591–612.

Larrain, J. 1989. *Theories of Development: Capitalism, Colonialism and Dependence*. Cambridge: Polity Press.

Lerche, J. 2013. "The Agrarian Question in Neoliberal India: Agrarian Transition Bypassed?" *Journal of Agrarian Change*, 13(3): 382–404.

Levien, M. 2012. "Special Economic Zones and Accumulation by Dispossession in India," *Journal of Agrarian Change*, 11(4): 454–83.

Lewis, W. A. 1954. "Economic Development with Unlimited Supplies of Labour," *Manchester School of Economic and Social Studies*, 22(May): 139–91.

Li, T. M. 2011. "Centering Labor in the Land Grab Debate," *Journal of Peasant Studies*, 38 (2): 281–98.

Marx, K. 1958–1971. "Brief an die redaktion der 'Otetschestwennyje Sapiski'." In Werke, 39 Bde. und 2 Erg. Bde. Edited by K. Marx and F. Engels, Berlin: Dietz.

Mezzadra, S. 2011. "The Topicality of Prehistory: A New Reading of Marx's Analysis of So-Called Primitive Accumulation," *Rethinking Marxism*, 23(3): 302–21.

National Commission for Enterprises in the Unorganised Sector (NCEUS) 2009. *The Challenge of Employment in India: An Informal Economy Perspective*, Volume I, Main Report, New Delhi.

Nielsen, K. B. 2010. "Contesting India's Development? Industrialisation, Land Acquisition, and Protest in West Bengal," *Forum for Development Studies*, 37(2): 145–70.

Patnaik, U. 2007. *The Republic of Hunger and Other Essays*. Gurgaon: Three Essays Collective.

Piketty, T. 2014. *Capital in the Twenty-first Century*. Cambridge, MA: Harvard University Press.

Polanyi, K. 2001. *The Great Transformation: The Political and Economic Origins of Our Time*. 2nd edn, Boston, MA: Beacon Press.

Razavi, S. 2009. "Engendering the Political Economy of Agrarian Change," *Journal of Peasant Studies*, 36(1): 197–226.

Saith, A. 1985. "Primitive Accumulation, Agrarian Reform and Socialist Transitions: An Argument," *Journal of Development Studies*, 22(1): 1–48.

Sampat, P. 2010. "Special Economic Zones in India: Reconfiguring Displacement in a Neoliberal Order?" *City and Society*, 22(2): 166–82.

Sanyal, K. 2007. *Rethinking Capitalist Development: Primitive Accumulation, Governmentality and Post-Colonial Capitalism*. New Delhi: Routledge.

Sud, N. 2014. "Governing India's Land," *World Development*, 60(August): 43–56.

Walker, K. L. M. 2008. "Neoliberalism on the Ground in Rural India: Predatory Growth, Agrarian Crisis, Internal Colonization, and the Intensification of Class Struggle," *Journal of Peasant Studies*, 35(4): 557–620.

Wallerstein, I. 1979. *The Capitalist World Economy*. Cambridge: Cambridge University Press.

Whitehead, A. and Tsikata, D. 2003. "Policy Discourses on Women's Land Rights in Sub-Saharan Africa: The Implications of the Return to the Customary," *Journal of Agrarian Change*, 3(1–2): 67–112.

Wily, L. A. 2012. "Looking Back to See Forward: The Legal Niceties of Land Theft in Land Rushes," *Journal of Peasant Studies*, 39(3–4): 751–75.

World Bank. 2007. *India: Land Policies for Growth and Poverty Reduction*. New Delhi: Oxford University Press.

Part I

Primitive and Contemporary Accumulation

2

From Primitive Accumulation to Regimes of Dispossession

Theses on India's Land Question

Michael Levien

2.1 Introduction

Since the mid-2000s, the dispossession of land has become unprecedentedly central to India's political economy. While tens of millions of people were dispossessed of their land in the period of state-led development (Fernandes 2008), the scale of land dispossession has accelerated since liberalization. Its causes, moreover, have profoundly shifted as privatized forms of industry, infrastructure, real estate, and extraction have eclipsed public sector projects as causes for forcibly taking land from farmers. What has also changed, particularly in the last decade, is the breadth and strength of anti-dispossession politics. While dozens of movements, most notably the Narmada Bachao Andolan (Save the Narmada Movement), fought tenaciously against dams and other public sector development projects from the 1980s onwards, since roughly 2005 so-called "land wars" have become more widespread and *successful* in stopping and stalling projects than ever before. This opposition has been occasionally aided (although typically not started) by political parties as land dispossession has become electorally salient for the first time in India's history. This opposition has, in the aggregate, reached a point where it now poses a significant obstacle to accumulation: by one estimate, in 2011, capital investments worth 3 percent of India's gross domestic product (GDP) were stalled due to land acquisition problems (CMIE 2012). Stalled investments have been singled out—rightly or wrongly—as a cause of India's recent economic slowdown and have become a key object of concern for both state and capital. Overcoming these "roadblocks" was the goal of a historic revision of India's Land Acquisition Act (LAA) in 2013 and a top priority of the Bharatiya Janata Party (BJP) government that assumed power in 2014. The Indian state remains caught, however, between the land requirements of capitalist growth and the political compulsions of electoral democracy; whether, how, and for whom this contradiction gets resolved constitutes India's "new" land question.

Like the other contributions to this volume, this chapter seeks to understand this conjuncture. It suggests that doing so requires theoretical reconstruction and comparative historical perspective. It argues that existing theoretical perspectives, whether Marx's primitive accumulation or Harvey's accumulation by dispossession, are not adequate for understanding the relationship between land dispossession and capitalism—in India or elsewhere. They are inadequate because they do not help us understand how dispossession changes over time and space, and they do not illuminate the politics of dispossession. While building on this theoretical tradition, I advance the concept of "regimes of dispossession" as an alternative way of understanding the relationship between land dispossession and capitalism. Dispatching with the obfuscating assumption that dispossession is a "necessary" cost of development or stage in the development of capitalism, I argue that dispossession is fundamentally a social relation of coercive redistribution. While this relationship exists in probably all social formations, it is driven by different forms of accumulation and class interests under different historical phases of capitalism. We can thus think of dispossession as being organized into socially and historically specific regimes. Under different regimes of dispossession, states seek to redistribute resources to different classes (or class fractions) for different economic purposes. Given these purposes and interests, regimes of dispossession have different combinations of means—force, legitimacy, and material concessions—available for making people comply with their dispossession. The concept of "regimes of dispossession" is useful, I argue, for two reasons. First, it helps us to compare and thus critically interrogate the specific economic purposes that states, at any given time, seek to legitimize as "development." Second, it helps us to understand the conditions under which dispossession is most likely to encounter noncompliance and be effectively stopped. Regimes of dispossession thus provide the basis for a political economy of dispossession without the assumptions of economic progress or political inevitability.

Applying this concept to land dispossession in post-independence India, I argue that beginning in the early 1990s, India shifted from a regime that dispossessed land for state-led projects of industrial and agricultural transformation (the production of commodities) to one that dispossesses land for private and decreasingly productive investments (the commodification of land itself). This radical transformation of Indian states into land brokers for private capital is entirely missed by economists who insist that today's land grabs are a necessary cost of "development" (Banerjee et al. 2007; Bardhan 2011; Chakravorty 2013) as well as those theorists who see them as part of India's passage through the (presumably inevitable) stage of "primitive accumulation" (Chatterjee 2008). Both formulations naturalize dispossession by turning contemporary forms of "class robbery" (Thompson 1963: 218) into historical necessity.[1] The land grabs of the neoliberal period do not represent simply more "development-induced displacement" or the inexorable march of "primitive accumulation," but rather the emergence of a new regime of redistributing landed wealth upwards. This new regime of dispossessing land for increasingly financialized private purposes is arguably less "developmental" than its Nehruvian predecessor. It is certainly far more politically tenuous. Its continuance depends on the Indian state's questionable ability to substitute land prices for nationalist legitimacy.

2.2 Six Theses on the Indian Land Question

In the following six theses, I develop the concept of regimes of dispossession as a way of understanding the relationship between land grabs and specific historical phases of capitalism, and demonstrate its ability to illuminate India's contemporary land question.

Thesis 1: Contemporary land grabs are not "primitive accumulation."

While the concept is much beloved and applied generously to land grabs and many other social phenomena that smell of theft and fraud, it is time to recognize that so-called "primitive accumulation" obscures more than it illuminates about contemporary forms of land dispossession. In the final chapters of Volume 1 of *Capital*, Marx advanced the concept to explain how the origins of capitalist social relations lay in "the historical process of divorcing the producer from the means of production" (1977: 875). Basing his analysis partly on the English enclosures, Marx argued that the bloody and violent process of expelling peasants from the land generated the *preconditions* for capitalism by effecting the "two transformations, whereby the social means of subsistence and production are turned into capital, and the immediate producers are turned into wage-laborers" (Marx 1977: 875). In a process stretching over several centuries and culminating in the Parliamentary Enclosures Acts of the eighteenth century, the English common lands (along with state and church lands) were turned into sheep walks and capitalist farms while commoners were turned into a proletariat. Mocking Adam Smith's "idyllic" fairy tale of capitalism's origins, Marx observes that capital came into the world "dripping from head to toe, from every pore, with blood and dirt" (1977: 926).

It is important to recognize that there is ambiguity in Marx's own conception of primitive accumulation. Fundamentally, this section of *Capital* offers an origin story about how the preconditions for capitalism came into being.[2] In this sense, primitive accumulation is defined by its *function* in the development of capitalism: it is those historical processes that establish capitalist social relations. On the other hand, Marx's theory of primitive accumulation also appears to make a *means*-specific distinction between those processes that gave rise to a capitalist mode of production in which "conquest, enslavement, robbery, murder, in short, force, played the greatest part" (1977: 874), and a mature capitalist system that, once developed, can dispense with extra-economic coercion and rely on the "silent compulsion of economic relations" (1977: 899). In the first possible conception, it is the *function* that distinguishes primitive accumulation from capitalist accumulation proper, and in the second it is its extra-economic *means*. This leaves two possible ways of understanding primitive accumulation as a general concept: 1) it is any process that creates the preconditions for capitalism; or 2) it is any process that creates capitalism's preconditions through the use of force. There is also a third possibility: primitive accumulation is any process of accumulation that involves force. Only in this last option would primitive accumulation be ongoing under capitalism. Marx, however, confined primitive accumulation to an early chapter in the development of capitalism, calling it, "an accumulation which is not the result of the capitalist mode of production but its point of departure" (1977: 873).[3] To the extent that it continues, it does so *serially* as different countries *transition* to capitalism.[4]

Within the functional conception, there is another ambiguity. How much emphasis should one place on each of the "two transformations": turning the land into capital, and turning the peasantry into a proletariat? In his analysis of the enclosures, Marx devotes little attention to the economic forces motivating the gentry to enclose the commons for sheep walks (the rising price of wool generated by the booming textile industry of Flanders) compared to its main consequence: creating a class of wage laborers "freed" from their means of production. This is because Marx was more interested in providing a historical account of the emergence of a proletariat rather than a political economy of land dispossession. But by focusing on dispossession's consequences rather than its causes, Marx's concept of primitive accumulation is based on a kind of post hoc functionalism that prevents it from becoming a theory of ongoing land dispossession under capitalism. Merely calling primitive accumulation "ongoing" does not resolve these conceptual problems.

Subsequent applications of primitive accumulation, particularly in the literature on agrarian transition, have compounded the initial ambiguities in Marx's conception. First, analysts of the transition to capitalism in England have argued that it was not necessarily the enclosures but rather reduction to "market dependence" that signaled the rise of capitalist social relations in the countryside (Brenner 2001; Wood 2002), raising the question of whether coercive land expropriation is necessarily a part of primitive accumulation. Decades of research on the "agrarian question" outside of England demonstrated that the widespread enclosure of land constitutes only one historically specific pathway to capitalist development (Byres 1991). Primitive accumulation, consequently, came to mean one of two things. First, it could refer to any process that proletarianized peasants and forged a class of agrarian capitalists. This often proceeded not through wholesale eviction from land but through a gradual process of class differentiation (Adnan 1985: 57; Kautsky 1988: 17; Lenin 1967). Economic processes like debt came to be seen as equally effective levers of primitive accumulation as the extra-economic enclosure of land (Bhaduri 1983). Second, primitive accumulation could refer to the general process of diverting agricultural surpluses for capitalist industrialization (Byres 1991: 11). In this usage, primitive accumulation did not necessarily mean establishing capitalist social relations, but transferring surpluses from agriculture to industry.[5] Such primitive accumulation was seen as a necessary condition for economic growth in all "backward" countries, including socialist ones (Byres 1991: 11; Ka and Selden 1986; Preobrazhensky 1965).[6]

The vast research on the origins of capitalism and agrarian transitions in Europe and the Global South has thus amplified the original ambiguities in Marx's concept of primitive accumulation. It is no longer clear if primitive accumulation refers to the dispossession of land through extra-economic means, to any process that generates the preconditions of capitalism, or even to any process that provides surpluses for industrialization. My purpose here is not to question the utility of primitive accumulation for understanding transitions to capitalism,[7] but to simply point out that it is not adequate for explaining the persistence of land dispossession and the various forms it assumes within capitalism. Agrarian transitions and land grabs for capitalist development are distinct phenomena in need of different concepts.

Today's dispossessions cannot be understood in the old functional sense of creating the preconditions for capitalism: dispossessing farmers for special economic zones (SEZs) or factories or mines does not necessarily inaugurate capitalist social relations (which often already exist) and it does not mark transitions between modes of production. Dispossessions today often have little to do with agriculture and are not about resolving the "agrarian question." They involve dispossessing land—sometimes already held within capitalist social relations—for *nonagricultural* development and are thus entirely orthogonal to the slow and uneven development of capitalist agriculture. They are, moreover, geared toward capitalizing land with little regard for what happens to the dispossessed farmer whose labor power is largely superfluous (both to specific projects and to accumulation as a whole). They reflect not an early stage of capitalism but advanced capitalist demands on land and natural resources. They take a myriad of sector-specific forms—whether dams, highways, mines, steel plants, SEZs, or housing—that do not resemble the "classic" pattern of the enclosures. They must be understood within a new problematic: not what is the role of land dispossession in transitions to capitalism, but what is the role of land dispossession within capitalism? This requires, I suggest, shifting attention from the transition *between* modes of production to variation in regimes of dispossession *within* the capitalist mode of production. Before developing this framework, however, we must turn to the strengths and limitations of Harvey's more recent formulation of "accumulation by dispossession."

> *Thesis 2*: While contemporary land grabs may be considered accumulation by dispossession, we should understand this as a political process of state redistribution and not a functional response to over-accumulation.

In *The New Imperialism* (2003), Harvey argues that "accumulation by dispossession" has become the predominant mode of accumulation under neoliberalism and that the traditional Marxist focus on labor must be complemented with greater attention to proliferating struggles over the dispossession of various forms of public and private wealth. Harvey's examples of accumulation by dispossession include the expropriation of land and natural resources from peasant populations; the conversion of common or state property into private property; the extraction of rents from intellectual property rights; the privatization of collective social assets (such as pensions, health care, and other social entitlements); and the various predations of finance capital. Harvey sees the resurging importance of these mechanisms relative to "expanded reproduction" as a response to the problem of over-accumulated capital in the core economies of the Global North.

To its merit, Harvey's accumulation by dispossession explicitly recognizes that dispossession is driven by advanced capitalism. Accumulation by dispossession unmoors the notion of primitive accumulation from the interstices of modes of production, making it applicable to very diverse forms of dispossession that have little do with agriculture and that do not necessarily inaugurate capitalist social relations: for example, dispossessions for factories, dams, SEZs, slum redevelopment, mining, privatized infrastructure, and real estate. Accumulation by dispossession marks a significant advancement on primitive accumulation precisely in its ability to capture diverse contemporary dispossessions that take sector-specific and geographically dispersed forms, and whose significance for capital lies more in the expropriated asset (e.g., the land) than in the labor power of the dispossessed owner.

But what exactly defines accumulation by dispossession? While freeing primitive accumulation from its function of generating capitalist social relations,

Harvey redefines the concept through another, more contemporary function: namely, absorbing over-accumulated capital in the global economy. We might expect Harvey to further specify accumulation by dispossession as those processes that provide outlets for capital through *extra-economic* means—as in Marx's characterization of primitive accumulation as involving, "conquest, enslavement, robbery, murder, in short, force" (1977: 874). However, Harvey does not provide a clear definition of accumulation by dispossession, and explicitly claims that it is "primarily economic rather than extra-economic" (Harvey 2006a: 159). Without this means-specific distinction, it is not clear what these different processes share, nor what separates accumulation by dispossession from other "fixes" to the problem of over-accumulation or from the "normal" expanded reproduction of capital (see Brenner 2006; Levien 2011, 2012). The concept's boundary is vitiated, undermining its analytical utility.

The consequence is that Harvey underemphasizes the most significant aspect of land dispossession: namely, that it is a deeply political process in which owners of the means of coercion transparently redistribute assets from one group to another.[8] Although he recognizes that state force can be central to dispossession (Harvey 2003: 154), he overextends the concept to include so many diverse phenomena—including those such as finance capital that have an indirect or unclear link to state force—that he cannot explicitly incorporate the state into his theory. As a result, Harvey's framework misses the crucial and distinguishing feature of accumulation by dispossession as it applies to land: its fundamental and transparent reliance on state force. This prevents him from grasping the distinctive *politics of dispossession*.

As a process of transparent and coercive redistribution, land dispossession is a political process whose outcome is determined by class struggle and not merely the circuits of capital (Brenner 1976; De Angelis 2007). There is nothing automatic about capital (over-accumulated or not) finding outlets in land or in any other asset; by reading every instance of dispossession as a result of the global impulses of capital, Harvey fails to answer the question of why impulses toward accumulation translate into *dispossession* in a particular context? Why do capitalists need states to dispossess land for them rather than simply purchasing it on the market? And why do states do it for them? Dispossession requires a state that is willing and able to use its monopoly of the means of violence to expropriate land from certain classes for the benefit of others. It often arises, as in the case of land dispossession in India, from the desire of states to help capitalists overcome barriers to accumulation, such as rural land markets dominated by smallholding peasantries. But, in different times and places, we find that states have been willing and able to dispossess land for *different* classes (or class fractions) for *different* economic purposes. In India, the colonial state dispossessed land for railroads and natural resource extraction to benefit metropolitan capital; the Nehruvian state dispossessed land for public sector industry and infrastructure; and the neoliberal state dispossesses land for all forms of private accumulation including real estate. Looking across countries, dispossession in India and China today is predominantly driven by urban–industrial purposes,[9] while in Africa and Latin America it is largely for agricultural plantations. Harvey's concept offers little basis for understanding this variation. While it may be true, as he argues, that the quantity of dispossession is increasing with the institution of neoliberal economic policies, it is unclear and probably impossible to show that over-accumulated capital in the Global North is the cause (much of the capital for India's SEZs, for example, has come from India itself).[10] More fundamentally, what we must understand—and what Harvey overlooks—is how and why states restructure themselves to dispossess land for different purposes and different classes in different periods.

One must also explain how states *succeed* in dispossessing people of their land for any given set of purposes and what accounts for variation in this success. The transparent use of state coercion to dispossess land creates an immediate antagonism between the would-be dispossessed and the state. Unable to rely on the mystification common to the exploitation of labor under capitalism, states must justify land dispossession with ideological claims to be serving "a public purpose" or the "national interest"; in the past century, they have typically done this through the language of "development." The persuasiveness of these claims rests considerably on whether the economic purpose driving dispossession can be aligned with a widely accepted concept of national development. Where ideological justifications and/or material compensations prove inadequate and resistance emerges, the ability of the state to dispossess is decided by the balance of political forces.

With accumulation by dispossession, then, the state, politics, and ideology are—as Perry Anderson argues for *precapitalist* modes of production (1974: 403–4)—*internal* and *constitutive* features of accumulation itself. This is the very significant implication of the fact that extra-economic coercion, which is thought to subside with the transition to capitalism, is an intrinsic and ongoing aspect of accumulation under advanced capitalism. It means that the character and outcome of dispossession in different times and places will be shaped by heterogeneous and nationally specific political, economic, and ideological factors that cannot be read off of global imperatives of capital. Their specific configuration in any given place and time can be understood as a regime of dispossession.

> *Thesis 3*: Dispossession is organized into socially and historically specific regimes.

When owners of the means of coercion consistently dispossess land for particular classes for a fairly coherent range of purposes, we can call this a regime of dispossession. If a regime of production is an institutionalized way of extracting surplus value from workers (Burawoy 1985), a regime of dispossession is an institutionalized way of expropriating land from current owners or users. It has two essential components: a state willing to dispossess for a particular set of economic purposes that are tied to particular class interests, and a way of generating compliance to this dispossession. These two aspects are intrinsically linked.

At any given time, states have a set of economic purposes for which they are willing and able to dispossess land. With respect to private property, this is partly a legal matter of what is considered a "public purpose" under relevant eminent domain statutes and domestic case law (common and state land are typically dispossessed with little legal impediment). However, it remains primarily a political question since, as we have seen in India, states amend (and sometimes flout) these laws to suit their purposes at any given time, and, further, because states might not choose to dispossess land for all the purposes allowed to them by law.

Some might consider this a *non*-question under the assumption that states are always instruments of the dominant class and are therefore always willing to dispossess whatever that class desires. However, assuming that state willingness to dispossess flows automatically from the "needs" of capitalism elides important variation and also renders dispossession inevitable. First, many states have for long periods in their history limited the forcible dispossession of land from their own citizens for "public purposes," construed narrowly as projects of the state (Reynolds 2010). While some classes inevitably profit more than others from state projects, one must distinguish between dispossessing land for public infrastructure and dispossessing land directly for private capital. The latter characterizes distinct historical regimes (such as India's present one), whose emergence must be explained. Second, dominant classes and class fractions change during different phases of capitalism; as such, the kinds of accumulation they propose to undertake with dispossessed land change along with them. This variation shapes the developmental consequences of dispossession as well as its politics.

Every regime of dispossession must have a way of compelling people to relinquish their land. This is a different power relation than that involved in the exploitation of labor: whereas exploitation requires that workers continuously work, dispossession requires a one-time expulsion of people from their land. *Whereas exploitation requires enduring consent at the point of production, dispossession only requires one-time compliance at the point of enclosure.*

Dispossessing people of their land is ultimately a question of political authority, or the ability of states to make people comply with their orders (Weber 1978: 212–15). Because dispossession cannot be mystified, there are three basic means available for doing this: coercion, material compensation, and normative persuasion (e.g., legitimacy). All regimes of dispossession rely to some degree on the actual or threatened use of force. When a state notifies farmers that it seeks to acquire their land, the potential use of violence backs this intent; regardless of any compensation that might be offered, unless farmers have the right to refuse, land acquisition is fundamentally coercive. If farmers do ultimately refuse to vacate their land, the threat of coercion becomes actual violence. Given its political costs, however, states typically hope to economize on violence and rely on other means. Regimes of dispossession vary in the extent to which they must rely completely on coercion or can additionally draw on normative and material inducements to separate people from their land.

States almost always find it necessary to justify their use of coercive power to redistribute property. This is not simply the case in liberal democracies; even if they can use relatively more coercion, authoritarian governments still typically feel compelled to justify dispossession as serving the public or national interest (as we see in China today). A major determinant of the success and stability of a regime of dispossession is the extent to which its appeals to the public interest are convincing to the dispossessed and to would-be supporters. This hinges crucially on the economic purposes justifying dispossession and their anticipated beneficiaries.

The other basis for compliance is appealing to the material interests of the dispossessed. While there were scant incentives for English peasants to relinquish the commons—they were driven off without compensation and transformed into a destitute proletariat—most contemporary societies require that those forcibly dispossessed of their private property be given some form of compensation. States sometimes base this compensation on an assessment of a property's "fair market value" (though determining this in the absence of a voluntary transaction poses a problem), and, in recent decades, they have sometimes augmented this compensation with explicit policies for the "resettlement and rehabilitation" of the dispossessed. While historically such concessions have been meager, states almost always make some sort of material promises to the dispossessed, trying to convince them that they will be given some stake in the "development" that their dispossession makes possible. Both the types of promises made—irrigated land, jobs, real estate values, and so on—and the ability of states to deliver on these promises depend crucially on the particular kind of accumulation that is driving dispossession.

Different regimes of dispossession are thus able to offer different kinds and degrees of incentives to get farmers to comply with their dispossession. This makes some regimes of dispossession politically formidable and others quite tenuous. If seeing dispossession as a necessary aspect of capitalism can only disempower political struggles against it, regimes of dispossession helps to identify why "accumulation by dispossession" can sometimes be halted by the noncompliance of rural people.

In the remainder of this chapter, I argue that the framework of regimes of dispossession helps to illuminate the changing political economy of dispossession in postliberalization India.

> *Thesis 4*: Since 1991, India has passed from a regime that dispossessed land for state-led industrial and infrastructural expansion to one that dispossesses land for private—and increasingly financial—capital.

Between 1947 and 1991, the Indian state largely dispossessed land for public sector projects to expand the industrial and agricultural productivity of the country. The main forms of this dispossession were public sector dams, steel towns, industrial areas, and mining. It is important to note that while India's Land Acquisition Act (LAA) allowed for the acquisition of land for private companies, in practice this was relatively rare before economic liberalization. While companies could receive plots of land in industrial areas/estates, the state maintained ownership and collected ground rent from companies. States did not acquire land for private companies to re-sell on the market; rather, the rent from dispossessed land accrued to the state not private capitalists, who were allotted leased land on the condition of engaging in industrial production.

This is not to say that dispossession in this period actually served the "public interest" or "development." Such terms are always obfuscations of particular class interests. Dispossessing land for these projects made possible accumulation that disproportionately benefited the industrial bourgeoisie, dominant agrarian classes, and public sector elites (Bardhan 1984; Dwivedi 2006; Nilsen 2010). Although this accumulation had some beneficial effects for other classes, these projects were distributionally regressive as they came at the cost of the impoverishment and proletarianization of approximately tens of millions of rural people who were typically already among the country's poorest and most marginalized (Fernandes 2008). Many were also ecologically destructive and failed on their own terms.

Nevertheless, these projects resonated with the Nehruvian vision of national development that had widespread public legitimacy in the postindependence years. This is not to say that the dispossessed found their expropriation and impoverishment legitimate, but that the broader legitimacy of postcolonial developmentalism made resisting these projects difficult. This legitimacy was inextricably linked to the fact that large numbers of people believed they would benefit from the irrigation of large dams or receive jobs in public sector industries. When farmers resisted, they were quickly suppressed and this suppression was itself legitimized as in the national interest. Land acquisition was not politically salient and farmers only occasionally found allies to take up their cause, usually to demand more compensation (cf. Brass 2011; Struempell 2014). It was only in the 1970s and 1980s that strong opposition movements began to challenge this model of development and the "displacement" it required. Movements such as the Narmada Bachao Andolan challenged the legitimacy of "destructive development" and pioneered anti-dispossession politics in India. Tragically, they were rarely successful in stopping their own dispossession, underscoring the enduring strength of the developmentalist regime of dispossession. A combination of coercion and powerful ideological appeals to the nation ensured that the development model would itself give way before these political movements could significantly impede its ability to dispossess land.

In the early 1990s, however, economic liberalization initiated the genesis of a new regime of dispossession driven by increasingly privatized, decreasingly productive, and less labor-absorbing purposes. While there are many ways to periodize India's liberalization (cf. Kohli 2012), the early 1990s was the key turning point for land acquisition.

First, economic liberalization after 1991 unleashed increasing private demand for land for industry, infrastructure, and real estate. With the decision to liberalize private investment and to diminish the public sector's role in the economy (Chandrasekhar and Ghosh 2002), the share of private capital in the economy steadily increased and became dominant by the end of the decade (Kohli 2012: 45). While liberalization did not substantially spur the manufacturing sector, it unleashed dramatic growth in India's service sector, particularly information technology (IT) and business process outsourcing (BPO) (Bardhan 2010; Kohli 2012). This IT/BPO boom contributed to a dramatic growth in demand for commercial real estate that could not be accommodated within the confines of older cities (Searle 2010), leading to the growth of Hi-Tech parks on the peri-urban frontier. This growth of the IT/BPO sector dovetailed with a second generation of reforms that liberalized the real estate and infrastructure sectors (the privatization of which is first considered in the Eighth Five-Year Plan formulated in 1992). The central government introduced a series of policies to attract private investment in power (1992), roads (1997), and ports (1997) as public–private partnerships (PPP) became the preferred method for building infrastructure. By the 2000s, 37 percent of infrastructure investment was coming from the private sector (Gulati 2011: 380). Crucially, compensating private infrastructure investors with excess land and/or development rights became an increasingly popular method of cost recovery in these arrangements—whether for roads, airports, or affordable housing (Ahluwalia 1998; IDFC 2008, 2009). Infrastructure investment thus became a vehicle for private real estate accumulation, culminating with special economic zones (SEZs) in the mid-2000s.

The privatization of infrastructure has thus been synergistic with the liberalization of India's real estate sector. While urban development was a state-dominated affair before liberalization, in the 1990s, the government loosened restrictions on bank lending to private developers. By 2002, private developers were allowed to obtain financing (for construction, though not land purchases) from foreign investors. This policy was further liberalized after 2005 (Searle 2010: 30). This credit expansion, combined with growing demand for housing and office space, precipitated a dramatic real estate boom by the mid-2000s.

In addition to real estate, economic liberalization unleashed a mineral rush across central and eastern India. Beginning in 1993, India began to liberalize its largely nationalized mining industry. The New National Mineral Policy of 1993 and subsequent revisions over the next decade facilitated private and foreign investment in prospecting, extracting, and processing of most minerals. After changes in 1997 and 2000, foreign private equity investments of up to 100 percent were granted automatic approval. By the 2000s, mineral prices were increasing with the Indian and Chinese construction booms driving up demand. While the mining sector has not grown exceptionally fast, the number of memorandums of understanding (MoUs) signed with private mining companies has accelerated and several states (Jharkhand, Chattisgarh, and Orissa) have been particularly proactive in handing over large mining concessions to private investors.

The process of economic liberalization that began in the 1990s has thus created "a voracious appetite for space to meet the demands of industrialization, infrastructure building, urban expansion and resource extraction" (Ghatak and Ghosh 2011: 65). This demand grew steadily over the 1990s, receded somewhat during the East Asia financial crisis-triggered slowdown between 1997 and the early 2000s, and ascended new heights in the mid-2000s as India's growth rate reached 9 percent and a liberalized real estate market entered a dramatic boom.

This demand could not be met by the ordinary operation of land markets. Most of the land available for such projects lay in the hands of India's large smallholding farmers and there are several well-known obstacles to consolidating large chunks of rural land. First, very small holdings make negotiations difficult and holdouts likely. Second, legal problems are almost guaranteed for large projects given titling issues and ambiguity in ownership. Third, many farmers remain reluctant to sell land for multiple reasons, including the lack of attractive exit options from agriculture. The growing demand for land initiated by liberalization in India thus confronted the supply barrier of rural land markets that do not provide "an open field" for the circulation of capital (Harvey 2006b: 271). If large land-consuming private investments were to go forward, the state would have to dispossess land for them.

There were two main incentives for states to help capital overcome this obstacle. First, liberalization also unleashed fierce interstate competition for investment (Jenkins 1999; Rudolph and Rudolph 2001). What has been underappreciated is the critical role land dispossession has assumed in this competition.[11] Given the barrier to amassing large chunks of land through market purchase, companies making large enough investments insist on a government commitment to acquiring it. Such a commitment is an invariable item in the MoUs signed between companies and state governments. According to both government officials and industry representatives that I interviewed in seven states, the ability of state governments to dispossess peasants has become of critical importance in the interstate competition for investment. As Nirupam Sen, then Industries Minister of West Bengal, explained:

> Providing land is one of the most important things, you see, because until and unless you get a hold of the land, there is no question of setting up industry.... That is the most important input. If you don't get it, if you cannot provide it, then they will seek somewhere else where they can get that land. (Interview, February 28, 2011, Kolkata)

The flight of Tata Motors from West Bengal to Gujarat after meeting stiff opposition in Singur powerfully illustrated the centrality of land to the interstate competition for footloose capital.

The second incentive for states to move into dispossessing land for private and increasingly real estate-driven purposes was the enormous licit and illicit rents this would make possible. Acquiring land and selling it to private investors became a significant source of revenue for industrial development corporations, urban development authorities, and other parastatals. Moreover, this increasing state role in land acquisition and allocation expanded the opportunities for illicit graft, which takes the form of government officials and politicians buying land in advance of projects being announced (or selling that information to others) or demanding bribes for land allocation, land conversion, or shifting project boundaries. This nexus of government officials, politicians, brokers, and developers generates "corruption" and black money on a scale that makes the "license raj" look quaint.

The upshot is that state governments restructured themselves into land broker states. No longer confining themselves to dispossessing land for state-led projects of productive material expansion, states turned to dispossessing peasants for any private economic purpose that constitutes "growth," including real estate speculation. As a shorthand, we can characterize this as a shift from a regime of "land for production" to one of "land for the market." No longer is land dispossessed just for the production of commodities, but increasingly for its own commodification. This began in the early 1990s as industrial development corporations and other agencies began acquiring land

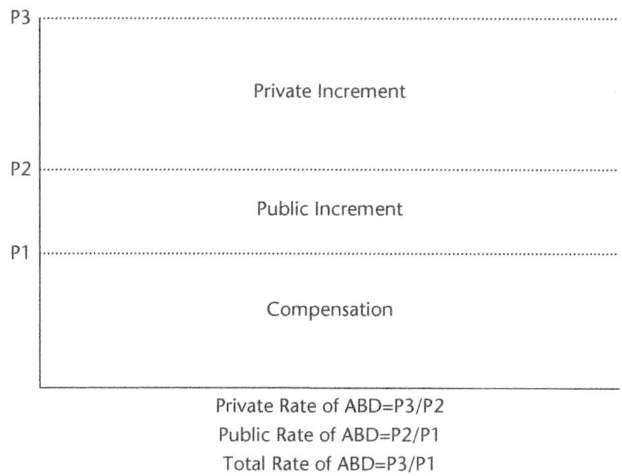

Figure 2.1. Rate of accumulation by dispossession
Source: Compiled by author.

for private and decreasingly industrial purposes (Levien 2013), and reached maturity in the mid-2000s with SEZs, other forms of PPP infrastructure, and the auctioning by urban development authorities of peri-urban farmland to private investors. Rather than simply dispossessing land for production, states began dispossessing land so that private companies could capture the differential between the price paid to farmers and its ultimate market value. I call this ratio between the compensation and market prices the *rate of accumulation by dispossession* (illustrated in Figure 2.1). In the case of the Mahindra World City SEZ in Rajasthan, I have used documents obtained through the Right to Information (RTI) Act to conservatively estimate this rate as 439 percent—including a profit of over Rs. 25 million per acre on land turned to residential use—based on 2011 prices, which have more than doubled since (see Levien 2012: 947–8). In peri-urban Delhi, the Greater Noida Industrial Development Authority has been documented acquiring land at Rs. 820 per square meter and reselling it to developers at Rs. 35,000 (Sood 2011). The key point is that under India's neoliberal regime maximizing such windfalls has itself become the purpose of dispossession, devoid of any broader considerations of "development."

Thesis 5: Under this neoliberal regime, dispossession has become less developmental.

The old term "development-induced displacement" always assumed what needed to be examined: that it was in fact "development" for which people were being displaced. This assumption, which resonates with the state's claim that anything it does is development, is reproduced by recent commentators who, in analyzing contemporary land struggles in India, insist that acquiring land is a necessary cost of development and that the main issue is the terms of compensation (cf. Banerjee et al. 2007; Bardhan 2011; Chakravorty 2013). In his recent book, Sanjoy Chakravorty argues, for example, that "Development cannot take place without changing the use of some land. . . . The problem is not with acquisition per se, but with unjust acquisition" (2013: 173). By unjust acquisition Chakravorty means acquisition compensated at an artificially low price; the solution, according to Chakravorty, is to "get the prices right." This perspective, however, misses the historical transformation in the economic purposes driving land dispossession from the developmentalist to neoliberal periods, and simply accepts that these purposes constitute "development" of a sufficiently "public" nature. Partha Chatterjee (2008) accomplishes a similar reduction when, following Sanyal (2007), he chalks up today's land grabs to a vaguely specified "primitive accumulation" whose political fallout must be managed by the state. While focusing on the "governmentality" he sees involved in the latter, he ironically accepts an ultra-orthodox concept of "primitive accumulation" that renders today's land grabs into the necessary birth pangs of capitalism when they are in fact simply new forms of class redistribution mediated by the state. These accounts make it appear as if the privatized and speculative forms of dispossession in the neoliberal era—a Reliance SEZ, a DLF Mall, a high-end residential colony, a Formula 1 racetrack—represent generic and laudable "development" (Chakravorty) or a necessary stage in the development of capitalism (Chatterjee).

While development is an inescapably normative concept, by most definitions many of today's land grabs fail the test. Indeed, there is a strong case to be made that the projects for which the Nehruvian state acquired land had much greater claims to development than those of today. Consider India's steel towns, a huge source of dispossession in the early decades after independence (Parry 1999; Parry and Struempell 2008; Struempell 2014). India's public sector steel towns provided relatively high-quality employment to tens of thousands of people, including many of those whom they displaced. They allowed India to autonomously manufacture the key building block of an industrial economy. They had huge linkages with the domestic economy and generated ancillary industries in their vicinity. And their revenues were captured by the state. Even dams, for all their many problems, at least produced electricity and provided irrigation to hundreds of thousands of farmers. Now consider today's private SEZs. These zones are driven—and their location determined—by the interests of their developers in reselling land at a profit. Aside from real estate speculation, the production that exists in them is oriented towards non-labor-intensive industries; almost two-thirds are in the IT/ITES (information technology enabled services) sector (Government of India n.d.). The tax breaks afforded to them represent a huge drain on government revenue. My own study of a multipurpose (but IT/ITES-focused) SEZ in Rajasthan found that it employed one individual in only 18 percent of dispossessed families (all in temporary subcontracted positions), created few productive linkages to the surrounding areas, and that its main effect was real estate speculation that dramatically exacerbated preexisting class, caste, and gender inequalities. When the dust settled, my random-sample survey of ninety-four households in four villages found that 65 percent of dispossessed families and 88 percent of Scheduled Caste/Scheduled Tribe (SC/ST) families reported having less income; 50 percent of families and 75 percent of SC/ST families reported having less food; and 76 percent of families and 88 percent of SC/ST families reported receiving more "loss" than "benefit" from the project (see Figure 2.2). And this was within water-scarce villages of Rajasthan that received a better-than-average compensation package for an SEZ that contains some of India's largest IT companies and is considered a "success." One can imagine how much less impressive are the "developmental" results of

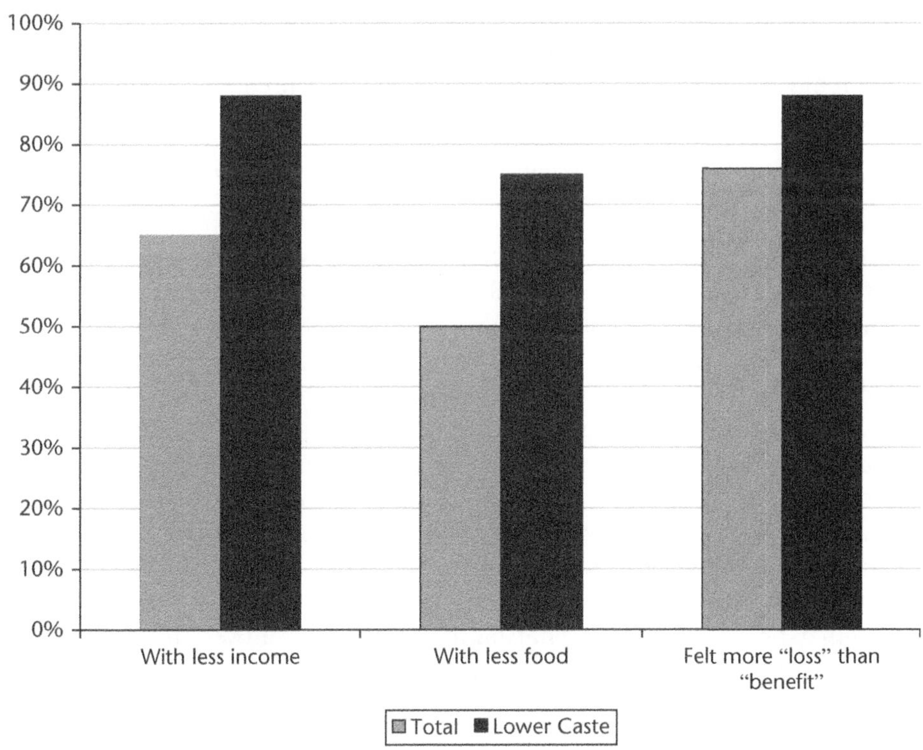

Figure 2.2. Impoverishment by dispossession for Mahindra World City SEZ in Rajasthan

Source: Based on author's fieldwork.

SEZs in more agriculturally prosperous regions, not to mention the residential colonies, golf courses, racetracks, and malls for which land is being expropriated in so many other parts of India.

It is no longer plausible to *assume* that dispossession is a necessary price of development or stage of capitalism. Instead, we must begin to take seriously—and empirically study—the variation in the socioeconomic consequences of dispossession across kinds of projects and diverse agrarian milieu. Understanding such variation helps us to interrogate rather than quietly accept the justifications for dispossession put forward by states and capital in the name of "development." The concept of regimes of dispossession helps to highlight this variation, illuminating the different forms of accumulation and associated class interests that dispossession serves in different times and places. Recognizing that dispossession is organized into qualitatively different regimes helps us to throw off the yoke of teleology and historical determinism. If the economic purposes driving dispossession change over time and place, then the dispossession of direct producers is not a necessary stage of economic development: it is *neither a stage nor necessarily generative of development*. The concept of regimes of dispossession invites us to see dispossession as a form of coercive redistribution that serves different purposes and class interests in different periods but whose contribution to economic change and the "public interest" is not only open to question, but subject to political contestation.

> *Thesis 6*: India's neoliberal regime is politically tenuous and is unlikely to be rescued by higher compensation.

Just as India's neoliberal regime of dispossession was reaching scale in the mid-2000s, it began to encounter increasingly widespread and militant land wars. The dimensions of these struggles are by now well known and need not be repeated. It is enough to say that since the mid-2000s, they have become endemic and uncommonly successful. Beneath the more high-profile struggles—Nandigram, Singur, Raigad, Kalinganagar, Jagatsinghpur, Bhatta-Parsaul, Niyamgiri—are endemic wars of attrition across the country. In addition to organized opposition are the more prosaic legal challenges that, according to my interviews, besiege industrial development corporations, urban development authorities, and other parastatals in almost all states. Cutting across the vicissitudes of individual struggles is an aggregate picture of remarkable obstruction. That land acquisition is now a top concern of Indian capital and the central government is testament to this newfound success.

What explains this remarkable success? There are several factors, which prove challenging to isolate. The first is the previous groundwork laid by movements against Nehruvian-era projects, such as the Narmada Bachao Andolan and the many movements gathered under the National Alliance of

People's Movements (NAPM). The second is the spread of NGOs and mass media into rural areas, and the rise of opposition parties. The precise effects of these changes remain to be investigated, and it is important to remember that journalists, NGOs, and parties typically arrive on the scene only once opposition to a project is well underway. But more fundamental than these background factors, I would suggest, are several features intrinsic to the neoliberal regime of dispossession itself. As we have seen, the shift toward dispossessing land for the private sector, and particularly for real estate purposes, has greatly magnified the discrepancy between compensation prices and market prices. Not only has the rate of accumulation by dispossession increased, but it is also more visible under the neoliberal regime than it was before. Instead of dispossession subsidizing the internal rate of return of a large infrastructure project, it is now a mere instrument for maximizing the profit of private companies—and farmers need only know the market price of land to estimate the extent to which they are being ripped off. The second and related factor, as I have already suggested, is the increasingly exclusionary growth that land dispossession is facilitating in the neoliberal era (cf. D'Costa 2014; Dreze and Sen 2013; Kohli 2012). Inseparable from these material considerations is the dubious legitimacy of dispossessing land for private and often speculative purposes. Few farmers can be convinced that giving their land to a private corporation for real estate constitutes a "public purpose" justifying their forcible removal. Even the media, political parties (at least when they are in opposition), and courts have become increasingly skeptical of forcible land acquisition for real estate speculation, as seen in the uproar over SEZs (see Levien 2013). Normative persuasion is simply not one of the resources available to India's neoliberal regime for producing compliance. This leaves material compensation and brute force as the main mechanisms for getting farmers off their land.

After 2007, we saw concerted attempts by state governments and the central government to overcome "land wars" by increasing material compensation. Uttar Pradesh, Haryana, and Rajasthan were among the early states that put policies into place that went beyond the usual below-market compensation price (the "circle rate") typically given to farmers. The overthrow of the Left Front government by Mamata Banerjee's Trinamool Congress (TMC) after the former's brutal treatment of farmers in Singur and Nandigram demonstrated that land acquisition had arrived as an electorally salient issue and that it would henceforth be politically difficult to continue shoving farmers off their land with minimal compensation. For the first time, offering a better deal to farmers became a matter of interparty competition. Nevertheless, there are limits to state-level policies as capital can always flee to states that are better able to politically manage with low prices—Andhra Pradesh, Tamil Nadu, and Gujarat, for example, all continued acquiring land without giving particularly high compensation (though not without problems). This generated a race to the bottom. As an official with the Haryana Industrial and Infrastructure Development Corporation told me, after Haryana put into place its unprecedentedly remunerative compensation policy at the behest of powerful farmers' groups, high land prices become a deterrent to investors who started going to other states (Interview, February 21, 2011). The only way to overcome this race to the bottom was national legislation to uniformly raise compensation rates.

After six years of wrangling, and much dilution at the behest of capital, the Right to Fair Compensation and Transparency in Land Acquisition, Rehabilitation and Resettlement Act (henceforth LARRA) was passed in September 2013. Explicitly framed as a compromise between the needs of industry and the interests of farmers, the legislation was intended to orchestrate a class compromise that would overcome endemic land wars and ensure a predictable supply of land for capital by giving farmers higher prices. Most essentially, the act allowed the government to continue acquiring land for private companies but required consent thresholds—80 percent of affected farmers for purely private projects and 70 percent for PPP projects—and social impact assessments.[12] In turn, it offered farmers increased compensation and resettlement and rehabilitation benefits. It sought, in other words, to ensure compliance to an ideologically tenuous neoliberal regime of dispossession by making material concessions in the rate of accumulation by dispossession, coupled with some procedural safeguards. It was Rahul Gandhi who apparently came up with the idea of turning land acquisition into a new right—the right to be fairly dispossessed, as it were.

Most farmers' groups were critical of LARRA, arguing that it allowed land acquisition for private companies to continue, did not sufficiently increase compensation to farmers and had many loopholes in its consent and resettlement and rehabilitation clauses. Capital argued that the bill would make land too expensive and delay projects, bringing economic growth to a grinding halt. Within days of Narendra Modi coming to office on an election platform that included fast-tracking large private investments, industry associations began a lobbying offensive to amend the law. On December 30, 2014, the government issued an ordinance—a temporary executive order that must ultimately be approved by parliament—that overturned key provisions of LARRA, including the consent and social impact assessment requirements. But immediate and widespread opposition—led by a reinvigorated Congress and supported by factions of the Sangh Parivar (family of Hindu nationalist organizations) itself—forced the BJP, with an eye toward upcoming state elections, to back down seven months later. Again, India's neoliberal regime of dispossession ran up against the political weight of farmers in India's democracy. LARRA remains the law of the land and the question is now whether it will significantly dissipate India's land wars. I believe this is unlikely, for three reasons.

First, the compensation rates under LARRA are unlikely to appease most farmers. While early versions of the bill called for a uniform fixing of compensation levels at six times the government assessed value of the land, or "circle rate" (always much below the market level), the final bill reduced this to effectively four times the rate in rural areas and two times the rate in urban areas (a distinction it leaves to state governments to establish). Given the rates of accumulation by dispossession described earlier in this section, this will hardly bring compensation up to the level of market value. The relentless lobbying of industrialists and builders to reduce compensation levels in the drafting of LARRA illustrates Marx's observation that capitalists have a difficult time seeing beyond their noses (i.e., short-term profits) to coordinate their long-term class interests. While some worried that the bill made prices too high (Chakravorty 2013, 2015), this amounts to little more than a defense of using eminent domain to generate corporate super-profits. LARRA did not bring compensation prices up to market prices and is thus unlikely to significantly reduce "land wars."

Second, and more fundamentally, the idea of utilizing the exchange value of land to build a class compromise between urban capital and farmers assumes what needs to be explained, which is how farmers come to value their land at its exchange value in the first place. In other words, it is not clear that all farmers can be bought, or at least that their "reservation price" is in a range acceptable to state and capital. Many farmers' struggles on the peripheries of expanding cities are, no doubt, struggles over prices. Nevertheless, it is a mistake to collapse, as Partha Chatterjee (2008) does, all of dispossession politics into ad hoc negotiation over its terms. There remains a second category of anti-dispossession movement that has shown absolutely no interest in compensation. From Nandigram to Niyamgiri, Raigad, Singur, and Jagatsinghpur, we find numerous examples of farmers unwilling to even consider compensation. One should notice that not all of these are in *adivasi* (indigenous or tribal) areas, where such attitudes are considered to be more prevalent. By refusing to value their land at its exchange value, these farmers cannot be brought into a class compromise on the terrain of commodification.

Finally, even among those who are willing to consider compensation, a final obstacle to the Indian state's ability to pacify anti-dispossession struggles lies, as pointed out earlier, in the developmental contradictions of the neoliberal growth model itself. The non-labor-intensive and real estate driven growth that characterizes India's SEZs—and India's growth model as a whole—has little to offer most farmers aside from land values. If the Nehruvian state dispossessed land for economic purposes that farmers could at least be *theoretically* included in—irrigated agriculture and public sector industrialization—the neoliberal regime of dispossessing land for real estate and non-labor-intensive growth does not even offer this possibility. For India's vast semi-proletariat, this makes holding on to even small pieces of land to supplement wage labor far preferable to pure landlessness. While economists justify forcible land acquisition by the macro-level efficiency of transferring land from agriculture to "higher value" land uses (Bardhan 2011; Chakravorty 2013), what they fail to understand is the micro-level rationality that is often behind a farmer's opposition to this transfer.

If the neoliberal regime is short on legitimacy and faces several obstacles to building compliance on a material basis, there remains the time-honored route of intimidation and violence. Undoubtedly, the neoliberal regime of dispossession has, like its predecessor, relied significantly on brutal state violence. There are simply too many cases of states using police, party cadre, and thugs to kill, beat, rape, and loot protesting farmers. Nevertheless, this violence often backfires, and the land wars of the past seven years have shown the limitations of brute coercion. The spread of media and the relentless work of movements and non-governmental organizations (NGOs) have made it more difficult to hide this violence than in the past (although, to be sure, it is still far easier to get away with it in the remote forested areas of central India than in the peri-urban plains). It is also much harder to legitimize this violence, as I have suggested, when it is being used to transfer land to a private corporation than to the state itself. Where anti-dispossession struggles have in recent years succeeded in attracting significant outside attention, it has often prevented the massive use of violence to definitively break their blockades—witness the ability of farmers in Jagatsinghpur to hold up the largest proposed foreign direct investment in India's history with bamboo gates and human chains. What the Maoist insurgency shows, moreover, is how this state violence can backfire on a massive scale.

This is not to say that state violence against rural people who refuse to be dispossessed will cease any time soon. The current government remains committed to fast-tracking private investments and infrastructure projects, and doing this with inadequate material incentives implies greater reliance on state force. Nevertheless, it is unclear how much an aggressive central government can accomplish; acquiring land remains primarily the responsibility of state governments, and it is they who most directly face the political fallout of violently dispossessing farmers for the Reliances of the world. There will certainly, therefore, be great subnational variation in how this plays out (Sud 2014). But the most likely aggregate scenario is that states will fumble towards a less-than-fully-successful class compromise with politically influential farmers in the plains, and continue to use substantial force against marginalized groups in remote areas. It is far from clear, however, that the sum total will be anything resembling a stable regime of dispossession.

2.3 Conclusion

By the latter half of the 2000s, the Indian state had to confront the fact that the land requirements of its neoliberal growth model were difficult to fulfill under conditions of electoral democracy. India's vast rural population, the second largest in the world, controls the land that is desired in ever greater quantities for the private investment that is central to the country's neoliberal phase of capitalist expansion. Whereas in the period of state-led development, states were able to mobilize the substantial legitimacy of Nehruvian developmentalism to transfer land from farmers to public sector projects, the neoliberal regime lacks the normative power to compel farmers to "sacrifice for the nation." Its remaining hope, aside from increasingly illegitimate and counterproductive violence, is to offer farmers some share of the unprecedented real estate values that dispossession is being used to manufacture and capture. But land values are a bad substitute for inclusive growth, both developmentally and politically. With land dispossession increasingly being put to use by predatory state governments in the service of an exclusionary development model, it is unlikely that this neoliberal regime, even with recent legislative amendments, will be able to accomplish what it seeks: the easy and predictable transfer of land from farmers to capital.

The upshot is that farmers are making themselves the most significant obstacle to capitalist growth in India. While the peasantry has long been seen as an obstacle to progress, this specific roadblock is as unexpected for modernization theory and classical Marxism as it is for "India Inc." These land struggles are shaping the trajectory of Indian capitalism in unforeseen and unpredictable ways, and have firmly lodged the "land question" at the center of Indian politics. Whether these anti-dispossession movements can open the way for more equitable paths of development and social change remains to be seen. But their chances will be far greater if they can ally with a reinvigorated left that has abandoned old theories of dispossession's role in history.

Notes

1. On the need for "denaturalizing dispossession," see Hart (2006).
2. As Perelman (2000: 25) points out, Marx translated Adam Smith's "previous" into German as *ursprünglich*, which was subsequently retranslated into English as "primitive."

3. In the more Hegelian idiom of the *Grundrisse,* Marx states, "The conditions and presuppositions of the *becoming,* of the *rising,* of capital presuppose precisely that it is not yet in being but merely in *becoming;* they therefore disappear as real capital arises, capital which itself, on the basis of its own reality, posits the conditions for its realization" (1973: 459).
4. Others have argued that a theory of ongoing primitive accumulation can be found in Marx (De Angelis 2007). I think, at best, one can argue that Marx's account of primitive accumulation *implies* that it would be ongoing (why would capitalism's thirst for land come to an end?), but Marx himself relegates it to the "prehistory" of capital. For the distinction between a serial and recursive conception of primitive accumulation, I thank Michael Watts.
5. The two were connected to the extent that capitalist agriculture generated larger surpluses that could be appropriated for subsidizing industrialization.
6. For a critique of this argument, see Gerschenkron (1962).
7. However, the problematic of transition faced its own difficulties and its relevance may anyway be declining along with the importance of agriculture to national projects of capitalist development (Bernstein 2012; Lerche 2013). See also the introduction to this volume.
8. I use group here rather than class because the dispossessed often belong to multiple classes (especially in the countryside), while the recipients of the land are typically of the capitalist class (though they can be from different fractions of that class).
9. Although dispossession for agribusiness is also increasing in China.
10. See Levien (2012).
11. On the broader significance of land liberalization and new forms of urban governance, see Sud (2012) and Goldman (2011).
12. It seems likely that states can get around this provision by allocating private companies land from their "land banks." Most states already acquire land well in advance of demand for particular projects; since they are not explicitly for private projects when acquired, these acquisitions would arguably not come under the 70–80 percent approval clause—though this may be decided by the courts.

References

Adnan, S. 1985. "Classical and Contemporary Approaches to Agrarian Capitalism." *Economic and Political Weekly,* 20(30): PE53–PE64.

Ahluwalia, I. J. 1998. "The Contribution of Planning to Indian Industrialisation." In *The State, Development Planning and Liberalisation in India,* edited by Terence J. Byres, New Delhi: Oxford University Press, pp. 254–97.

Anderson, P. 1974. *Lineages of the Absolutist State.* London: New Left Books.

Banerjee, A. V., Bardhan, P., Basu, K., Chaudhury, M. D., Ghatak, M., Guha, A. S., Majumdar, M., Mookherjee, D., and Ray, D. 2007. "Beyond Nandigram: Industrialisation in West Bengal." *Economic and Political Weekly,* 42(17): 1487–9.

Bardhan, P. 1984. *The Political Economy of Development in India.* New Delhi: Oxford University Press.

Bardhan, P. 2010. *Awakening Giants, Feet of Clay: Assessing the Economic Rise of China and India*. Princeton, NJ: Princeton University Press.

Bardhan, P. 2011. "Industrialization and the Land Acquisition Conundrum." *Development Outreach*, 13(1) (April): 54–7.

Bernstein, H. 2012. "Agrarian Questions from Transition to Globalization." In *Peasants and Globalization: Political Economy, Rural Transformation, and the Agrarian Question*, edited by A. Haroon Akram-Lodhi and Cristobal Kay, London: Routledge, pp. 239–61.

Bhaduri, A. 1983. *The Economic Structure of Backward Agriculture*. London: Academic Press.

Brass, P. 2011. *An Indian Political Life: Charan Singh and Congress Politics, 1937–1961*. New Delhi: Sage.

Brenner, R. 1976. "Agrarian Class Structure and Economic Development in Pre-Industrial Europe." *Past and Present*, 70: 30–75.

Brenner, R. 2001. "The Low Countries in the Transition to Capitalism." *Journal of Agrarian Change*, 1(2): 169–241.

Brenner, R. 2006. "What Is, and What Is Not, Imperialism?" *Historical Materialism*, 14(4): 79–105.

Burawoy, M. 1985. *The Politics of Production: Factory Regimes Under Capitalism and Socialism*. London: Verso.

Byres, T. J. 1991. "The Agrarian Question and Differing Forms of Capitalist Agrarian Transition: An Essay with Reference to Asia." In *Rural Transformation in Asia*, edited by Jan Breman and Sudipto Mundle, New Delhi: Oxford University Press, pp. 3–76.

Chakravorty, S. 2013. *The Price of Land: Acquisition, Conflict, Consequence*. New Delhi: Oxford University Press.

Chakravorty, S. 2015. "Improving an Unworkable Law." *The Hindu*, January 7.

Chandrasekhar, C. P. and Ghosh, J. 2002. *The Market That Failed: Neoliberal Economic Reforms in India*. New Delhi: LeftWord Books.

Chatterjee, P. 2008. "Democracy and Economic Transformation in India." *Economic and Political Weekly*, 43(16): 53–62.

CMIE (Center for Monitoring Indian Economy). 2012. "Sharp Increase in Projects Getting Shelved." Available at: <http://www.cmie.com/kommon/bin/sr.php?kall=wclrdhtm.php&cmienvdt=20120704101759176&pc=099000000000&type=INSIGHTS> (accessed August 15, 2012).

D'Costa, A. P. 2014. "Compressed Capitalism and Development: Primitive Accumulation, Petty Commodity Production, and Capitalist Maturity in India and China." *Critical Asian Studies*, 46(2): 317–44.

De Angelis, M. 2007. *The Beginning of History: Value Struggles and Global Capital*, London: Pluto Press.

Dreze, J. and Sen, A. 2013. *An Uncertain Glory: India and Its Contradictions*. Princeton, NJ: Princeton University Press.

Dwivedi, R. 2006. *Conflict and Collective Action: The Sardar Sarovar Project in India*. London: Routledge.

Fernandes, W. 2008. "Sixty Years of Development-Induced Displacement in India." In *India Social Development Report 2008: Development and Displacement*, edited by H. M. Mathur, New Delhi: Oxford University Press, pp. 89–102.

Gerschenkron, A. 1962. *Economic Backwardness in Historical Perspective: A Book of Essays*. Cambridge, MA: Harvard University Press.

Ghatak, M. and Ghosh, P. 2011. "The Land Acquisition Bill: A Critique and a Proposal." *Economic and Political Weekly*, 46(41): 65–72.

Goldman, M. 2011. "Speculative Urbanism and the Making of the Next World City." *International Journal of Urban and Regional Research*, 35(3): 555–81.

Government of India. n.d. "Special Economic Zones in India." Ministry of Commerce and Industries. Available at: <http://sezindia.nic.in/> (accessed February 1, 2011).

Gulati, M. 2011. "The Infrastructure Sector in India 2010–2011." In *India Infrastructure Report 2011: Water: Policy and Performance for Sustainable Development*, edited by Infrastructure Development Finance Corporation. New Delhi: Oxford University Press, pp. 379–96.

Hart, G. 2006. "Denaturalizing Dispossession: Critical Ethnography in the Age of Resurgent Imperialism." *Antipode*, 38(5): 977–1004.

Harvey, D. 2003. *The New Imperialism*, Oxford: Oxford University Press.

Harvey, D. 2006a. "Comment on Commentaries." *Historical Materialism*, 14(4): 157–66.

Harvey, D. 2006b. *The Limits to Capital*, London: Verso.

IDFC (Infrastructure Development Finance Corporation). 2008. *India Infrastructure Report 2008: Business Models of the Future*. New Delhi: Oxford University Press.

IDFC. 2009. *India Infrastructure Report 2009: Land—A Critical Resource for Infrastructure*, New Delhi: Oxford University Press.

Jenkins, R. 1999. *Democratic Politics and Economic Reform in India*. Cambridge: Cambridge University Press.

Ka, C. and Selden, M. 1986. "Original Accumulation, Equity and Late Industrialization: The Cases of Socialist China and Capitalist Taiwan." *World Development*, 14(10): 1293–310.

Kautsky, K. 1988. *The Agrarian Question*. London: Zwan Publications.

Kohli, A. 2012. *Poverty Amid Plenty in the New India*. Cambridge: Cambridge University Press.

Lenin, V. I. 1967. *The Development of Capitalism in Russia*. Moscow: Progress Publishers.

Lerche, J. 2013. "The Agrarian Question in Neoliberal India: Agrarian Transition Bypassed?" *Journal of Agrarian Change*, 13(3): 382–404.

Levien, M. 2011. "Special Economic Zones and Accumulation by Dispossession in India." *Journal of Agrarian Change*, 11(4): 454–83.

Levien, M. 2012. "The Land Question: Special Economic Zones and the Political Economy of Dispossession in India." *Journal of Peasant Studies*, 39(3–4): 933–69.

Levien, M. 2013. "Regimes of Dispossession: From Steel Towns to Special Economic Zones." *Development and Change*, 44(2): 381–407.

Marx, K. 1973. *Grundrisse*. London: Penguin.

Marx, K. 1977. *Capital*,vol.1, New York: Vintage.

Nilsen, A. G. 2010. *Dispossession and Resistance in India: The River and the Rage*. London: Routledge.

Parry, J. P. 1999. "Two Cheers for Reservations: The Satnamis and the Steel Plant." In *Institutions and Inequalities: Essays in Honor of Andre Beteille*, edited by R. Guha and J. P. Parry. New Delhi: Oxford University Press, pp. 128–69.

Parry, J. P. and Struempell, C. 2008. "On the Desecration of Nehru's 'Temples': Bhilai and Rourkela Compared." *Economic and Political Weekly*, 43(19): 47–57.

Perelman, M. 2000. *The Invention of Capitalism: Classical Political Economy and the Secret History of Primitive Accumulation*. Durham: Duke University Press.

Preobrazhensky, E. 1965. *The New Economics*. Oxford: Clarendon Press.

Reynolds, S. 2010. *Before Eminent Domain: Toward a History of Expropriation of Land for the Common Good*. Chapel Hill, NC: The University of North Carolina Press.

Rudolph, L. I. and Rudolph, S. H. 2001. "Iconisation of Chandrababu: Sharing Sovereignty in India's Federal Market Economy." *Economic and Political Weekly*, 36(18): 1541–52.

Sanyal, K. 2007. *Rethinking Capitalist Development: Primitive Accumulation, Govermentality, and Post-colonial Capitalism*. New Delhi: Routledge.

Searle, R. 2010. "Making Space for Capital: The Production of Global Landscapes in Contemporary India." Unpublished PhD Dissertation, University of Pennsylvania.

Sood, J. 2011. "Road to Disaster." *Down to Earth*. June 15.

Struempell, C. 2014. "The Politics of Dispossession in an Odisha Steel Town." *Contributions to Indian Sociology*, 48(1): 45–72.

Sud, N. 2012. *Liberalization, Hindu Nationalism and the State: A Biography of Gujarat*. New Delhi: Oxford University Press.

Sud, N. 2014. "Governing India's Land." *World Development*, 60: 43–56.

Thompson, E. P. 1963. *The Making of the English Working Class*. New York: Vintage.

Weber, M. 1978. *Economy and Society*, Berkeley, CA: University of California Press.

Wood, E. M. 2002. *The Origin of Capitalism: A Longer View*. London: Verso.

3

Land Grabs, Primitive Accumulation, and Resistance in Neoliberal India

Persistence of the Self-Employed and Divergence from the "Transition to Capitalism"?

Shapan Adnan

3.1 Introduction

Land grabs involving separation of non-capitalist producers from their lands have been interpreted in terms of the theoretical framework of "primitive accumulation" and the "transition to capitalism" (Marx 1976). This transition schema envisages that the labor of the dispossessed groups and their erstwhile lands would be potentially deployed in production processes run by capitalists, resulting in the progressive capitalist transformation of the economy. However, the consequences of land alienation in India in the context of neoliberal globalization appear to deviate partially from this schema because significant sections of the dispossessed remain self-employed rather than being absorbed in capitalist wage employment. The stability of this pattern in recent times, despite continuing land grabs and primitive accumulation, has raised questions about the inevitability and applicability of the schema of "transition to capitalism" in India.

In this chapter I argue that such outcomes come about when dispossessed producers cannot find adequate wage employment in the formal and informal capitalist sectors and hence have limited avenues of survival other than self-employment. This pattern is reinforced by the predominance of unproductive capital controlled by merchant–usurers and the appropriation of precapitalist ground rent by landowners, all of which are predicated upon extraction of surplus from self-employed producers, and hence are systematically oriented against the emergence of incremental wage labor.

In Marx's (1976: 873–6) analysis of the origins of capitalism, the notion of "primitive accumulation" is used to explain how the separation of non-capitalist classes from their lands and productive assets enabled such resources and the labor of the dispossessed to become available, in the first place, for deployment in capitalist production. On the one hand, the expropriated lands (and other inputs) were concentrated among actual or potential capitalist producers, or landlords willing to lease them to capitalist tenants. On the other hand, the dispossessed groups faced circumstances in which they had no better avenue of survival than taking up wage employment in capitalist production. Thus, a double transformation followed primitive accumulation, whereby land became part of productive capital and the dispossessed were converted into wage laborers in the workforce (Marx 1976: 874). It is this sequence of interlinked processes leading from primitive accumulation to capitalist production, taking place on a social-historical scale, which constitutes the classic Marxian schema of "transition to capitalism." The *simultaneity* of these two transformations was integral to Marx's construct of primitive accumulation, which was primarily based on the historical experience of enclosures in Britain (Marx 1976). However, it constituted something of a special case that need not hold in other historical instances (Adnan 2013, 2014: 23; Levien 2012).

Historically, primitive accumulation has not been limited to a one-shot event at the origin of capitalism, but has constituted an ongoing process that has continued to make available land and other inputs to capitalist production (Marx 1976: 876; Perelman 1984: 8–12). However, the *institutional mechanisms and processes* mediating primitive accumulation have varied with evolving socioeconomic and political conditions. In the context of neoliberal globalization, primitive accumulation has been characterized by distinctive processes such as financialization, speculation, and the orchestration of crises. Harvey (2005) has incorporated these novel features in the notion of "accumulation by dispossession" (ABD), in addition to updating Marx's original concept of primitive accumulation in the context of "neoliberal imperialism" (Ashman and Callinicos 2006: 117; Brenner 2006: 102; Fine 2006: 143–4).[1] Given limits of space, however, it has not been presently possible to provide a more comprehensive review of the theoretical issues pertaining to primitive accumulation and ABD.[2]

When lands are grabbed from non-capitalist classes in an involuntary and/or unjust manner, such acts can trigger resistance from the actually or potentially dispossessed groups. These constitute *struggles against primitive accumulation/ABD*, as contrasted to working class struggles against capitalism or peasant struggles against landlords.[3] Resistance to land alienation by the dispossessed groups can take different forms: covert or overt, violent or nonviolent (Adas 1986; Adnan 2007: 210–11, 2013: 97–8; Scott 1985, 1986). When lands are appropriated by powerful groups and agencies, particularly the state, resistance usually takes nonviolent legal–constitutional forms, such as public meetings and demonstrations. However, under limiting conditions, resistance can be transformed into violent forms, including armed struggle (Adas 1986: 82–3; Adnan 2007). The *modalities* of resistance can also change in terms of scale, organization, and collective action, involving coalition formation and coordinated activities in multiple arenas (Tilly 1978: 7, 78). Such shifts in the forms and modalities of resistance can critically affect the outcomes of land alienation and primitive accumulation.[4]

While land has traditionally served as the means of production in agriculture as well as the key marker of power and status in social and political relations, in recent times it has become increasingly valued as a commodity in the wider economic and social arena (Akram-Lodhi 2012; Hall 2013). During the immediate postcolonial decades, the Indian state undertook large-scale land acquisition for massive projects of infrastructure constructions and development, which were justified in terms of "public purpose." However, with the advent of neoliberal globalization, there was a qualitative shift in the nature and purpose of state acquisition of land.[5] While the process continued, its purpose shifted increasingly towards making land available to private corporations and elite groups. Moreover, a variety of non-state actors, including multinational corporations, political parties, private armies, and criminal groups, became involved in large-scale land grabs (Banerjee 2014; Walker 2008).

Viewed in historical perspective, the role of land alienation in India has undergone a striking change in the context of neoliberal globalization. Consequently, the contemporary "land question" has also taken on novel dimensions beyond its classic formulation, as elaborated in the following paragraphs.

Trends in the composition of the Indian workforce in terms of employment status raise problematic issues in relation to the schema of transition to capitalism. First, while the Indian economy has grown significantly during the neoliberal era, expansion of wage employment has been uneven and limited. Symptomatic of this trend is the phenomenon of "jobless growth," the extent of which has fluctuated due to multiple factors.[6] Second, while people have been displaced from their farms and settlements by land grabs, the dispossessed groups have not necessarily been absorbed in wage employment—remaining unemployed or self-employed. In the Indian discourse, such outcomes have been dubbed "dispossession without proletarianization" (Basu and Das 2009: 158; Levien 2012: 954; U. Patnaik 1990; Sanyal and Bhattacharyya 2009: 35).

These trends in the Indian economy do not entirely accord with the schema of transition to capitalism. Since the conversion of expropriated land into productive capital has not been matched by the transformation of the dispossessed into wage workers, only one of the "two transformations" postulated by Marx has taken place substantively compared to the other (Adnan 2014; Levien 2012; Marx 1976: 834). The prospect of primitive accumulation being followed by the "transition to capitalism" remains uncertain to the extent that land grabs are not followed by the employment of the dispossessed in capitalist production.[7] Moreover, questions have arisen as to whether the Indian economy is heading toward an alternative trajectory that diverges from the transition to capitalism.[8]

In order to resolve these theoretical issues, it is necessary to examine empirical evidence on the extent to which land grabs have been followed by the absorption of the dispossessed in *capitalist wage employment*. However, while some micro-level studies are available, there do not appear to be any national-level data on the specific contribution of land alienation to the composition of the Indian workforce in terms of the relative proportions of the self-employed and wage workers. To the extent that national-level data are available on the composition of the workforce by employment status, these reflect the overall outcomes of multiple factors (Himanshu 2011), and hence cannot be explained by the employment impacts of land alienation alone, which constitutes just one of the contributory factors.

Given these data limitations, an indirect approach combining two complementary methods has been used in Sections 3.4 and 3.5. First, the national-level survey data are interpreted with suitable assumptions to assess inter-temporal trends in the employment status of the Indian workforce (divided into the self-employed, regular employees, and casual workers). Second, evidence from a case study of land acquisition (Levien 2012) is reviewed to trace the causal processes underlying the observed outcomes in terms of the relative weight of the self-employed and wage employees. While the findings of the micro-level analysis are not representative and cannot explain national-level employment trends, they do provide grounded evidence on the *causal processes* shaping the outcomes of land alienation, particularly in terms of impacts on employment status. Until more appropriate macro-level data are available, these micro-level insights can be utilized for interpreting the national-level trends. In Section 3.6, the results of these different methods have been put together to formulate a set of hypotheses to explain the factors underlying "dispossession without proletarianization" in neoliberal India.

Following from these considerations, the objectives of this chapter are formulated in terms of the following key questions:

1. What are the diverse mechanisms of land grabs mediating primitive accumulation in neoliberal India?

2. How have different forms and modalities of resistance by dispossessed groups influenced the outcomes of land alienation and primitive accumulation?
3. What have been the trends in the composition of the total Indian workforce in terms of the self-employed and wage employees?
4. What does micro-level analysis tell us about changes in employment status brought about by land alienation and its consequences?
5. What factors help to explain inter-temporal trends in the relative weight of the self-employed and wage employees in the Indian workforce?
6. What are the implications of the analysis for prospects of the "transition to capitalism" in India?

The rest of the chapter is laid out as follows. Section 3.2 summarizes the salient features of the neoliberal policy regime in India and the ways in which these have influenced different types of land alienation mechanisms. Section 3.3 analyzes the forms and modalities of resistance to land grabs and their varying outcomes. Inter-temporal trends in the relative proportions of self-employed and wage workers in the Indian workforce are reviewed in Section 3.4. Section 3.5 provides a critical review of existing interpretations of trends in the relative weight of wage workers and the self-employed and puts forward an explanatory hypothesis. In conclusion, Section 3.6 summarizes the inferences of the preceding analysis, formulates possible explanations of the trends in employment status, and assesses the prospects of the "transition to capitalism" in India.

3.2 Mechanisms of Land Grabs

The contemporary Indian economy has witnessed massive expropriation of land and natural resources through a range of mechanisms influenced by neoliberal policies, as specified in Section 3.2.1 (Basu 2007; P. Patnaik 2011, 2012: 31–2; U. Patnaik 2012: 240–1; Sebastian 2012: 21–7; Walker 2008: 580–9).

3.2.1 *The Neoliberal Policy Regime and Its Influence*

Following the adoption of neoliberal "reforms" during the 1990s, the Indian government undertook "massive programmes of privatisation and structural adjustment" (Chandrasekhar and Ghosh 2002; U. Patnaik 2012: 234; Walker 2008: 559). The "license-permit raj" characterizing the post-independence regulatory framework was replaced by a policy package combining deregulation, trade liberalization, financial sector reforms, and devaluation (Sanyal 2007: 242–5; Shah 2008: 80). Under this neoliberal policy regime, there were systematic cuts in public expenditure that led to reductions in redistributive welfare benefits and social services, affecting the most vulnerable sections of the population including the unemployed and the dispossessed (Basu and Das 2009: 159; Baviskar and Sundar 2008: 87–9; Bhaduri 2008: 12–13; P. Patnaik 2012; Shah 2008: 80; Walker 2008: 562).

Such policies of the Indian central government were aimed at opening up the domestic economy and gaining the confidence of investors and financial institutions regulating global capital flows in order to attract foreign direct investment (FDI) and prevent capital flight (Bhaduri 2008: 11–12; P. Patnaik 2008: 109). Lifting of restrictions on private corporate investment triggered a "race to the bottom" among the regional state governments of the Indian Union, which vied with each other to offer concessionary terms to finance capital for large-scale construction and development projects. Even traditionally left-wing governments in the states of West Bengal and Kerala joined the competition, reflecting a "virtual consensus among all major political parties about the priorities of rapid economic growth led by private investment" (Bhaduri 2008: 10; Chatterjee 2008: 57; Levien 2012: 944–6; P. Patnaik 2011: 3).

As elaborated in Section 3.2.2, these varied strands of the neoliberal policy regime and their consequences served to open up novel avenues of primitive accumulation in India through land alienation.

3.2.2 *Land Acquisition by Force and Manipulation*

The central Indian government set up special economic zones (SEZs) to serve as "hyperliberalized economic enclaves" for the purposes of "promoting exports, attracting FDI, developing infrastructure, and generating employment" (Levien 2012: 934). The territorial dimension of the SEZs made a substantive bounded area indispensable and hence their legal–institutional framework was designed to enable *systematic* acquisition of land (Adnan 2016). The rapid approval and establishment of many such zones all over India led to large-scale land grabs with varying outcomes.[9]

The key legal instrument used to obtain land for SEZs and other large projects was the archaic Land Acquisition Act of 1894 embodying the legal principle of *eminent domain* which empowered the state to acquire lands for *public* purposes, overriding objections by those dispossessed in the process, as also discussed by Malabika Pal in Chapter 6 in this volume (Adnan 2014: 29; Levien 2012: 944–6; Vasudevan 2008: 41; Walker 2008: 589). Ironically, however, while the state authorities invoked the rhetoric of *public purpose* to acquire lands, they went on to transfer these to *private profit-making* corporations, making the state serve effectively as "land broker" to the private corporate sector (Bhaduri 2008: 12; Levien 2012: 941–6; Walker 2008: 580).

Even though the law provided for compensation payments and rehabilitation arrangements for those whose lands had been acquired, these were implemented in discriminatory ways. While landowners holding formal ownership rights or titles were considered eligible for payment of compensation, the common or customary "rights of use" of others to the acquired areas were not recognized, although their livelihoods could depend upon access to these resources (Banerjee 2014: 276–7; Bhaduri 2008: 13; Walker 2008: 582–90). Even when compensation was paid, the rates offered by the administration to those losing land were typically considerably below the prevailing market price. This kind of compulsory undervaluation of land during acquisition, juxtaposed with the higher rate charged when it was transferred to private agencies, allowed the state to make inflated profits in a manner characteristic of ABD (Harvey 2005: 149–50; Levien 2012: 947–8).

Moreover, even the laws and regulations were often manipulated or violated in varying degrees during the processes of state acquisition, compensation, and rehabilitation (Banerjee 2014: 295). Various state agencies were found to use illegal means to circumvent legal restrictions on acquisition of communal lands of *adivasis* (indigenous peoples) without their consent.[10] In addition to state violence by security forces, a key trend in Indian land grabs has been "the use of officially sanctioned criminal violence at the level of the state and by private landholders against the rural poor" (Banerjee 2014: 282–3, 293–7; Nigam 2007: 6–7; Walker 2008: 559, 591–3).

3.2.3 *Land Alienation without the Use of Force*

In contrast to the points discussed in Sections 3.2.1 and 3.2.2, land alienation in neoliberal India has also taken place *without* the use of force or violence. In such instances, alternative mechanisms were at play, such as informal mediation, negotiations, persuasion, temptation, fraud, or forgery to orchestrate transfer of the lands (Adnan 2014, 2016).

For instance, the Tamil Nadu state government made use of a set of strategies to acquire land by agreement and minimize resistance. It set up official committees for obtaining land through private negotiations rather than compulsory state acquisition and utilized market mechanisms streamlined by supportive legislation (Vijayabaskar 2014: 303–20). In some cases, acquiescence to land acquisition was encouraged by material incentives such as allotment of housing plots and jobs with training facilities as part of a package deal. In a similar vein, the Mahindra World City (MWC) SEZ in Rajasthan offered compliant landowners smaller but developed, high-value commercial and residential plots in exchange for their acquired holdings (Levien 2012: 953).

In certain instances, land and natural resources were appropriated through sheer technological superiority or market bargaining power (Adnan 2014: 30).

For instance, private corporations producing soft drinks extracted "pure groundwater as a free raw material" from deep aquifers using expensive capital-intensive technology that could not be matched by ordinary water users of the locality (Bhaduri 2008: 13; cf. Walker 2008: 564–5).

3.2.4 *Indirect Mechanisms of Land Alienation*

There were also indirect mechanisms of land alienation that were mediated by processes geared to quite different objectives, but which nonetheless led to similar outcomes. For instance, measures of fiscal contraction in India led to severe income deflation as well as reduction in public expenditure involving cuts in public investment, rural development, welfare facilities, and employment generation programs (P. Patnaik 2008: 109, 2012: 35–7; U. Patnaik 2012: 249; Shah 2008: 80). While in the short run such neoliberal policy measures reduced the real incomes and purchasing power of the poor, in the long run their cumulative impacts undermined the viability of small-scale peasant and artisanal producers, constraining them to part with their lands and assets (P. Patnaik 2008: 113, 2012: 37; Walker 2008: 573–9). Such outcomes were similar to those of direct land enclosure but corresponded to *indirect* forms of primitive accumulation because they constituted *unintended* consequences of policy and development interventions with very different objectives (Adnan 2014: 26–33, 2016).

3.3 Resistance and Contestations

As land grabs have intensified in neoliberal India, these have also been confronted with different forms and modalities of resistance from dispossessed groups (Levien 2012; Walker 2008). In particular, the sheer lack of legitimacy of the Indian state's role as "land broker" to the private corporate sector has triggered a politics of resistance against primitive accumulation (Adnan 2014: 30–9). Among "the key markers" of the struggle were protests and social movements against construction of "special economic zones, expressways and other infrastructure projects," involving large-scale population displacements (Das 2012: 33–5). The need for protecting the land and livelihoods of those confronted with dispossession typically involved the *entire population* of the affected area, subsuming multiple classes and social groups. Those involved were not only landholders with property rights, but also landless peasants, *adivasis* ("tribes" or indigenous peoples), and other occupational groups without such rights but whose livelihoods depended crucially upon access to the acquired areas, for example farmlands, forests, wetlands, pastures, and various common property resources (Kunnath 2012: 104–7; Sanyal and Bhattacharyya 2009: 42–3).

The *forms* and *modalities* of such struggles against primitive accumulation have evolved over time and space (cf. Adnan 2007: 183–6; 2013: 97–8). The initial response of groups facing dispossession in different parts of India was to form local-level organizations to mobilize support for protests and collective action (Tilly 1978: 78; Walker 2008: 585–91). In subsequent stages, overarching federations such as the National Alliance of People's Movements (NAPM) were formed to interconnect local-level and issue-specific movements, as well as to contest, at the national level, policies and legislation designed to facilitate land grabs (Walker 2008: 590). Such shifts in the scale and modalities of struggle led to growing coalescence and consolidation among the localized movements and their supporting coalitions, making possible coordinated collective action in multiple arenas (Das 2012: 35–7; Levien 2012: 933–4, 2016; Vasudevan 2008: 43; cf. De Angelis 2001: 3).

Resistance to land grabs and lack of compensation by weak social groups usually made use of nonviolent and *socially accepted* forms of protest in India, such as petitioning, demonstrations, and blockades (*gherao*) of the concerned offices and installations. However, such protests typically proved futile against the forces repressing them. Consequently, in some instances, the concerned social groups and organizations embarked upon more active and overt forms of resistance. In limiting cases, the affected groups took up armed struggles against primitive accumulation caused by "displacement and dispossession from land-based livelihoods" (Harriss-White 2012: 142–3). Such instances also highlight the conditions under which *nonviolent* and/or *covert* forms of resistance to land alienation were transformed into *violent* and/or *overt* ones (cf. Adnan 2007: 214–15, 2013: 118–19; 2014: 31, fn 8).

For instance, in mineral-rich Chhattisgarh, the state authorities signed dozens of MOUs with foreign and domestic corporations and issued licenses enabling them to acquire minerals and forest resources at very low costs, corresponding to ABD (Bharadwaj 2009; Walker 2008: 593–5). Confronted by land grabs and forced displacements, the local *adivasis* had little option but to take up armed resistance in alliance with units of the Maoist Communist Party of India (CPI–ML) (Jha 2013; Kunnath 2012: 104–7; Sebastian 2012: 19–20).

Overall, land grabs for SEZs and other projects have triggered widespread struggles against primitive accumulation in neoliberal India, involving explicitly political contestations (Levien 2012; Walker 2008: 590). The outcomes of such contentions have varied considerably, depending upon the balance of forces between expropriation and resistance in particular localities (Adnan 2014: 30–9, 2016; Banerjee 2014: 275–6, 281–4). On the one hand, many SEZs have been set up successfully through state acquisition of land, forced displacement, and crushing of resistance of the evicted groups. On the other hand, the political chain reaction following violent repression of protesters in West Bengal's Nandigram SEZ "led to a cancellation of the project, a temporary moratorium on SEZs and a reduction in their maximum allowed size" (Levien 2012: 933–4). Comparable struggles have resulted in the cancellation, postponement, or downsizing of SEZs or other land-acquiring projects in different parts of India (Levien 2012; Sampat 2015). To the extent that such resistance has been able to reduce the pace and extent of land grabs, it has also succeeded in imposing social and political *limits* upon the inroads of primitive accumulation in neoliberal India.

3.4 Trends in the Composition of the Indian Workforce

In order to assess the outcomes of primitive accumulation mediated by land alienation, it is necessary to ascertain what kinds of livelihoods have been taken up by the dispossessed groups. Have they been transformed into wage workers in capitalist production, as postulated in the schema of transition to capitalism? Or, have their livelihoods shifted toward other forms of wage or self-employment? In either case, what are the long-term implications for the "transition to capitalism" in India?

While these issues are raised within the *theoretical* framework of the transition to capitalism, what trends have actually characterized neoliberal India is an *empirical* question. Its resolution therefore requires assessment of evidence on changes in the composition of the Indian workforce in terms of the relative shares of wage workers and the self-employed, as well as the factors explaining them.

3.4.1 *Trends in the Composition of the Indian Workforce by Employment Status*

Inter-temporal trends in the composition of the total Indian workforce from 1999–2000 to 2011–12 can be derived from the repeat sample survey data provided by the National Sample Survey Office (NSSO). These have been summarized by Shaw (2013: 25, table 4), with all Indian workers classified into three broad types of "employment status": (i) self-employed, (ii) regular salaried and wage workers, and (iii) casual wage workers.[11]

However, the concepts of employment status used by NSSO need to be critically appraised before the data can be accurately interpreted. Both regular and casual wage workers can be employed in either profit-making capitalist enterprises, or non-capitalist production units such as peasant farms to supplement family labor. Those participating in public works programs such as the National Rural Employment Guarantee Act (NREGA) correspond to casual workers, who are hired by state or local authorities for redistributive welfare or employment-generation objectives as contrasted to working for profit-seeking capitalist enterprises (Thomas 2012: 46–8). Correlatively, the category of the self-employed is comprised of "own account" operators running their enterprises as well as their family workers who are typically not paid formal wages, but receive "payments" or benefits in terms of sharing in the pooled resources of the household for subsistence needs (Hirway 2002, 2012). While, in theory, some of these could conceivably operate as capitalist enterprises, the vast majority of self-employed producers are not in a position to do so because they cannot meet capitalist conditions of reproduction requiring them to earn normal profit, over and above meeting subsistence and input costs (Adnan 1985: PE–56; U. Patnaik 1979: 401).

It is evident that the "employment status" in terms of which the NSSO workforce data are classified do not allow clear distinction between those who are wage workers in *capitalist* employment and those who are not. Consequently, I have made some stylized assumptions regarding the NSSO data when analyzing them in order to *estimate* the relative shares of the self-employed and wage workers.

As a first approximation, it is assumed that all regular and casual workers are "wage employees working in capitalist enterprises." This definition would obviously provide an *overestimate* since significant proportions of wage employment are generated by agencies different from profit-making capitalists, as noted above. Second, all the self-employed, consisting of operators running their own units and their family labor, are assumed to be working in non-capitalist petty commodity production (PCP) units. However, this is likely to provide an *underestimate* of the share of the self-employed in the workforce because of idiosyncrasies of the data collection procedures regarding the varied activities undertaken by family labor. "The NSSO concepts and methods are not able to capture... unpaid work, which consists of (i) unpaid family work, (ii) free collection of goods of different kinds, and (iii) household production for self-consumption" (Hirway 2012: 69–71).[12] Such survey procedures, combined with the overestimation of wage employees noted earlier, are likely to result in systematic underestimation of the self-employed workforce derived from the NSSO data.[13]

Bearing these qualifications in mind, the following inferences can be drawn about trends in the composition of the Indian workforce between 1999–2000 and 2011–12. The proportion of the self-employed has fluctuated in the range of 49–52 percent of the total workforce during this period, accounting for just over 50 percent in both the terminal years 1999–2000 and 2011–12, with a rise to 52 percent in 2004–5 and a fall to 49.5 percent in 2009–10 (Shaw 2013: 23–5, table 4). Since the self-employed are likely to have been *underestimated*, it may be warranted to infer that they constituted at least half, if not more, of the Indian workforce during 1999–2012, despite mild fluctuations.

Correlatively, the combined weight of the regular and causal wage employees according to the NSSO survey data has been relatively stable despite mild fluctuations, accounting for just under half of the Indian workforce: from 49.8 percent in 1999–2000 to 49.6 percent in 2011–12.[14] However, this aggregate figure is likely to be an *overestimate* of the *wage workers employed by profit-seeking capitalist enterprises*, since some of them were hired by petty commodity producers to supplement family labor, or public agencies for local area development, employment generation, and poverty reduction programs with redistributive objectives (Rangarajan, Iyer Kaul, and Seema 2011: 71). Moreover, there may be a certain degree of reversibility in employment status involving temporary and contingent shifts between self-employment (family labor) and casual work ("distress" wage employment), depending upon changes in economic conditions (Himanshu 2011: 53–4; Hirway 2012: 68; Kannan and Raveendran 2012: 79–80; Rangarajan, Iyer Kaul, and Seema 2011: 69–70).[15] These considerations suggest that wage workers *employed by capitalist enterprises* are likely to have constituted considerably *less than half* of the Indian workforce over 1999–2012.

Viewed in disaggregated terms, the share of regular wage/salaried employees in the Indian workforce increased from 15.8 percent in 1999–2000 to 20.3 percent in 2011–12, while that of casual wage workers fell correspondingly from 34.0 percent to 29.3 percent. Since the proportion of the self-employed has been stable over the same period, the growth of the regular wage-employed must have taken place at the expense of the casual wage workers (Rangarajan, Seema, and Vibeesh 2014: 118). Such steady growth in the proportion of regular employees (Rangarajan, Iyer Kaul, and Seema 2011: 71) and the corresponding decline of casual wage workers suggests that *institutionalized capitalist production relations* may have expanded and become more consolidated in India over 1999–2012. Nonetheless, the share of regular wage employees remained relatively small at one-fifth or less of the total Indian workforce over this period.

Overall, it can be concluded that, while the proportion of the self-employed in the Indian workforce accounted for a slightly larger share compared to that of total wage employees, it undoubtedly constituted a much greater proportion than that of the *subset of workers employed in capitalist enterprises*. Moreover, the proportionate composition of the total workforce in terms of employment status remained stable during 1999–2012, with only minor fluctuations.

3.5 Interpreting Trends in Wage Work and Self-Employment

These trends in the composition of the Indian workforce over 1999–2012 flag a number of issues for further consideration. The fact that the self-employed constituted the single largest component and accounted stably for half of the workforce over the entire period, raises questions about what happens to the livelihood and employment status of dispossessed groups after the loss of their lands.

3.5.1 *Divergence from the "Transition to Capitalism"?*

While attempting to interpret (earlier) trends in wage and self-employment, Sanyal (2007: 40) questions the very applicability of the classic Marxian schema of "transition to capitalism" to contemporary India. Instead, he contends that the dynamics of capitalist development generates contradictory trends: "while on the one hand, the process of primitive accumulation... leads to the destruction of the precapitalist sectors, on the other, it simultaneously produces a space that necessitates the recreation of those sectors" (Sanyal 2007: 39). In other words, Sanyal postulates a two-way process such that the breakdown and "re-creation" of precapitalist or non-capitalist production units take place in parallel, entailing a complex outcome that diverges partly from the schema of transition to capitalism. However, although he refers to primitive accumulation and "the internal logic of the expanded reproduction of capital," Sanyal does not provide a clear causal explanation of how and why precapitalist (or non-capitalist) units are "re-created" after their disintegration through mechanisms of primitive accumulation.

While noting the significance of self-employment in petty commodity production (PCP) in the Indian economy, Barbara Harriss-White (2012: 117–18) observes: "[PCP] is *not transitional.* If it is but a stage in the differentiation of individual capitals, it is *constantly being replenished* and *reproduced*... Its existence *does not imply a teleology of development*" (emphases added). This view may be paraphrased as follows: while individual PCP units may disintegrate through differentiation processes, they are also being continuously replenished, such that PCP as an aggregate category continues to be reproduced over time. However, the existence of PCP does not imply that it is a "transitional" category in a teleological framework of (capitalist) development.

Despite their differences, the views of Sanyal and Harriss-White are suggestive of a two-way process through which PCP units are being simultaneously undermined and re-generated in contemporary India. I have built upon these ideas to formulate a complex hypothesis to explain the employment outcomes of land grabs as the resultant of a two-way process, as follows. On the one hand, ongoing land alienation separates pre-existing landowners from their lands, including petty commodity producers, continuing to feed the ranks of the dispossessed. On the other hand, since the dispossessed do not find adequate wage employment, their alternative avenues of survival are confined to forms of self-employment. In the case of erstwhile producers who have lost their land, this involves reconstituting themselves into other kinds of self-employed enterprises that operate at a lower level of income and security (compared to the situation when they still had access to land).[16] Taken together, this two-way process results in the *net persistence* of the self-employed in the Indian workforce compared to wage workers.

3.5.2 *Self-Employed Producers and Circuits of Capital*

Sanyal and Bhattacharyya (2009: 39) propose a "three-tier" classification of the Indian labor force in which, apart from "direct wage-workers in capitalist production," they distinguish between two types/subcategories of the self-employed. Type I consists of those working "in informal enterprises linked to capitalist enterprises in the formal economy" through subcontracting, outsourcing, contract farming, and/or putting out arrangements (Sanyal and Bhattacharyya 2009: 38–9). Given their direct linkages with the capitalist sector of the economy, this category of self-employed producers are integrated with the circuit of capital.

In contrast, Type II consists of "independent" self-employed producers working "in informal enterprises ... with no direct linkages to the formal economy." Sanyal and Bhattacharyya (2009: 38–9) regard them as constituting "an *overwhelming share of the growing labour force* [that] has been absorbed mainly ... in rural non-farm and urban informal activities ... [and are located in] a *production space outside the circuit of capital*" (emphases added).

This typology becomes relevant in explaining Sanyal and Bhattacharyya's (40) conclusion that "There is ... a growing informalisation of the Indian economy with a clear trend towards self-employment as the main source of livelihood of the informal labour force." In particular, the limited growth of wage employees, coupled with the presence of both categories of the self-employed, one of which (Type II) holds an "overwhelming share of the growing labour force," could possibly account for the greater weight of the self-employed compared to wage workers.[17]

However, the claim that the Type II self-employed are "excluded from the circuit of capital" is open to question. In the contemporary Indian economy, there is little scope for autarkic production and even petty commodity producers are significantly involved in the market, as indeed noted by Sanyal and Bhattacharyya (2009: 38). Consequently, to the extent that the Type II self-employed transact in input and output markets, they cannot avoid interacting with other classes of economic agents who also participate in the same markets, as well as the circuit of capital mediated by these "common markets" (cf. Adnan 1985: PE-53–5). Therefore, the claim that this component of the self-employed is "excluded from the circuit of capital" is not warranted.

Moreover, Sanyal and Bhattacharyya (2009: 38) do not distinguish between the circuits of *productive* and *unproductive* capital, or their qualitatively different relations with self-employed producers. The circuit of productive capital is generated by capitalist production based on inputs purchased from the market including hired wage labor, leading to market sale of the products for realization of profit (surplus value) (Marx 1976). In contrast, circuits of unproductive capital are generated by usurers' and merchants' capital through manipulation of price differences in non-labor input and product markets (Marx 1981: 728–45). The mechanisms of surplus extraction used by unproductive capital are *systematically predicated upon self-employed producers* who are exploited through the credit and/or product markets, as contrasted to the employment of wage workers through the labor market (Adnan 1985: PE–59–60). The same applies to the extraction of precapitalist ground rent through the lease market. Consequently, rather than presupposing that a category of self-employed producers (Type II) is "excluded from" the circuit of capital, it is more realistic to explore the *differential outcomes* of their interactions with the circuits of *productive* and *unproductive* capital, as taken up in Section 3.5.3.

3.5.3 *Outcomes of Land Acquisition for a Special Economic Zone*

The ground-level processes activated by land grabs and their impacts can be traced with the micro-level evidence from Levien's (2012: 948–65) ethnographic study of the outcomes of land acquisition for the MWC SEZ in rural Rajasthan. Significantly, the households whose lands had been acquired for the SEZ did not show any lack of market linkages indicative of "exclusion from the circuit of capital" (as postulated by Sanyal and Bhattacharyya for the self-employed Type II). On the contrary, the extent of their market participation had *increased* after acquisition of their lands, because they now had to buy the goods that they had formerly produced themselves. The local economy as a whole displayed increased levels of commoditization because of the ways in which the money received as compensation for land acquisition had been utilized, as well as associated changes in production relations involving the land-losing households (Levien 2012: 953–60).

First, the households that had received large amounts of compensation money typically invested to a greater extent in circuits of *unproductive* capital compared to productive investment involving capitalist production with wage labor (Levien 2012: 954). In many cases, they became involved in extraction of surplus as merchant–usurers, collection of precapitalist ground rent through purchase and leasing out of new lands, the "eating" of commission as land brokers, and/or making gains from real estate speculation (Levien 2012: 956–63). Second, the land-losing households with less compensation money resorted to nonfarm self-employment, involving sale of a variety of products and services in the village economy, becoming *transformed from one kind of PCP* (landholding peasant) to *another* (non-farm enterprise). Moreover, some of them were subjected to market-based surplus extraction by usurers' and merchants' capital or payment of precapitalist rent. Third, very few of the dispossessed households gained wage employment in the high-tech capitalist enterprises established on their former lands in the SEZ. Even those who were employed had to work in low-paying and insecure jobs such as guards, janitors, gardeners, and drivers, given their lack of skills and education (Levien 2012: 949).

Overall, the households whose lands were acquired for the high-tech MWC SEZ experienced multiple and divergent outcomes. Paradoxically, even though land acquisition was propelled by the needs of advanced capitalism, it did not generate significant wage employment for those dispossessed in the process. Most of them resorted to different forms of self-employment and/or surplus extraction based on unproductive capital and land rental, as contrasted to involvement in the circuit of productive capital as either capitalist producers or wage workers. It is evident that, in this particular instance, the short-run outcomes of primitive accumulation through land acquisition deviated significantly from the Marxian schema of "transition to capitalism."

While the findings of this case study of a single SEZ are obviously not generalizable for all India, they provide insights into some of the critical factors shaping the employment outcomes of primitive accumulation through land acquisition. In particular, the following hypothesis is suggested: the predominance of unproductive capital and (precapitalist) rent extraction systematically constrains the dispossessed to take up forms of self-employment for survival, as contrasted to the option of wage employment made available by productive capital.

3.6 Concluding Remarks

In this concluding section, I integrate the arguments developed in previous sections and draw out their implications for the mechanisms and determinants of primitive accumulation/ABD in neoliberal India. A set of hypotheses is put forward to explain trends in the employment status of the workforce in the Indian economy (Section 3.6.3). The domestic and global factors influencing the prospects of the transition to capitalism in India are briefly indicated.

3.6.1 *Mechanisms of Primitive Accumulation and ABD*

As evident from the preceding discussion, the concrete mechanisms of land alienation in neoliberal India display striking diversity. They involve market or nonmarket agencies, as well as the application or absence of force; they operate directly or indirectly, entailing intended or unintended consequences respectively (Adnan 2013: 92–3, 2014: 33, 2016; cf. De Angelis 2004: 78–80). Taken together, these varied options constitute a diverse *repertoire* from which *alternative mechanisms and strategies* of primitive accumulation/ABD can be chosen flexibly to suit specific social-historical circumstances (Harvey 2006: 159; Burawoy 1985: 214–19; Tilly 1978: 151–8).

In neoliberal India, ongoing primitive accumulation through land alienation has been facilitated by policies of deregulation, privatization, and austerity, as well as the growing significance of speculation by finance capital. These features correspond to Harvey's notion of ABD, and demonstrate its applicability to land alienation in the context of neoliberal globalization. Viewed in this perspective, primitive accumulation can be regarded as a *generic capitalism-facilitating* process, which can assume particular forms such as ABD at specific sociohistorical conjunctures (Adnan 2013: 123, 2014: 39, 2016; Hall 2013; Sampat 2008).

3.6.2 *Factors Shaping Primitive Accumulation and Its Outcomes*

The varied mechanisms of primitive accumulation/ABD in neoliberal India were driven or constrained by powerful antecedent forces and mediating institutions. In particular, the state exercised its power of eminent domain as well as its repressive capabilities to enforce land acquisition and crush instances of social and ideological opposition (Adnan 2014: 38, 2016; Jenkins, Kennedy, and Mukhopadhyay 2014) (see Kenneth Bo Nielsen and Alf Gunvald Nilsen in this volume (Chapter 5) for this dual role of the state). Moreover, private corporations and finance capital grabbed new territories and resources, making use of force or market transactions as necessary (De Angelis 2004: 75; Sebastian 2012: 13–16; Walker 2008: 583–95). Taken together, the Indian state, as well as domestic and transnational corporations and financial organizations, served as key drivers of primitive accumulation/ABD in the context of neoliberal globalization (Sampat 2008; Sebastian 2012: 13–16; Walker 2008).

Correlatively, land alienation in neoliberal India was constrained by resistance in diverse forms, as well as in different modalities, ranging from stand-alone movements to coordinated collective action in multiple arenas (Bharadwaj 2009; Levien 2012; Sampat 2015).

The contestations between these factors driving and constraining primitive accumulation/ABD in contemporary India have cut both ways, with varying outcomes. In certain instances, opposition to land grabs has been suppressed or defused by mechanisms ranging from violent repression to co-option and negotiated settlements, facilitating primitive accumulation. In other cases, however, resistance movements coordinated at local, national, and/or international levels, have been able to delay or cancel SEZs and other land-acquiring projects (Levien 2012; Sampat 2015; Walker 2008). Wherever such struggles have succeeded in stopping or slowing down land grabs, these have also served to impose social and political limits on the pace and extent of primitive accumulation/ABD in neoliberal India (Adnan 2014: 39).

Moreover, instances of resistance to primitive accumulation also *constrained capitalist production indirectly*, by limiting the supply of its required inputs such as land.[18] It is such *interactions* between the mechanisms of dispossession and resistance that have shaped the trajectory and contested limits of primitive accumulation and capitalist development in neoliberal India.

3.6.3 *Factors Sustaining Self-Employment in the Indian Workforce*

The arguments developed in Sections 3.6.1 and 3.6.2 have been placed within the broader framework of the Indian and global economies to formulate the following hypotheses to explain the trends in the employment status of the workforce, noted earlier.

First, the formal and informal capitalist sectors of the Indian economy have not been generating sufficient incremental wage employment to absorb the vast labor force available. This is in part due to the highly capital-intensive (and labor-displacing) technologies that firms in India's formal sector have been constrained to adopt in order to remain competitive in the global market (Bhaduri 2008; Sanyal 2007). While regular wage work has increased compared to casual wage labor over 1999–2012, the scale of their total labor absorption has not been such as to make a significant dent in the weight of the self-employed in the Indian workforce.

Second, faced with the lack of wage employment after loss of their lands, most dispossessed groups have become involved in circuits of unproductive capital, or invested the money received as compensation payment or sale revenue in buying new land for extracting precapitalist ground rent. The production relations generated by the circuits of unproductive capital controlled by merchant–usurers, as well as the appropriation of rent by landlords, are *systematically predicated upon extraction of surplus from self-employed producers* rather than wage workers (Adnan 1985: PE–59–60). Consequently, the predominance of unproductive capital and rent extraction over productive capital has systematically constrained the transformation of dispossessed groups into wage labor. This is likely to have been a critical factor contributing to the higher proportion of the self-employed compared to wage workers in capitalist employment in the Indian workforce, as well as the stability of this pattern over 1999–2012.

Third, the fact that half or more of the Indian workforce has continued to survive through self-employment over 1999–2012 cannot be uncritically attributed to the purported "welfarist governmentality" and redistributive "development" programs of the state, donor agencies, nongovernmental organizations, and so on (Adnan 2014: 36–7), despite claims to the contrary (Chatterjee 2008: 55–6; Sanyal 2007: 36–7, 59–61). While redistributive public works programs such as NREGA have generated incremental employment and raised wage rates across all segments of the labor market, their overall contributions to reducing poverty and unemployment have been at best marginal (Hirway 2012: 68; Kannan and Raveendran 2012: 80; Rangarajan, Iyer Kaul, and Seema 2011: 71). More generally, cutting down on redistributive and welfarist programs has been a key feature of the neoliberal policy regime characterizing contemporary India (Basu and Das 2009: 159; Baviskar and Sundar 2008: 87–9; Shah 2008: 80). Instead, the continued survival of the self-employed is to be largely attributed to their *resistance* to land alienation and primitive accumulation/ABD, as well as their *resourcefulness* in finding avenues of survival based on their own efforts, inclusive of undertaking "distress employment"—typically involving overwork and underconsumption (Himanshu 2011: 53–4; Hirway 2012: 68; Kannan and Raveendran 2012: 79–80; Rangarajan, Iyer Kaul, and Seema 2011: 69–70).

Taken together, these three hypotheses help to explain why at least half of the Indian workforce remained in various forms of self-employment over 1999–2012, rather than being increasingly absorbed in wage employment in capitalist production. It is possible that the continuation and accentuation of such causal processes in the long run could potentially make the trajectory of the Indian economy diverge from the Marxian schema of "transition to capitalism."

3.6.4 *Prospects of the Transition to Capitalism in India*

It should be noted that the analysis so far has been focused on the impacts of land alienation on the workforce *within the bounds of the Indian economy*. However, if the framework of analysis is broadened to take account of the wider *global* economy, other significant processes come into view. In particular, even during the neoliberal era, a fraction of those evicted from land in India has been joining the stream of international migrant workers who have found employment in capitalist enterprises in other countries of the world.

The paradox is that both the growth of international employment of migrant workers and the stagnation in domestic wage employment constitute divergent outcomes of common or comparable primitive accumulation processes taking place inside India. Given the *parallel operation of these opposed trends*, the direction of transformation of the Indian economy in the long run remains an open question. While the prospects of the "transition to capitalism" in India continues to remain uncertain, it cannot also be ruled out at this stage. Indeed, any teleological and deterministic presupposition or prediction about the longer-term outcome would be unwarranted at this juncture (Harriss-White 2012: 117–18; cf. Adnan 1985: PE–53).

Acknowledgments

I appreciate comments on the chapter made by the participants of the conference. I am particularly grateful for critical feedback from Amiya Bagchi, Barbara Harriss-White, Judith Heyer, and Pranab Basu. Comments from an anonymous reviewer were also helpful in revising the paper. I am obliged to Anthony D'Costa for his patience and editorial assistance. I thank Achin Chakraborty and Himanshu for helping with access to source documents and Sushil Khanna and Mritiunjoy Mohanty for discussions in Kolkata. This chapter was also presented at a national seminar at the Department of Economics of Visva-Bharati in Santiniketan on March 21, 2014, thanks to arrangements by Sudipta Bhattacharya.

Notes

1. Despite its innovative aspects, Harvey's concept of ABD has been subject to a range of theoretical criticisms, including conflation of its mechanisms with the normal workings of capitalism (Ashman and Callinicos 2006: 119; Brenner 2006: 98–102; Fine 2006: 143–6; Harvey 2006; Levien 2012).
2. However, critical reviews of the theoretical debates on primitive accumulation and ABD are available in my earlier publications (Adnan 2013: 91–9, 2014: 24–7, 2016).
3. See Adnan (2014: 38–9); Borras and Franco (2010); Das (2012: 33–5) Harvey (2005: 162–9); Levien (2012); Sanyal and Bhattacharyya (2009: 41–3); and Walker (2008).
4. See Adnan (2013: 97); Brenner (1976: 70–1, 1977: 73–5); De Angelis (2004: 79); Moore (1966: 13, 28); and Perelman (1984: 46).
5. See Levien's (2013) persuasive attempt to construct a periodization of modern Indian economic and political history in terms of "regimes of dispossession."
6. See Bhaduri (2008: 11); Chandrasekhar and Ghosh (2011); Chowdhury (2011: 23); Himanshu (2011: 43–57); Sanyal (2007: 245–7); Sanyal and Bhattacharyya (2009: 36–7); and Thomas (2012: 48).
7. D'Costa (2014) has attempted to explain this situation with the idea of "compressed capitalism."
8. See Harriss-White (2012); Levien (2012); Sanyal (2007); and Sanyal and Bhattacharyya (2009). A critical review of these interpretations is available in Adnan (2014: 33–7).
9. As of February 2016, there were an estimated 187 SEZs in operation throughout India. Available at: <http://en.wikipedia.org/wiki/Special_economic_zone#India> (accessed April 28, 2016).

10. For instance, by violating or bypassing the Panchayats (Extension to Scheduled Areas) Act, 1996 (PESA) (Bhaduri 2008: 13).
11. The sources of the data cited by Shaw (2013: 25, table 4) are the NSSO Employment and Unemployment Surveys of 1999–2000, 2004–5, 2009–10, and 2011–12. The data are computed according to the Usual Principal Status (UPS) measure based on "workers who are employed for a major part of the year" (Shaw 2013: 23).
12. In particular, a range of social-occupational groups are excluded from the labor force by definition (particularly code numbers 91–3), such that their possible economic role and contributions are not included in the NSSO data (Hirway 2012: 69–71). In addition, activities for household self-consumption involving "processing of primary products" and craft and artisanal activities are not even considered by NSSO surveys, while "activities related to agricultural production," though formally considered, are treated as negligible and hence those undertaking them are treated as non-workers (p. 69). However, Hirway argues that these assumptions of NSSO surveys are refuted by time use survey data (pp. 69–71).
13. The problematic issues pertain to the "missing labor force" and the "withdrawal" and "very low workforce participation" of particular social groups, such as women, the elderly, and the disabled (Chandrasekhar and Ghosh 2011; Himanshu 2011: 48–9; Hirway 2012: 68–71; Kannan and Raveendran 2012: 79, table 5; Rangarajan et al. 2014: 121).
14. Recomputed from the data provided by Shaw (2013: 25, table 4).
15. Himanshu (2011: 53–8) observes that there are fluctuations and reversible shifts between self-employment and casual wage work among sections of the Indian workforce, resulting from changes in economic conditions and policies.
16. Reviewing macro-level employment trends in the contemporary Indian economy, Himanshu (2011: 53) notes that cultivating households faced with limits to land are likely to either take up casual wage employment or "seek employment in the non-agricultural sector where this can take the form of self-employment."
17. The articulation of this argument by Sanyal and Bhattacharyya (38–40) is not very clear and coherent. This is the best interpretation that I could make of the text.
18. Historically, resistance to primitive accumulation by peasants and other non-worker classes have also resulted in indirect opposition to capitalism. I am indebted to Amiya Bagchi for comments elucidating this point.

References

Adas, M. 1986. "From Footdragging to Flight: the Evasive History of Peasant Avoidance Protest in South and South-East Asia." *Journal of Peasant Studies*, 13(2): 64–86.

Adnan, S. 1985. "Classical and Contemporary Approaches to Agrarian Capitalism." *Economic and Political Weekly*, 20(30): PE–53–64.

Adnan, S. 2007. "Departures from Everyday Resistance and Flexible Strategies of Domination: The Making and Unmaking of a Poor Peasant Mobilization in Bangladesh." *The Journal of Agrarian Change*, 7(2): 183–224.

Adnan, S. 2013. "Land Grabs and Primitive Accumulation in Deltaic Bangladesh: Interactions between Neoliberal Globalization, State Interventions, Power Relations and Peasant Resistance." *The Journal of Peasant Studies*, 40(1): 87–128.

Adnan, S. 2014. "Primitive Accumulation and the 'Transition to Capitalism' in Neoliberal India: Mechanisms, Resistance, and the Persistence of Self-Employed Labour." In *Indian Capitalism in Development*, edited by Barbara Harriss-White and Judith Heyer, London: Routledge, pp. 23–45.

Adnan, S. 2016. "Alienation in Neoliberal India and Bangladesh: Diversity of Mechanisms and Theoretical Implications." *South Asia Multidisciplinary Academic Journal (SAMAJ)* (online), 13: 1–20. Available at: <http://samaj.revues.org/4130> (accessed September 9, 2016).

Akram-Lodhi, H. 2012. "Contextualizing Land Grabbing: Contemporary Land Deals, the Global Subsistence Crisis and the World Food System." *Canadian Journal of Development Studies*, 33(2): 119–42.

Ashman, S. and Callinicos, A. 2006. "Capital Accumulation and the State System: Assessing David Harvey's *The New Imperialism*." *Historical Materialism*, 14(4): 107–31.

Banerjee, P. S. 2014. "Orissa and West Bengal: The SEZ Imbroglio." In *Power, Policy and Protest: The Politics of India's Special Economic Zones*, edited by R. Jenkins, L. Kennedy, and P. Mukhopadhyay, New Delhi: Oxford University Press, pp. 273–304.

Basu, P. K. 2007. "Political Economy of Land Grab." *Economic and Political Weekly*, 42(14): 1281–7.

Basu, D. and Das, D. 2009. "Political Economy of Contemporary India: Some Comments." *Economic and Political Weekly*, 44(22): 157–9.

Baviskar, A. and Sundar, N. 2008. "Democracy versus Economic Transformation?" *Economic and Political Weekly*, 43(46): 87–9.

Bhaduri, A. 2008. "Predatory Growth." *Economic and Political Weekly*, 43(16): 10–14.

Bharadwaj, S. 2009. "Gravest Displacement, Bravest Resistance: The Struggle of Adivasis of Bastar, Chhattisgarh against Imperialist Corporate Landgrab." *Sanhati*, June 1: 1–39. Available at: <http://sanhati.com/excerpted/1545/> (accessed February 19, 2013).

Borras, S. M. Jr. and Franco, J. 2010. "Towards a Broader View of the Politics of Global Land Grab: Rethinking Land Issues: Reframing Resistance." Initiatives in Critical Agrarian Studies (ICAS), Land Deal Politics Initiative (LDPI), and Transnational Institute (TNI). ICAS Working Paper Series No. 001.

Brenner, R. 1976. "Agrarian Class Structure and Economic Development in Pre-Industrial Europe." *Past and Present*, 70: 30–75.

Brenner, R. 1977. "The Origins of Capitalist Development: A Critique of Neo-Smithian Marxism." *New Left Review*, 104: 25–92.

Brenner, R. 2006. "What Is, and What Is Not, Imperialism?" *Historical Materialism*, 14(4): 79–105.

Burawoy, M. 1985. *The Politics of Production: Factory Regimes under Capitalism and Socialism*. London: Verso Books.

Chandrasekhar, C. P. and Ghosh, J. 2002. *The Market That Failed*. New Delhi: LeftWord Books.

Chandrasekhar, C. P. and Ghosh, J. 2011. "Latest Employment Trends from the NSSO." *Business Line*, July 12.

Chatterjee, P. 2008. "Democracy and Economic Transformation in India." *Economic and Political Weekly*, 43(16): 53–62.

Chowdhury, S. 2011. "Employment in India: What Does the Latest Data Show?" *Economic and Political Weekly*, 46(32): 23–6.

D'Costa, A. P. 2014. "Compressed Capitalism and Development: Primitive Accumulation, Petty Commodity Production, and Capitalist Maturity in India and China." *Critical Asian Studies*, 46(2): 317–44.

Das, B. 2012. "Resisting Development, Theorizing Resistance." *Economic and Political Weekly*, 47(33) (August 18): 33–7 (Review of Alf Gunvald Nilsen. 2010. *Dispossession and Resistance in India: The River and the Rage*, London and New York: Routledge).

De Angelis, M. 2001. "Marx and Primitive Accumulation: The Continuous Character of Capital's 'Enclosures'." *The Commoner*, No. 2 (September). Available at: <http://www.rebelnet.gr/files/De_Angelis,_M.,_2001._Marx_and_primitive_accumulation_The_continuous_character_of_capital%5C's_enclosures,.pdf> (accessed September 16, 2016).

De Angelis, M. 2004. "Separating the Doing and the Deed: Capital and the Continuous Character of Enclosures." *Historical Materialism*, 12(2): 57–87.

Fine, B. 2006. "Debating the 'New' Imperialism." *Historical Materialism*, 14(4): 133–56.

Hall, D. 2013. "Primitive Accumulation, Accumulation by Dispossession and the Global Land Grab." *Third World Quarterly*, 34(9): 1582–604.

Harriss-White, B. 2012. "Capitalism and the Common Man: Peasants and Petty Production in Africa and South Asia." *Agrarian South: Journal of Political Economy*, 1(2): 109–60.

Harvey, D. 2005. *The New Imperialism*. Oxford: Oxford University Press.

Harvey, D. 2006. "Comment on Commentaries." *Historical Materialism*, 14(4): 157–66.

Himanshu. 2011. "Employment Trends in India: A Re-Examination." *Economic and Political Weekly*, 46(37): 43–59.

Hirway, I. 2002. "Employment and Unemployment Situation in 1990s: How Good Are NSS Data?" *Economic and Political Weekly*, 37(21): 2027–36.

Hirway, I. 2012. "Missing Labour Force: An Explanation." *Economic and Political Weekly*, 47(37): 67–72.

Jenkins, R., Kennedy, L., and Mukhopadhyay, P. (eds) 2014. *Power, Policy and Protest: The Politics of India's Special Economic Zones*. New Delhi: Oxford University Press.

Jha, P. S. 2013. "Mr PM Think Twice before Jumping the Gun." *Tehelka*, 24(10) (June 15). Available at: <http://tehelka.com/mr-pm-think-twice-before-jumping-the-gun/> (accessed September 9, 2016).

Kannan, K. P. and Raveendran, G. 2012. "Counting and Profiling the Missing Labour Force." *Economic and Political Weekly*, 47(6): 77–80.

Kunnath, G. 2012. *Rebels from the Mud Houses: Dalits and the Making of the Maoist Revolution in Bihar*. New Delhi: Social Science Press.

Levien, M. 2012. "The Land Question: Special Economic Zones and the Political Economy of Dispossession in India." *Journal of Peasant Studies*, 39(3–4): 933–69.

Levien, M. 2013. "Regimes of Dispossession: From Steel Towns to Special Economic Zones." *Development and Change*, (44)2: 381–407.

Marx, K. 1976. *Capital: A Critique of Political Economy*, vol. 1. Harmondsworth: Penguin Books in association with New Left Review.

Marx, K. 1981. *Capital: A Critique Of Political Economy*, vol. 3. Harmondsworth: Penguin Books in association with New Left Review.

Moore, B. Jr. 1966. *Social Origins of Dictatorship and Democracy: Lord and Peasant in the Making of the Modern World*. Boston, MA: Beacon Press.

Nigam, A. 2007. "NAPM Report: To Nandigram via Singur: A Preliminary Report on the Struggle and Violence in Nandigram." Available at: <http://kafila.org/2007/02/27/napm-report-to-nandigram-via-singur/> (accessed September 9, 2016).

Patnaik, P. 2008. "The Accumulation Process in the Period of Globalisation." *Economic and Political Weekly*, 43(26–7): 108–13.

Patnaik, P. 2011. "Things, Not People." *Frontline*, 28: 21. Available at: <http://www.frontline.in/navigation/?type=static&page=archive> (accessed February 9, 2013).

Patnaik, P. 2012. "The Peasant Question and Contemporary Capitalism: Some Reflections with Reference to India." *Agrarian South: Journal of Political Economy*, 1(1): 27–42.

Patnaik, U. 1979. "Neo-Populism and Marxism: The Chayanovian View of the Agrarian Question and Its Fundamental Fallacy." *Journal of Peasant Studies*, 6(4): 375–420.

Patnaik, U. (ed) 1990. *Agrarian Relations and Accumulation: The Mode of Production Debate in India*. Mumbai: Oxford University Press.

Patnaik, U. 2012. "Some Aspects of the Contemporary Agrarian Question." *Agrarian South: Journal of Political Economy*, 1(3): 233–54.

Perelman, M. 1984. *Classical Political Economy: Primitive Accumulation and the Social Division of Labour*. Totowa, NJ: Rowman & Allanheld.

Rangarajan, C., Iyer Kaul, P., and Seema. 2011. "Where Is the Missing Labour Force?" *Economic and Political Weekly*, 46(39): 68–72.

Rangarajan, C., Seema, and Vibeesh, E. M. 2014. "Developments in the Workforce between 2009–10 and 2011–12." *Economic and Political Weekly*, 49(23): 117–21.

Sampat, P. 2008. "Special Economic Zones in India." *Economic and Political Weekly*, 43(28): 25–9.

Sampat, P. 2015. "The 'Goan Impasse': Land Rights and Resistance to SEZs in Goa, India." *Journal of Peasant Studies*, 42(3–4): 765–90.

Sanyal, K. 2007. *Rethinking Capitalist Development: Primitive Accumulation, Governmentality and Post-Colonial Capitalism*. London, New York, and New Delhi: Routledge.

Sanyal, K. and Bhattacharyya, R. 2009. "Beyond the Factory: Globalisation, Informalisation of Production and the New Locations of Labour." *Economic and Political Weekly*, 44(22): 35–44.

Scott, J. C. 1985. *Weapons of the Weak: Everyday Forms of Peasant Resistance*. New Haven, CT and London: Yale University Press.

Scott, J. C. 1986. "Everyday Forms of Peasant Resistance." *Journal of Peasant Studies*, 13(2): 5–35.

Sebastian, G. 2012. "The Big U-Turn: New Forms of Primitive Accumulation and the Shift in Patriotic Sensibilities in India." Paper Presented at Seminar on "Parliamentary Democracy and Neoliberalism in India," October 18–19, 2012, organized by Institute of Parliamentary Affairs, Government of Kerala, in association with Department of Economics, University College, Thiruvananthapuram (mimeo).

Shah, M. 2008. "Structures of Power in Indian Society: A Response." *Economic and Political Weekly*, 43(46): 78–83.

Shaw, A. 2013. "Employment Trends in India: An Overview of NSSO's 68th Round." *Economic and Political Weekly*, 48(42): 23–5.

Thomas, J. J. 2012. "India's Labour Market during the 2000s: Surveying the Changes." *Economic and Political Weekly*, 47(51): 39–51.

Tilly, C. 1978. *From Mobilization to Revolution*. Reading, MA: Addison-Wesley Publishing Company.

Vasudevan, R. 2008. "Accumulation by Dispossession in India." *Economic and Political Weekly*, 43(11): 41–3 (Review of The Perspectives Team. 2007. *Abandoned: Development and Displacement*, New Delhi: Perspectives).

Vijayabaskar, M. 2014. "The Politics of Silence." In *Power, Policy and Protest: The Politics of India's Special Economic Zones*, edited by R. Jenkins, L. Kennedy, and P. Mukhopadhyay, New Delhi: Oxford University Press, pp. 305–30.

Walker, K. Le Mons. 2008. "Neoliberalism on the Ground in Rural India: Predatory Growth, Agrarian Crisis, Internal Colonization, and the Intensification of Class Struggle." *Journal of Peasant Studies*, 35(4): 557–620.

4

Agrarian Crisis and Accumulation in Rural India

Locating the Land Question within the Agrarian Question

Arindam Banerjee

4.1 Introduction

Historically, in global experiences of the development of capitalism, the "land question" has been intricately linked with the resolution of the "agrarian question." The emergence of capitalism within the old structures of feudalism, originally in Western Europe and later in other advanced capitalist countries, has also been a process of structural transformation of economies from a largely land-dependent mass of agricultural producers to the emergence of an industrial workforce, and subsequently to a sizeable service sector. This process of transition to capitalism has clearly not been a linear and homogenous one even within the developed world and it is crucially shaped by the manner in which the agrarian question was resolved in different *times* and *spaces* (Byres 2003a). The heterogeneity of this transition is underscored further when one looks at the experience of late developers (D'Costa 2014).

The agrarian question (AQ) has evolved historically based on the experience of different regions with capitalism. In that sense, AQ is a continually developing concept and one need not deem the classical versions as unchangeable or universally applicable. Undoubtedly, the classical AQ ideas regarding class alliances, development of productive forces, and mobilization of surplus for industrialization that were held as necessary conditions and outcomes of a successful resolution of the AQ have some relevance in contemporary developing countries. However, for a realistic understanding of the challenges before capitalist transition, these ideas should be applied in a manner that is amenable to accommodating the historical and regional specificities within which developing countries are situated.

In the context of developing countries, including India, there has been new thinking whether the classical agrarian question, whereby agricultural surplus is mobilized for industrialization, remains relevant under conditions of globalization or is bypassed without its resolution (Bernstein 1996; Lerche 2013). The AQ is held to be closed without its resolution in the classical sense because the role of agrarian accumulation of capital in structural transformation is supposedly undermined. However, the adoption of this position makes it difficult to answer some questions that recent development experiences have prompted. The increasing demand for land in rural areas to extend urbanization and industrialization has led to significant contradictions that rural communities face vis-à-vis land acquisition. A major question that needs to be answered is why, even under conditions where income from agricultural land increasingly fails to sustain livelihoods, the resistance of rural communities to land acquisition has intensified over the last few years.

Is the agrarian question passé for developing countries today? If so, then the historical role of the peasantry is reduced to simply abdicating claims on land and other natural resources for the purpose of industrialization. Should then one categorize peasant resistance to land acquisitions as a Luddite, anti-progressive, and anti-capitalist position? Will such a corporate-led growth process form a feasible path to development? The answer to these questions will help us comprehend the land question in India, which in turn, requires an examination of the patterns of contemporary agrarian accumulation in India and the status of the agrarian question. Both land acquisition, and the resistance to it, are intricately linked with the changed production conditions in Indian agriculture under neoliberal reforms.

In this chapter, I examine the relevance of the agrarian question in contemporary developing countries such as India and the feasibility and contradictions of a neoliberal corporate-led growth process, which is necessarily accompanied by significant displacement of petty producers from land and large-scale livelihood destruction. The chapter is organized as follows. In Section 4.2, I briefly discuss the broad evolution of India's agrarian question as it pertains to land. Section 4.3 analyzes the conditions of agriculture and the symptoms of the agrarian crisis based on primary data and secondary evidence. Based on an understanding of the agrarian crisis, Section 4.4 delves into some recent hypotheses on the agrarian question in developing countries under globalization and outlines the current juncture of the agrarian question in India. Section 4.5 locates the land question within the agrarian question and presents the conclusions.

4.2 Historical Prelude to the Agrarian Question

The role of colonialism in the early capitalist transitions is significant, to a large extent primary, in more than one way. The external sources of accumulation via the colonial drain of wealth and direct (European out-migration) and indirect (via export of manufactured commodities and colonial deindustrialization) export of unemployment to the colonies played a crucial role in the capitalist transition. The land question was often settled beyond the national boundaries through the forcible eviction of natives in America or Australia, in the "New World," from the best agricultural lands (Bagchi 1982: ch. 3; Magdoff 1978). No country striving for development in the postcolonial world has the advantage of accessing such momentous advantages, specific to the history of colonialism.

The agrarian question in developing countries, and the land question located within it, have always been different, both qualitatively and in terms of dimensions. The colonial experiences, which mostly preserved feudal economic institutions and sometimes established and ossified them, also added new dimensions to the agrarian question in the Global South. The integration of colonies such as India into the circuits of world trade without any significant transformation of the modes of production toward capitalism led to situations where the surplus was often accumulated in the hands of local ruling classes, subordinated by the metropolis, and mostly in precapitalist form (Patnaik 1999).

In the case of colonial India, there was a significant export orientation of agriculture where large masses of peasants were producing for the world market but under onerous tenancy conditions coupled with extra-economic control of their labor by a landlord class. The latter, while accumulating surplus, lacked the incentive to invest in agriculture and improve the productive capacities of land, since there were always other precapitalist ways of stepping up surplus appropriation.[1] Nevertheless, the landlord class, as the agent of land revenue collection, was a critical agent for the mobilization and transfer of tribute from the colony to the metropolis under British rule (Bagchi 2010).

The colonial experience meant that after independence redistribution of land emerged as the most important agrarian question that was crucial for the emancipation of the Indian peasantry from the oppressive stranglehold of the landlord class that also collaborated with imperialism. However, the landlord–bourgeoisie alliance that came to dominate the Indian state after 1947 prevented any meaningful land reforms in most parts of the country, except in a few pockets where protracted peasant struggles could not be suppressed by the ruling classes (Harriss 2013). Rather, capitalist development in agriculture was triggered more through the technological intervention of the green revolution strategy in the mid-1960s, and not through a radical transformation of the rural feudal society. Further helped by the development of public interventions like the crop management system, credit provisioning, seed research, and so on, and stepping up of public investment in agriculture, capitalist landlords (transition from above) and rich peasants (transition from below) emerged through a process of peasant differentiation.[2]

Land concentration remained high during this capitalist transition and the barrier of ground rent restricting investments in agriculture (Patnaik 1983) continued to prevail in most parts of the country. The unevenness of this capitalist development of Indian agriculture was also well noted in the literature. The landmark study by Bhalla and Singh (1997) revealed that agricultural growth in the 1960s and 1970s was mainly concentrated in the wheat-growing northern states such as Punjab, Haryana, and Western Uttar Pradesh, though it later spread to the eastern and southern regions in the 1980s. Institutional changes such as land redistribution and tenancy reforms in West Bengal played a major role in a late green revolution in the state (Saha and Swaminathan 1994).

A shift to neoliberal policies in the early 1990s led to the emergence of new conditions for Indian agriculture and consequently the agrarian question. Trade liberalization in agricultural commodities and reductions in public expenditure in agriculture posed new constraints on Indian farmers, leading to the precipitation of an agrarian crisis in many parts of the country (Patnaik 2002; Reddy and Mishra 2009; see also Chapter 9 in this volume). Since around 2004, a high-growth regime in the Indian economy, accompanied by a languishing agricultural sector, has also raised questions regarding the disjunction between the agricultural sector and the rest of the economy.

4.3 Agrarian Crisis and Accumulation under Neoliberalism

The agrarian crisis and stagnation is predominant now in different parts of the country, largely as a result over the last two decades of an export-oriented strategy, state withdrawal from agricultural investments, institutions, and services, and an overall neoliberal outlook toward small-scale petty production. Driven by fiscal fundamentalism, there was a veritable withdrawal of the state from economic operations, especially from agriculture (Banerjee 2009; Patnaik 2002). A host of policies adopted, such as the rationalization of input subsidies, the downsizing of public interventions including incentive pricing, a further decline in public investments, and the contraction of institutional credit availability in rural areas, all contributed to a deflation in real farm incomes and the emergence of indebtedness among the peasantry (Ramachandran and Swaminathan 2005; Reddy and Mishra 2009).

Further, the push for export-oriented agriculture implied that millions of small and marginal farmers in the country were exposed to global competition from other developing country farmers, particularly in certain export crops such as cotton and palm oil. This resulted in declining crop prices in the late 1990s and significant income deflation for petty producers in agriculture (Patnaik 2002). The enhanced volatility of crop prices in the world markets and rising food prices meant that the effects of higher export crop prices in the new century were not always unambiguously positive, in real terms, for the Indian farmers, resulting in a prolonged agricultural stagnation. The resultant agrarian crisis is primarily a crisis of accumulation among sections of the upper peasantry and a crisis of simple reproduction for large sections of the peasantry. A rich rural elite comprised of capitalist landlords, thin sections of the rich peasants, and other collaborative agents of organized capital within the rural areas may have found ways and means for continued accumulation even within this larger crisis. The primary data analysis later in the section substantiates this situation in agriculture.

To understand the various facets of this crisis, it is useful to examine the patterns of accumulation within the peasantry under the neoliberal economic regime. Based on a rural field study in the states of West Bengal and Telengana (Andhra Pradesh at the time of the survey) in 2006, we will look at some characteristics of the recent accumulation processes in rural areas and their implications for the land question. Of the two states studied, one is a food-growing region and the other a cultivator of commercial crops. This helps us to understand the accumulation question under two different conditions. Major results of this study have already been published (Banerjee 2009), where various dimensions of the agrarian crisis have been outlined. Here, we look at specific results from the study, which is relevant for our inquiry into the agrarian question.

4.3.1 *Setting and Methodology*

The question of accumulation is approached with the assumption of a heterogeneous peasantry, comprising various classes. The peasant classification is based on the capacity of farm households to appropriate surplus directly through use of hired labor and indirectly through secondary means of exploitation in land or credit markets. Among several methods of studying peasant classes (see Thorner 1982), this classification method based on hired labor use helps us to separate accumulation occurring in the "sphere of production" and the "sphere of exchange," allowing us to gain certain insights on the agrarian question.

In the rural areas, the landless labor and the landlord classes do not perform manual labor on their own farms; for the former this is involuntary, as they do not have any access to land, whereas for the latter it is voluntary. Between these two classes there is a heterogeneous peasantry comprising the upper and lower peasantry. The upper peasantry is made up of net labor-exploiting classes, within which the rich peasants are expected to be more dynamic in terms of accumulation and depend primarily on hired labor for cultivation. The accumulation of capital historically in the hands of these classes increases the capital stock on their farms and enables them to appropriate increasing volumes of surplus produced by the farm labor. We have used the labor-exploitation criterion developed by Patnaik (1976) to identify and study the various peasant classes in these areas.[3]

The survey covered seventy-seven households in two villages in the Raina-2 Block, Bardhhaman district in West Bengal. These villages had an agro-climactic characteristic akin to the fertile alluvial Indo-Gangetic plains. The abundance of groundwater and high levels of irrigation allowed multiple cropping; usually two crops of rice and potato are sown but farmers with greater access to money capital could go for a third crop of *boro* (winter) rice or vegetables. The region in Telengana is located in the Saidapur Block in Karimnagar district in Andhra Pradesh. The three villages surveyed in the Saidapur consist of sixty households, primarily cotton growers. Around half of their operated area is irrigated. The black soil in the region produces notably high yields of cotton when cultivated under irrigated conditions.

4.3.2 *Peasant Classes and Labor Exploitation*

The survey results organized in accordance with our peasant class framework and land-size classes reveal the agrarian structure in these two regions. In the West Bengal region, a large proportion of the households (89.6 percent) had small holdings within 5 acres. The most notable feature of this region is the predominance of a rich peasant class, which forms the largest economic class, with around 39 percent. This is primarily due to the high demand for hired labor for the multiple cropping practiced in this region. Significant migration of labor from other parts of the state and also from neighboring Jharkhand state meets this high demand. The rich peasant households are distributed across all land-size classes, implying that the high labor demand is generated from both small and large landholdings. This reflects the relatively more broad-based capital accumulation experienced in West Bengal after the land reform measures of the 1980s (also see Bhalla and Singh 1997).[4]

In Telengana, the distribution of the households across land-size classes is more spread out. One-quarter of the households operate more than 5 acres (2 hectares). Unlike in West Bengal, the rich peasant class is not only thinner (13.33 percent) but is also confined to the two highest size-class brackets (5 acres and above). The small and middle peasant classes jointly comprise

Table 4.1. Labor hiring-in and hiring-out: West Bengal (WB) and Telengana (TL) (in labor-days)

Class	F per hh	Hi per hh	100*[Hi/(F + Hi)]	Ho per hh	100*[Ho/(F + Ho)]	100*[Hi/(F + Ho)]
West Bengal, Raina						
RL	0.00	0.00	0.00	190.00	100.00	0.00
PP	24.04	28.72	54.44	228.52	90.48	11.37
SP	154.25	113.76	42.45	180.00	53.85	34.04
MP	106.00	102.88	49.25	21.31	16.74	80.81
RP	121.70	448.57	78.66	11.17	8.40	337.61
LLD	0.00	0.00	0.00	0.00	0.00	0.00
Total	85.26	211.38	71.26	97.26	53.29	115.81
Telengana, Saidapur						
RL	0.00	0.00	0.00	258.57	100.00	0.00
PP	201.43	62.29	23.62	160.00	44.27	17.23
SP	292.00	91.30	23.82	181.44	38.32	19.28
MP	298.33	169.96	36.29	5.00	1.65	56.03
RP	265.00	613.51	69.84	0.00	0.00	231.51
LLD	0.00	1919.00	100.00	0.00	0.00	0.00
Total	240.17	193.09	44.57	125.43	34.31	52.81

Notes: RL: Rural labor; PP: Poor Peasants; SP: Small Peasants; MP: Middle Peasants; RP: Rich Peasants; LLD: Landlords; hh: household; F: Family labor; H_i: Labor days hired-in; H_o: Labor days hired-out.

Source: Primary Field Survey

more than 60 percent of the households, indicating that the use of family labor is at least as important as hired labor for a majority of households in this irrigated cotton-cultivating region.

The class-differentiated character of the peasantry is amply reflected by the patterns of exploitation of hired labor and self-exploitation of family labor. In both regions, the methodology of categorizing peasant households into economic classes is validated by the wide differences in the hiring-in and hiring-out of labor and their relative importance vis-à-vis the use of family labor. The upper peasantry classes in both regions, particularly the rich peasant households are much more dependent on hired labor for cultivation. This differentiation within the peasantry is well reflected by Table 4.1.

The percentage of labor days hired-out (H_o) within the total labor days (family labor and labor days sold to others) worked by a household ($F + H_o$) drops sharply as we move from the lower to the upper peasantry. Similarly, the total labor days hired-in (H_i) as a percentage of total labor days ($F + H_o$) worked by a household multiplies many times as we move up the hierarchy of peasant classes. The latter variable theoretically implies the ratio of surplus appropriated from employing others' labor by a peasant household relative to the surplus created by their own labor (partly appropriated by themselves and partly exchanged for wages in the labor market). This ratio [$H_i/(F + H_o)$] for the rich peasant class in West Bengal is around thirty times that for the poor peasant class and nearly ten times that for the small peasants. Similarly, in Telengana, this ratio for the rich peasants is more than that for the poor peasants by a multiple of more than 13.

The upper peasantry classes are in a position to appropriate and accumulate the surplus produced within agriculture purely on the basis of their historically evolved production techniques. The specific capital–labor relations that their production system embodies due to the historical accumulation of capital marks these classes as the more dynamic ones in the context of any agrarian transition.

4.3.3 *Asset Ownership and Incomes*

The distribution of assets within the peasant household reflects the historical process of capital accumulation within the peasantry, which is crucial for any capitalist transition of agriculture to occur. The ownership of physical assets, more specifically the control over the means of production by a peasant class, determines the production technique that a peasant household employs in cultivation, capacity to exploit labor on their farms, and consequently, the volume of surplus/deficit generated. The investment of that surplus back into agriculture remains crucial for capital accumulation in the hands of the dynamic peasant classes.

Table 4.2 indicates the near monopoly of the rich peasants over assets in West Bengal. This class, which forms 39 percent of the households in this region, owns 80.5 percent of the assets. The concentration of productive assets is also high. They control 69.2 percent of livestock and 87.3 percent of machinery and implements. A deeper inspection of the field data reveals that the topmost land-size class, with an operated area of 10 acres or more (3.9 percent of the sample and all rich peasant households), owns 44.1 percent of the assets. High levels of inequality in asset ownership exist even within the rich peasant class, implying an intraclass differentiation, the source of which we elaborate later in this discussion.

In Telengana, the concentration of assets is relatively more diffused compared to West Bengal, as the middle peasant class also has a significant share in the assets relative to their economic class size. The only landlord household in this region, the rich peasants, and the middle peasants together own 73.4 percent of livestock and 80.3 percent of machinery and implements, indicating strong control over productive capital by all the labor-exploiting classes.

The concentration of productive assets in both regions is compatible with the higher exploitation of hired labor by the upper peasantry classes. Does this translate into higher surplus appropriation and capital accumulation by these classes? For this purpose, we look at different indicators of income: farm labor income (FLI), farm disposable incomes (FDI), and the total household

Table 4.2. Distribution of assets across peasant classes (in %)

Class	Distribution of households	Total assets	Livestock	Implements and machinery
WB				
RL	2.6	0.2	0.7	0.0
PP	32.5	8.5	8.9	1.0
SP	5.2	1.3	1.6	4.2
MP	20.8	9.4	19.6	7.6
RP	39.0	80.5	69.2	87.3
LLD	0.0	0.0	0.0	0.0
Total	100.0	100.0	100.0	100.0
TL				
RL	11.7	2.2	0.0	0.1
PP	11.7	4.6	6.5	1.4
SP	41.7	22.5	20.1	18.1
MP	20.0	21.3	41.0	9.9
RP	13.3	33.2	29.2	28.5
LLD	1.7	16.3	3.2	41.9
Total	100.0	100.0	100.0	100.0

Notes: RL: Rural labor; PP: Poor Peasants; SP: Small Peasants; MP: Middle Peasants; RP: Rich Peasants; LLD: Landlords; WB: West Bengal; TL: Telengana.

Source: Primary Field Survey

income (THI).[5] This sheds light on the constraints facing different peasant classes as far as the realization and re-investment of surpluses in their farms are concerned.

The FLI represents the income of farm households from cultivation, prior to the payment of rent and interest, depicting the hypothetical economic situation where land and credit monopoly are absent. Correspondingly, FDI is the farm income after ground rent and interest is paid out. A peasant household has to reproduce family labor, save, and undertake farm investments from this disposable income. The divergence between the FLI and FDI represents the dimensions of secondary exploitation relations, that is, extraction of farm surplus through rent and usury, which exist in Indian agriculture in various degrees.

In West Bengal, the FLI of the rich peasants is more than twenty times that of the poor peasant class (Table 4.3). The thin, small peasant class in this region generates a negative FLI primarily due to very low yields of potato. This can be held as an exceptional case, although even with normal yields, their FLI would have still been much lower than the middle or rich peasants. In Telengana, the rich peasants and the landlord household have markedly higher incomes than the other three peasant classes. The average FLI for the rich peasants is as much as 45.1 percent of the gross value of output (GVO); the sole landlord household has an FLI more than 40 percent of the GVO. The direct surplus appropriation through labor exploitation by the topmost classes

Table 4.3. Income situation across peasant classes (in INR)

Class	FLI [1]	FLI as % of GVO [2]	Interest payment* [3]	Rent* [4]	FDI [5=(1−(3 + 4)]
WB					
RL	400	10.3	400 (100)	0 (0)	0
PP	1253	10.4	1264 (100.9)	154 (12.3)	−165
SP	−4146	−14.2	8669 (−)	750 (−)	−13,565
MP	12,990	27.5	3014 (23.2)	1759 (13.5)	8218
RP	28,964	21.5	8980 (31.0)	1057 (3.6)	18,928
LLD	0	0.0	0 (−)	0 (−)	0
Total	14,186	20.9	4996 (35.2)	866 (6.1)	8324
TL					
RL	−65	0.0	7336 (−)	0 (−)	−7401
PP	8330	31.1	5449 (65.4)	6157 (73.9)	−3275
SP	2776	9.0	4354 (156.8)	3792 (136.6)	−5369
MP	10,411	17.2	12,764 (122.6)	1483 (14.2)	−3837
RP	106,968	45.1	76,431 (71.5)	500 (0.5)	30,037
LLD	246,497	41.8	38,000 (15.4)	8000 (3.3)	200,497
Total	22,574	32.5	16,683 (73.9)	2795 (12.4)	3097

* The figures in parentheses in Columns 3 and 4 are interest payments and rent expressed as a percentage of FLI
Notes: FLI: Farm Labor Income; GVO: Gross Value Output; FDI: Farm Disposable Income; RL: Rural labor; PP: Poor Peasants; SP: Small Peasants; MP: Middle Peasants; RP: Rich Peasants; LLD: Landlords; WB: West Bengal; TL: Telengana.
Source: Primary Field Survey

is definitely way ahead of the remaining peasant households, as expected in a peasantry with capitalist differentiation.

When we look at the FDI, the income situation drastically worsens. All lower peasantry classes and the middle peasants in Telengana generate a negative FDI, signifying the difficulty of mere simple reproduction of most peasant farms (see also Chapter 9, this volume). Of more significance is the voluminous drain of the farm incomes of the upper peasant classes, primarily by interest payments. For rich peasants in West Bengal, 31.0 percent of the FLI was required to service interest payments and in Telengana, the same figure was a massive 71.5 percent. Relatively, land rents do not bleed a major portion of the farm incomes, except for poor and small peasants in Telengana.

What emerges from the above analysis of farm incomes is the stranglehold of usury and indebtedness over the peasant classes that produce significant surplus from cultivation and livestock. A further detailed inquiry into the sample data results reveals that the sole capitalist landlord household in Telengana and the three rich peasant households in West Bengal who cultivated the largest landholdings (some of them also owning the rice mills in the locality and doubled up as crop commission agents or intermediaries) were the rare households that did not experience any significant depletion of their surpluses through secondary channels of exploitation like usury. This also explains the intraclass differentiation within the West Bengal rich peasants.

Most rich peasant households, while appropriating surplus produced by wage labor on their farms, lost that surplus to various usurious interests in the village. The field enquiry also revealed that the informal credit sources were mostly seed and input dealers or village residents who had formal jobs in the service sector and less commonly the old landlord–moneylender that were earlier observed (Bhaduri 1973; Chandra 1974) in the 1970s in some parts of the country.

This escalated importance of usury in the study regions is not exceptional but is a more prevalent phenomenon in Indian agriculture in the reforms period. Rising indebtedness has become both a cause and effect of the pervasive agrarian crisis since the late 1990s. The Situation Assessment Survey of Farmers conducted by the National Sample Survey in 2002–3 revealed that more than 48 percent of farm households were indebted. Agricultural and professional moneylenders had gained importance as a source of credit to rural households during the reforms period, accompanied by a corresponding withdrawal of commercial banks from rural areas as part of the banking reforms in the 1990s (Banerjee 2008; Ramachandran and Swaminathan 2005).

Table 4.4 further underscores the fact that without any secondary appropriation of surplus and nonagricultural incomes, it is not possible to carry out either a simple reproduction of peasant farms or investments in agriculture. In West Bengal, while the FDI does not allow most households to meet a calorie-norm-based consumption poverty line,[6] the income from service charges and self-employment enables a thin section within the rich peasant class to supplement their incomes considerably. Three rich peasant households cultivating land between 5 and 10 acres are classified as non-poor mainly due to their nonfarm income. Another three rich peasant households cultivating more than 10 acres and also the major rice mill owners of the region actually end up with considerable investable surplus. However, the average rich peasant households barely meet the poverty line, implying that the remaining twenty-four households within this class have serious problems in even a simple reproduction of their farms. In Telengana, the landlord household has some nonfarm income due to other formal service employment while all other classes cannot meet the poverty line consumption levels.[7]

These findings are to a large extent corroborated by the Situation Assessment Survey of Agricultural Households in 2012–13. Table 4.5 (see Column 12) reveals that more than 86 percent of agricultural households in India have negative or paltry investments; their farm and nonfarm incomes taken together mostly fall short of the required consumption expenses.

Table 4.4. Total household income and poverty line per household by economic and land size-classes: West Bengal advanced region (in INR)

Class	FDI [1]	Income from paid employment [2]	Income from service charges, self-employment, etc. [3]	Total household income (annual) [4=1+2+3]	Annual household poverty line (2200 Kcal) [5]
RL	0	11,400	0	11,400	30,942
PP	−165	14,150	401	14,385	34,380
SP	−13,565	10,200	600	−2765	41,256
MP	8218	1279	0	9496	30,512
RP	18,928	670	17,933	37,531	48,361
LLD	0	0	0	0	0
Total	8324	5947	7148	21,419	39,291
0	0	11,400	0	11,400	30,942
0.01–1.0	−684	9200	244	8760	31,697
1.01–2.5	10,706	2100	126	12,932	31,485
2.51–5.0	12,505	2000	2222	16,727	64,176
5.1–10.0	47,362	0	11,333	58,695	48,132
10.1 and above	70,309	0	161,333	231,643	114,600
Total	8324	5947	7148	21,419	39,291

Note: Land size-classes are in acres; RL: Rural Labor; PP: Poor Peasants; SP: Small Peasants; MP: Middle Peasants; RP: Rich Peasants; LLD: Landlords.

Source: Primary Field Survey, Column 5 estimated by author

What the above assessment implies is that even the labor-exploiting rich peasants are unable to achieve social reproduction; undertaking productivity-enhancing farm investments is beyond their means. Reducing consumption levels, taking out consumption loans from informal sources, or defaulting on existing credit liabilities are the only ways in which they can drag themselves into the next production cycle. Indebtedness and immiserization has penetrated the upper peasantry under neoliberal conditions. Capitalist landlords and thin sections of the rich peasants who have established extended control over the value chain in cultivation, both on input and output ends, continue the accumulation process more vigorously within the conditions of agrarian crisis. It is evident that the base of capital accumulation has also shrunk in the process.

This transformation in the process of surplus appropriation in agriculture under the neoliberal policy regime has important implications for the agrarian question in the country. The new pattern of accumulation that has accompanied the agrarian crisis has hinged on secondary relations of exploitation through myriad exchange relations rather than primary labor relations within the sphere of production. Based on these findings, we discuss the classical agrarian question and how that changes under the twin conditions of neoliberalism and globalization in developing countries such as India.

4.4 Is the Agrarian Question Relevant under Globalization?

The conceptualization of the agrarian question within Marxist scholarship has evolved greatly over time and yet remains as contentious as ever. The classic agrarian question broadly has three different yet interrelated aspects (Bernstein 1996; Byres 2003a). Bernstein (1996) identifies these various facets respectively as AQ1, AQ2, and AQ3. The classical debates have progressively engaged with the issue of political class alliances (primarily between the working class and the peasantry) for a transition to socialism (AQ1; Marx/Engels), the development of productive forces by a dynamic rich peasantry that challenges landlord-dominated feudalism (AQ2; Lenin/Kautsky), and finally, on the question of accumulation of surplus and capital necessary for industrialization in backward economies (AQ3; Preobrazhensky debates, formalized by Byres). Byres (2003b) cautions that such a dissection of the agrarian question, while correct, cannot be applied mechanically given that the three components are necessarily interconnected during any transition experience.

Recently, scholars from the South (Moyo, Jha, and Yeros 2013) have pointed out that an overemphasis of AQ2 and AQ3 in the discourse has led to a reductionist view of the agrarian question, particularly when it concerns developing countries that were ex-colonies. According to this view, this reductionism largely occurs due to the fact that the agrarian question is often primarily informed by the conditions under which different paths of capitalist development traversed in the North (W. Europe, Japan, and the USA), accompanied by a certain degree of obfuscation of the role of colonial exploitation in the process.

In contrast, in most ex-colonies[8] often in collaboration with local landed classes, the emancipation of the peasantry was an integral part of national liberation movements and constituted the primary component of the agrarian question (Moyo, Jha, and Yeros 2013). This argument assumes significance in the context of contemporary globalization, which increasingly re-incorporates developing country peasantries within the network of global capitalism, often on unequal terms.

The growing dominance of global finance capital since the 1980s, leading to the hegemony of the twin paradigm of *free trade* and *free markets*, has necessitated a rethinking of the agrarian question and its relevance in developing countries. Within this re-engagement, Bernstein's (1996, 2006) hypothesis of the *bypassing of the agrarian question* is a particularly strong and widely accepted formulation.

Bernstein argues that under the globalization regime, with freer flows of capital, it is not necessary for the agrarian question to be resolved within national boundaries. The AQ gets bifurcated into the *AQ of capital* and the *AQ of labor*. The former is resolved through the inflow of capital into a developing nation as foreign direct investment. Bernstein argues that while

Table 4.5. Rural India: Agricultural household monthly income from all sources, consumption expenditure, and investment in productive assets (in INR), 2012–2013

Size-class (ha.)	Net income receipts					CE	Balance	IA	Surplus/deficit	% of HHs	Cum. % of HHs
	W	C	AF	NFB	Total						
1	2	3	4	5	6	7	8=(6–7)	9	10=(8–9)	11	12
<0.01	2902	30	1181	447	4561	5108	−547	55	−602	2.6	2.6
0.01–0.4	2386	687	621	459	4152	5401	−1249	251	−1500	31.9	34.5
0.41–1.00	2011	2145	629	462	5247	6020	−773	540	−1313	34.9	69.4
1.01–2.00	1728	4209	818	593	7348	6457	891	422	469	17.2	86.6
2.01–4.00	1657	7359	1161	554	10,730	7786	2944	746	2198	9.3	95.9
4.01–10.00	2031	15,243	1501	861	19,637	10,104	9533	1975	7558	3.7	99.6
>10.00	1311	35,685	2622	1770	41,388	14,447	26,941	6987	19,954	0.4	100
All sizes	2071	3081	763	512	6426	6223	203	513	−310	100	

Note: W=wage income; C=Income from cultivation; AF=Income from animal farming; NFB=Non-farm business income; CE=Consumption expenditure; IA=Investments in productive assets; HH is Households.

Source: NSSO (2014)

developing countries at the moment of decolonization did not have any access to external sources of accumulation, globalization of capital opens up such opportunities in the form of foreign investment. Simultaneously, this diminishes the importance of historically accumulated agrarian capital for industrialization. Consequently, the AQ2 and AQ3 need not be resolved within rural class dynamics but are determined by the agency of foreign capital. The question of structural transformation of the workforce, or the AQ of labor, may not get addressed under globalization due to competition leading to rising capital intensity, efficiency, and scale requirements.

This liquidation of the national–historical agrarian question seems inappropriate from the perspective of developing countries on at least three grounds. First, the bypassing of the agrarian question in developing countries is premised heavily on assuming a typified path of development akin to advanced country industrialization. Most developing countries have not experienced structural transformation similar to the advanced capitalist countries. The "death of the peasantry" that was predicted (Hobsbawm 1973, 1994) has clearly not happened in large parts of Asia and Africa. This obstinacy of the peasants to not disappear, lack of adequate industrial jobs, and the proliferation of the informal economy are held as an indication that the agrarian question cannot be resolved within developing countries. The globalized accumulation of capital, according to Bernstein, is therefore a way this transition can progress.

This line of reasoning is untenable when we read the history of capitalism not independent of the history of labor. The significant out-migration of population from the early industrializers of Western Europe into the "New World" led to a re-population of white colonies, where land and other productive resources were grabbed from the indigenous populations; the latter often concentrated into reserves or exterminated (Bagchi 1982; Frank 1967). This out-migration was further intensified during the nineteenth century, which witnessed the fastest strides of capitalism in Western Europe. Estimates of migration from Europe between 1821 and 1915 (based on Kenwood and Lougheed 1971) stand at nearly 44 million. The population of Europe increased by 129 million in this period: from 134 million in 1821 to 263 million in 1915 (Maddison 2007). For every two persons in Western Europe who were born and stayed back during this period, one had to emigrate from the continent in search of jobs and a livelihood. The international out-migration from developing countries in contemporary times is of relatively miniscule dimensions compared to the above.

Apart from this "direct" export of unemployment, early industrializers such as Britain also had the advantage of "indirectly" exporting the unemployed through the forcible imposition of "free trade" on colonies and the export of machine-made manufactures, causing deindustrialization and job losses in the latter (Bagchi 2010). Additionally, colonial transfer of "tribute" from colonies buttressed capital accumulation and financed industrialization both in Britain and the "New World" (Habib 1995; Mukherjee 2010). The sustained resettlement of labor in the "New World," indirect export of unemployment, and the "drain of wealth" from colonies form important linkages between colonialism and the "very specific" path of capitalist development in advanced countries.

Since such options were never available in relatively labor-abundant developing countries after their de-colonization, capitalist transformation cannot be expected to replicate that of the early industrializers. Such an expectation can lead to erroneous positions on the status of the agrarian question in developing countries. This is not to argue that the transition in developing countries is complete but to underscore that it is an ongoing process whose coordinates are at wide variance with that of earlier capitalist experiences.

Secondly, under liberalization, the entry of foreign investment is definitely freer with reforms of capital-control legislation. However, such foreign investment mostly comes with a price for poor developing countries. The subsequent repatriation of capital back to the metropolis mostly imposes an additional burden on the receiver of such capital. Further, freer capital mobility also opens up the possibility of "capital flight" from these countries, engulfing the latter in financial crises. At any cost, foreign investment does not substitute for *colonial tribute*,[9] which added to capital accumulation in core economies without any additional cost.

Finally, any foreign capital that enters, for example, agribusiness capital, must still negotiate the political economy of agrarian class structures in specific countries, unless there is an absolute corporate takeover of agriculture. The historical peasant differentiation will determine which classes of peasants can get incorporated within the global value chains, which demand high-quality production for world markets. The terms and conditions under which such incorporation/subordination takes place through institutions like contract farming, will also differ across peasant classes. Undoubtedly, with global capital incorporating local/national capital from different countries/regions through mergers and joint ventures, this negotiation between organized capital (not always foreign) and petty producers assumes a complex form; at no point, however, can it be said that local agrarian class structures are irrelevant.

In fact, with the growing dominance of global capital, a section of the erstwhile ruling classes discovers new modes of surplus appropriation. The modes of exploitation in upstream and downstream activities in agriculture that have gained prominence with the agrarian crisis in India reflect the collaboration of agrarian ruling classes with organized capital in direct and indirect ways. Even as international capital emerges stronger than ever, it may still not be plausible that Bernstein's *agrarian question of capital* gets entirely resolved outside the national boundaries. Bernstein's position would amount to "determinism of global capital" as termed by Byres (2003b).

Recently, Lerche (2013) has put forward a nuanced argument that India has bypassed the "classical agrarian transition" under the influence of neoliberal policies. He stops short of adopting Bernstein's argument of the agrarian question being entirely bypassed on the grounds of continuing accumulation in Indian agriculture: "... surplus value created in agriculture does feed into accumulation of traders-cum-moneylenders, agro industries and multinational companies operating in the up- and downstream agriculturally related industries ... " (Lerche 2013: 400).

This observation is similar to our findings wherein secondary means of exploitation determine the volume of accumulation (or dissipation) of capital by capitalist landlords/farmers or sections of rich peasants. Lerche, however, argues that this capital accumulation does not play the historic role of developing productive forces in agriculture (AQ2) or fueling the process of industrialization (AQ3): "agriculture does not appear to significantly support growth in Indian non-agricultural sectors, neither through capital transfers nor through the creation of a major rural market for industrial produce" (Lerche 2013: 400).

This hypothesis can only be partially correct since even during the agrarian crisis the agricultural input industries or crop marketing/retailing companies have not faced any significant strain. Rather, the agrarian crisis has evolved as a phenomenon where accumulation for more organized capital and their agents has proceeded at the expense of petty producers whose incomes are squeezed. One cannot argue that the rural capital accumulation is entirely delinked from the growth process in the country. There is evidence of investment by rural ruling classes in real estate, hospitality sector, financial instruments, and so on. This diversion of the accumulated agrarian surplus to profitable sectors of the economy may still fuel the neoliberal growth pattern in the country, although it does not lead to industrialization, the latter being contingent also on factors external to the agrarian question, such as specific industrial and trade policies or competing speculative bubbles.

Although Lerche does not adopt Bernstein's position regarding the agrarian question, he remains skeptical regarding political alliances (AQ1) between workers and peasants that can restore centrality to the agrarian transition. For him, the dividing line in the agrarian structure in India is between the "labor-exploiting and accumulating classes" and the rest of the agrarian population. Pitting the landless and semi-proletarian classes against all dynamic accumulating rural classes is difficult and thereby the resistance to neoliberal policies will necessarily be rendered weak. Citing the estimates of ceiling surplus land in India (Rawal 2008), he argues that redistributive land reforms would mean meager benefits (0.35 acres per landless household) for the laboring classes, although it is a novel idea.

There are two reasons why this argument needs to be revisited. First, while it is true that the average land size is far too small to harness economies of scale, it does not automatically nullify the case for redistributive land reforms. While one cannot escape the challenge of structural transformation by reverting to neopopulist reforms, alternative paradigms of agricultural organization like cooperatives, need to be explored through alliances of rural labor and poor and small peasants. This can potentially address the labor question and the state must be forced to respond to such aspects of the agrarian question.

Secondly, accumulation and labor exploitation in agriculture are not identical under neoliberal conditions. Our data analysis in the previous section strongly illustrates such patterns of accumulation. The exploitation of labor alone does not ensure accumulation, the latter depending crucially on secondary relations like rent/usury/commissions, which Lerche also refers to. This implies a possibility of broader resistance against neoliberal globalization by including sections of middle/rich peasants. Needless to say, capitalist landlords/farmers and small sections of rich peasants, who have accelerated their accumulation during the agrarian crisis using secondary exploitative relations, will be outside this alliance. The prominent dividing line between small-scale petty capital and large organized capital (and their rural agents) assumes importance and is identical to the dividing line of accumulation.

The incorporation of rich/middle peasants in the alliance for the emancipation of the peasantry from the stranglehold of neoliberalism is an integral part of the agrarian question in India. However, with some degree of capital accumulation in the hands of middle/rich peasants under the earlier *dirigiste* regime, the latter are also significant employers of rural labor. This can create a possible contradiction in the agenda for redistributive land reforms and resisting neoliberalism. This is where we need to examine the land question and its relation with the agrarian question in India.

4.5 Locating the Land Question and Conclusions

The agrarian crisis and depression in farm incomes have complex implications for the value of agricultural land. There is no doubt that the productive value of agricultural land diminishes with declining farm incomes. This creates the appropriate context for grabbing farmland at low prices by a variety of interests such as real estate developers, service providers, or industrial/mining companies. Also, agricultural land at low value can be bought out by landed ruling classes, furthering the accumulation of social and economic power. Increasing land concentration and land acquisition in rural India can be viewed as a continuous process of primitive accumulation that accompanies the contemporary, corporate-led neoliberal growth process, although not necessarily leading to a structural transformation of the workforce.

Table 4.6. Share of rural households in operated area

	1993–4	2011–12
Bottom 70%	18.9	14.6
Middle 20%	33.2	35.3
Top 10%	47.9	50.2

Source: Based on Rawal 2013: table 6

Landlessness in rural India has notably increased in recent decades. Rawal (2013), using the NSS (National Sample Survey) Employment–Unemployment Surveys shows that landlessness in terms of proportion of rural households having no access to operated areas increased from 38.7 percent in 1993–4 to 48.5 percent in 2011–12. From Rawal's findings, it is evident that the share of the bottom 70 percent of households (including the landless) in the total operated area declined from 18.9 percent in 1993–4 to 14.5 percent in 2011–12 (Table 4.6).

The middle households gained in terms of operated area during this period largely through leasing-in. The distribution of owned area in rural areas is more concentrated and has worsened for the middle, too. Based on NSS data on land and livestock surveys, we have estimated that the share of both the bottom 60 percent and the middle 30 percent of rural households decreased to 8.3 percent and 36.5 percent of total owned area by 2002–3. The share of the top 10 percent of households was historically high in the post-independence period, at 55.2 percent (Nair and Banerjee 2012). The larger point is that while the bottom 60 percent of rural population is virtually landless, the rural middle classes also find it increasingly difficult to maintain ownership rights on agricultural land under the pressure of the ensuing agrarian crisis.

Income deflation in agriculture creates common ground between rural workers, semi-workers with marginal land holdings, and sections of middle/rich peasants as far as land is concerned. Unless there is a successful struggle to increase profitability in agriculture, the value of agricultural land will remain low with much less incentive for the rural poor to wage any struggle for land. While resistance to land acquisition, often led by middle peasants, has been spontaneous and frequent in recent times, movements for redistribution of agricultural land has not been so vigorous, not discounting ongoing land movements led by Ekta Parishad/Left parties in some parts of the country.

The change of land use from agricultural to nonagricultural is marked by drastic changes in the value of land. When there is an immediate change in land use with acquisition, under influences of speculative real-estate interests, the escalation in land values is momentous. The dimensions of these changes are difficult to fathom in advance when compensation for acquisition is determined. Changes in land prices due to speculation inevitably lead to ex-landowners feeling that the compensation is grossly inadequate, a negligible fraction of post-acquisition land values. This leads to one stream of resistance to land acquisition led by rural middle/upper-middle landowning classes, with the latter feeling excluded from the massive accumulation process. The land-poor rural masses have no common interest in these concerns.

A qualitatively different stream of resistance to land acquisition originates from small/middle peasants, who stand to lose their modest livelihoods. The lack of a meaningful alternative livelihood and the absence of state support in that regard makes the land owned and cultivated by them an indispensable and non-alienable asset. The market land prices used for determining compensation rarely capture the costs of livelihood transformations or possible future unemployment. Apart from these landowners, land acquisition for large development/industrialization projects, livelihoods of tenants/agrarian labor and non-agricultural petty producers is also affected. Even within complex socio-economic structures, this presents another situation where a broad political alliance of rural workers, agrarian and non-agrarian petty producers can be forged to oppose land takeovers and protect livelihoods; Singur in West Bengal and Kalinganagar in Odisha show that this is already happening in the country.

It is erroneous to conclude that either the agrarian question or agrarian transition has been bypassed in India. Rather, the resolution of the agrarian question has been arrested under neoliberalism. The agrarian question, primarily its political aspects (AQ1), is emerging as a conflict in the countryside between rural workers and petty producers on one hand and the thin accumulating classes and big capital on the other. The vast masses of informal labor in urban areas, often with rural linkages, and the urban working class, constitute natural allies in the resistance against neoliberalism. Defending the claims of millions of petty producers in rural India on land, natural resources, and livelihoods emerges as the primary focus of the contemporary agrarian question. The resolution of this fundamental political aspect will determine the development of productive forces in agriculture (AQ2) and the path of industrialization (AQ3).

In this context, the role of the state in protecting agrarian petty producers and employment-generating domestic industries from global competition assumes critical importance. Continued pursuance with neoliberal policies will narrow the base of capital accumulation in agriculture but also intensify agrarian struggles against such policies. Although the state is presently under neoliberal hegemony, it is clear that any capitalist development accompanied by large-scale alienation of peasant masses from land and employment destruction will be counterproductive and wrought with an intensifying crisis of livelihoods. What happens to the agrarian question and the embedded land question will depend primarily on political struggles of potential multiclass alliances against neoliberalism in rural India and its consequent impact on the state.

Appendix 1

Various categories of income of peasant households:

Farm labor Income (FLI)= Gross Value of Output (GVO) − [Total Material Input costs + Paid-out labor costs] (1)

Farm Disposable Income (FDI)= FLI − Rent Payments − Interest payments (2)

Total Household Income (THI)= FDI + Income from Paid-employment + Income from non-farm Self-employment (3)

The GVO is the sum of Crop Output, By-Product and Livestock Product. The Total Material Input costs is determined by the following relation-

Total Material Input Costs = $[S_f$ + Livestock $\text{Feed}_f + M_f] + [S_p$ + Livestock $\text{Feed}_p +$ Livestock $\text{Maintenance}_p + M_p + F_p + \text{Fuels}_p +$ Irrigation $\text{charges}_p + P_p +$ Service $\text{charges}_p] +$ Amortization cost.

where "S," "M," "F," and "P" denotes seed, manure, fertilizers, and pesticides respectively. The suffixes "*f* and *p* denote farm-produced inputs and purchased inputs respectively. The total paid-out labor costs comprise wages paid in both cash and kind. The yearly amortization cost for productive assets has been derived by a straight-line depreciation exercise based on the data on expected life of these assets.

The household data was collected in the survey for each farm operation for labor and non-labor inputs and price and quantity for sales in order to arrive at money estimates for input and output.

Appendix 2

Table 4.A1. Percentage distribution of households across classes in the three regions

Class	WB-rn	TL-sdp
RL	2.6	11.7
PP	32.5	11.7
SP	5.2	41.7
MP	20.8	20.0
RP	39.0	13.3
LLD	0.0	1.7
Total	100.0	100.0
0	2.6	10.0
0.01–1.0	53.2	18.3
1.01–2.5	24.7	16.7
2.51–5.0	11.7	30.0
5.1–10.0	3.9	16.7
10.1 and above	3.9	8.3
Total	100.0	100.0

Notes: RL: Rural labor; PP: Poor Peasants; SP: Small Peasants; MP: Middle Peasants; RP: Rich Peasants; LLD: Landlords; WB-rn: West Bengal, Raina; TL-sdp: Telengana, Saidapur.

Source: Primary Field Enquiry

Table 4.A2. Total household income and poverty line income per household by economic and land size-classes (in acres): Telengana advanced region (in INR)

Class	FDI [1]	Income from paid employment [2]	Income from service charges, self-employment, etc. [3]	Total household income (annual) [4=1+2+3]	Annual household poverty line (2200 Kcal) [5]
RL	−7401	12,493	0	5092	27,758
PP	−3275	6414	0	3139	33,792
SP	−5369	8672	0	3303	35,482
MP	−3837	150	3200	−487	33,792
RP	30,037	0	1350	31,387	40,128
LLD	200,497	0	40,000	240,497	59,136
Total	3097	5849	1487	10,432	35,059
0	−8409	12,400	0	3991	26,752
0.01–1.0	−6836	7733	0	896	33,024
1.01–2.5	−10,047	6290	0	−3757	32,947
2.51–5.0	−319	5094	2133	6909	34,731
5.1–10.0	−6693	3689	0	−3004	41,395
10.1 and above	96,916	0	10,160	107,076	42,240
Total	3097	5849	1487	10,432	35,059

Notes: RL: Rural labor; PP: Poor Peasants; SP: Small Peasants; MP: Middle Peasants; RP: Rich Peasants; LLD: Landlords.
Source: Primary Field Survey, Column 5 estimated by author

Notes

1. The grand Land Revenue Settlements during colonial rule recognized the "right over land" of the landlords without any countervailing protection of tenancy rights for a long time. This rendered unrestricted power to the landlords whereby they could enhance the burden of surplus appropriation on the tenants using the credible threat of eviction. Landlords also enjoyed control over the labor of tenants and their families in the context of such a lopsided legal framework.
2. See Byres (2003a) for the conception of "transition from above and below."
3. The calculation of the Labor Exploitation Index (E) for each household has been done considering both direct labor exploitation through hiring-in and hiring-out as well as indirect exploitation of labor through leasing-out and leasing-in of land using the following equation- $E = X/F = [(H_i - H_o) + (L_o - L_i)]/F$, where, H_i, H_o are labor days hired-in and hired-out respectively, L_o, L_i are the total labor days on land leased-out and land leased-in respectively and F is the family labor on the operated farm. Apart from the landless labor and landlords, the four peasant classes are poor peasants ($E<-1$), small peasants ($-1<E<0$), middle peasants ($0<E<1$) and rich peasants ($E>1$).
4. See Appendix 2, Table 4.A1 for the class structure in the two regions.
5. See Appendix 1 for definitions of the various income indicators and the computing methodology.

6. The $AHPL_i$ for the ith peasant class is estimated using the poverty line "monthly per capita expenditure" and the average family size for the ith peasant class using the following relation:

$$\mathbf{AHPL_I = 12 * [Poverty\ line\ MPCE * Average\ Family\ Size_i]}$$

Based on the 61st round of NSS data from the Consumption Expenditure Survey (Report No. 508) and the Nutritional Intake in India (Report No. 513) for 2004–5, we estimate that the MPCE required for attaining a consumption basket that embodies a nutritional intake of 2200 KCal per capita per day is Rs. 704 and Rs. 573 for Telengana and West Bengal respectively.

7. See Appendix 2, Table 4.A2.
8. There are rare exceptions such as South Korea and Taiwan that could transcend this trend and achieve a structural transformation similar to early industrializers.
9. Irfan Habib (1995) estimated that as early as 1801, the tribute to Britain from India alone constituted 30 percent of the former's domestic savings. Again in the late nineteenth century, the Indian tribute covered 40 percent of Britain's balance of payment (BoP) deficit with the rest of the world (Hobsbawm 1968). The BoP deficit maintained by Britain at this juncture was crucial for its financial hegemony over the world.

References

Bagchi, A. K. 1982. *The Political Economy of Underdevelopment.* Cambridge University Press.

Bagchi, A. K. 2010. *Colonialism and Indian Economy.* Oxford University Press, New Delhi.

Banerjee, A. 2008. "Trends in the Growth of Capitalist Relations and Investment in Indian Agriculture (1970 to the present)." PhD Dissertation Submitted to Jawaharlal Nehru University, New Delhi.

Banerjee, A. 2009. "Peasant Classes under Neo-Liberalism: A Class Analysis of Two States." *Economic and Political Weekly*, 44(15): 49–57.

Bernstein, H. 1996. "Agrarian Questions Then and Now." *Journal of Peasant Studies*, 24 (1–2): 22–59.

Bernstein, H. 2006. "Is There an Agrarian Question in the 20th Century?" *Canadian Journal of Development Studies*, 27(4): 449–60.

Bhaduri, A. 1973. "A Study of Agricultural Backwardness under Conditions of Semi-Feudalism." *Economic Journal*, 86: 120–37.

Bhalla, G. S. and Singh, G. 1997. "Recent Developments in Indian Agriculture: A State Level Analysis." *Economic and Political Weekly* (Review of Agriculture), 32(13): A2–A18.

Byres, T. J. 2003a. "Paths of Capitalist Agrarian Transition in the Past and in the Contemporary World." In *Agrarian Studies: Essays on Agrarian Relations in Less-Developed Countries*, edited by V. K. Ramachandran and M. Swaminathan, New Delhi: Tulika Books, pp. 54–83.

Byres, T. J. 2003b. "Structural Change, the Agrarian Question and the Possible Impact of Globalization." In *Work and Well-Being in the Age of Finance*, edited by J. Ghosh and C. P. Chandrasekhar, New Delhi: Tulika Books, pp. 171–211.

Chandra, N. K. 1974. "Farm Efficiency under Semi-Feudalism: A Critique of Marginalist Theories and Some Marxist Formulations." *Economic and Political Weekly*, 9(32, 33, and 34): 1309–32.

D'Costa, A. P. 2014. "Compressed Capitalism and Development: Primitive Accumulation, Petty Commodity Production, and Capitalist Maturity in India and China." *Critical Asian Studies*, 46(2): 317–44.

Frank, A. G. 1967. *Capitalism and Underdevelopment in Latin America*. New York: Monthly Review Press.

Habib, I. 1995. "Colonisation of the Indian Economy." In *Essays in Indian History: towards a Marxist Perception*, New Delhi: Tulika, pp. 304–46.

Harriss, J. 2013. "Does 'Landlordism' Still Matter? Reflections on Agrarian Change in India." *Journal of Agrarian Change*, 13(3): 351–64.

Hobsbawm, E. 1968. *Industry and Empire*. Harmondsworth: Penguin.

Hobsbawm, E. 1973. "Peasants and Politics." *Journal of Peasant Studies*, 1(1): 3–22.

Hobsbawm, E. 1994. *Age of Extremes: The Short Twentieth Century 1941–1991*, London: Abacus.

Kenwood, A. G. and Lougheed, L. 1971. *The Growth of the International Economy 1820–1960*. London: Allen and Unwin.

Lerche, J. 2013. "The Agrarian Question in Neo-liberal India: Agrarian Transition Bypassed?" *Journal of Agrarian Change*, 13(3): 383–404.

Maddison, A. 2007. *Contours of the World Economy 1-2030 AD: Essays in Macro-Economic History*. Oxford: Oxford University Press.

Magdoff, H. 1978. *Imperialism: From the Colonial Age to the Present*. New Delhi: Aakar Books.

Moyo, S., Jha, P., and Yeros, P. 2013. "The Classical Agrarian Question: Myth, Reality and Relevance Today." *Agrarian South: Journal of Political Economy*, 2(1): 93–119.

Mukherjee, A. 2010. "Empire: How Colonial India Made Modern Britain." *Economic and Political Weekly*, 45(50): 73–82.

Nair, K. N. and Banerjee, A. 2012. "Structural Changes in Land Distribution and Its Implications for Improving Access to Land." In *Shaping India: Economic Change in Historical Perspective*, edited by D. Narayana and Raman Mahadevan, New Delhi: Routledge, pp. 46–66.

NSSO (National Sample Survey Organization). 2014. "Key Indicators of Situation of Agricultural Households in India, 2012–13," NSS Seventieth Round, Government of India, New Delhi.

Patnaik, U. 1976. "Class Differentiation within the Peasantry: An Approach to Analysis of Indian Agriculture." *Economic and Political Weekly* (Review of Agriculture), 11(39): 82–101.

Patnaik, U. 1983. "Classical Theory of Rent and Its Application to India: Some Preliminary Propositions, with Some Thoughts on Sharecropping." *Journal of Peasant Studies*, 10(2 and 3): 71–87.

Patnaik, U. 1999. "The Process of Commercialization under Colonial Conditions." In *The Long Transition: Essays on Political Economy*, New Delhi: Tulika, pp. 252–304.

Patnaik, U. 2002. "Deflation and Déjà vu: Indian Agriculture in the World Economy." In *Agrarian Studies: Essays on Agrarian Relations in Less-Developed Countries*, edited by V. K. Ramachandran and M. Swaminathan, New Delhi: Tulika, pp. 111–43.

Ramachandran, V. K. and Swaminathan, M. 2005. *Financial Liberalization and Rural Credit in India*. New Delhi: Tulika.

Rawal, V. 2008. "Ownership Holdings of Land in Rural India: Putting the Record Straight." *Economic and Political Weekly*, 43(10): 43–7.

Rawal, V. 2013. "Changes in the Distribution of Operational Landholdings in Rural India: A Study of National Sample Survey Data." *Review of Agrarian Studies*, 3(2): 73–104.

Reddy, D. N. and Mishra, S. 2009. *Agrarian Crisis in India*. New Delhi: Oxford University Press.

Saha, A. and Swaminathan, M. 1994. "Agricultural Growth in West Bengal in the 1980s: A Disaggregation by Districts and Crops." *Economic and Political Weekly*, 29(13): A2–A11.

Thorner, A. 1982. "Semi-Feudalism or Capitalism? Contemporary Debate on Classes and Modes of Production in India, Part 2." *Economic and Political Weekly*, 17(50): 1993–9.

Part II

Legal-Institutional Dimensions of "Regimes of Dispossession"

5

Law Struggles, Lawmaking, and the Politics of Hegemony in Neoliberal India

Toward a Critical Perspective on the 2013 Land Acquisition Act

Kenneth Bo Nielsen and Alf Gunvald Nilsen

5.1 Introduction

The liberalization of the Indian economy since the early 1990s has ushered in a new phase of conflict over land acquisition, where land is increasingly being acquired by state governments for private investors (Bedi 2013; Levien 2013; Sud 2014).[1] As a result, since the mid-2000s India has witnessed thousands of small "land wars" (Levien 2012, 2013) that have centered crucially on the forcible transfer of land from small and marginal farmers or indigenous groups to industrial or residential use, or to special economic zones (SEZ) (Jenkins, Kennedy, and Mukhopadhyay 2014).[2] The resulting new "multiscalar geopolitics of popular resistance" (Banerjee-Guha 2013: 167) is often seen pitting marginal rural populations against a "neoliberal" (Kapoor 2011) Indian "land-broker state" (Levien 2013) whose actions and policies point to certain fundamental changes in the political economy and structure of the Indian state in the context of economic reform (Chatterjee 2008; Gupta and Sivaramakrishnan 2011). In the current situation, the balance of power has ostensibly decisively shifted in favor of corporate capital, which now sets "the terms to which other political formations can only respond" (Chatterjee 2008: 61). Yet, as several (partially) successful land struggles testify, Indian politicians and state governments are not "immune to popular pressure" (Sampat 2013: 19) emanating from social movements resisting new forms of displacement and dispossession.

In this chapter we are specifically concerned with analyzing how law mediates India's current land struggles and to what effect. As Nandini Sundar (2011) has succinctly observed, many of India's land struggles have relied considerably on a relatively new form of activism, which Sundar calls "law-struggles," in which collective action is increasingly geared toward achieving the introduction of laws that safeguard the rights, entitlements, and needs of subaltern groups, including those threatened by displacement and dispossession. As a sign of the increasing politicization of "the masses" and the maturation of liberal democracy in India, "legal mobilization" by subaltern groups has, to Sundar, never before been as important as it is today (Sundar 2011: 177).

As several of the chapters in this volume show, the significance of "law struggles" is related to the state's blatant misuse of the prior existing land acquisition legislation, the Land Acquisition Act of 1894, which was in effect until January 1, 2014. This act has been a major source of agitation for civil society and subaltern movements over the past decade (Palit 2012: 6). One key grievance has been that, whereas the Indian state has arrogated to itself substantial legal powers to exercise the right of eminent domain, there has been a consistent failure to put in place a legal regime that protects the interests of marginal rural populations and that guarantees the right to resettlement and rehabilitation for displaced groups (Levien 2011: 69). Demands for modifications in the legal framework underpinning the exercise of eminent domain has thus been high on the agenda of the many grassroots organizations working to put a stop to the large-scale acquisition of land for the benefit of private sector enterprises.[3] These efforts were seemingly crowned with success when, after a prolonged law-making process, the Lok Sabha finally passed a new act in 2013, the Right to Fair Compensation and Transparency in Land Acquisition, Rehabilitation and Resettlement Act (LARR), which claimed to rectify this. According to its preamble, the new act would ensure:

> A humane, participative, informed and transparent process for land acquisition for industrialisation, development of essential infrastructural facilities and urbanisation with the least disturbance to the owners of the land . . . and provide just and fair compensation to the affected families whose land has been acquired . . . and [ensure] that the cumulative outcomes of compulsory acquisition should be that affected persons become partners in development. (Government of India 2013: 1)

Put in these terms, the aims of the act appear irreproachable. Yet, as we discuss later, and as Malabika Pal's contribution to this volume similarly brings out, the preamble may well be seen to promise a good deal more than what the letter of the law is capable of delivering.

In this chapter we map the emergence of this act to argue that, in the context of mushrooming and often violent ongoing conflicts over land acquisition, "law struggles" and lawmaking need to be understood dialectically. On the one hand, social movements mobilize to advance or implement legal frameworks in order to oppose the marginalizing ramifications of neoliberalization; on the other hand, political authorities design and introduce legal frameworks in order to contain opposition and construct "the material basis for consent to a regime of dispossession" (Levien 2013: 400) in a context of neoliberal economic and social restructuring. Drawing on Gramsci (1998: 161), we suggest that for political authorities, lawmaking serves as a modality through which a "compromise equilibrium" (see Section 5.2) can be established between dominant and subaltern groups, which ultimately enables the long-term advance of neoliberal restructuring. In other words, while the state via lawmaking offers concessions to the demands of social movements, and thus compels private investors to agree to a set of concessions that in some ways might militate against their immediate short-term interests, it *also* establishes foundations for a "compromise" that might provide a firmer basis for the onward march of neoliberalization of the Indian economy in the context of advanced global capitalism.

We proceed by first briefly clarifying the Gramscian underpinnings of our approach. To ground our analysis we then introduce the case of one of India's most talked-about land struggles as our empirical point of entry, namely, the movement in Singur, West Bengal, from 2006 onwards against the acquisition of 997 acres of agricultural land for a Tata Motors small-car factory. To shed light on how the dynamics of law struggle and lawmaking can be seen to produce compromise equilibria, we then devote the major part of the chapter to map the process—from mobilization to legislation—through which the Right to Fair Compensation and Transparency in Land Acquisition, Rehabilitation and Resettlement Act, 2013, came into being. We also offer a close reading of its final contents to further substantiate our argument about the links between law struggle, lawmaking, and compromise equilibria as outlined earlier. We then briefly examine the more recent (failed) attempt by the current Bharatiya Janata Party (BJP)-led government to amend the act, before, in the conclusion, we summarize the implications of our analysis for understanding how capitalism is unfolding in India today.

5.2 Law Struggles, Lawmaking, and Neoliberal Hegemony in India

The power of dominant groups, Gramsci (1998: 12) famously argued, finds its manifestation in two ways: on the one hand in "the spontaneous consent given by the great masses of the population to the general direction imposed on social life by the dominant fundamental group" and, on the other hand, in "the apparatus of state coercive power which 'legally' enforces discipline on those groups who do not 'consent' either actively or passively" to an extant social order. Consent was of course at the heart of his theorization of hegemony, and the construction of hegemony, in turn, was mediated through what Gramsci refers to as "the integral state" (Gramsci 1998: 239)—that is, the fusion of political and civil society—and the way in which it imbricates its institutions, discourses, and technologies of rule into the fabric of the everyday life of subaltern groups. However, the construction of hegemony should not be conceived of as a one-way and top-down process in which dominant groups simply impose ideologies of dominance upon subaltern groups. Rather, the process is a gradual and contentious one, which proceeds through an ongoing churning in which "unstable equilibria... between the interests of the fundamental group and those of the subordinate groups" (Gramsci 1998: 182) are continually formed and superseded. In other words, actually existing hegemony will never assume the form of a "monolithic ideological formation." To the contrary, it will only ever be manifest as "a problematic, contested, political *process* of struggle" (Roseberry 1995: 77; see also Mallon 1995: ch. 1).

Gramsci's keen awareness of how dominant groups were compelled to concede to subaltern demands in order to entrench and reproduce their hegemonic position was combined with an emphatic insistence that "subaltern groups are always subject to the activity of ruling groups, even when they rebel and rise up" (Gramsci 1998: 55). What this means is that when subaltern groups develop oppositional projects to protest or resist the rule of dominant groups, they do not do so at a distance from the institutional and symbolic modalities through which hegemony is constructed, but rather in and through these modalities (Green 2002). There are two implications that follow from this. First, the institutions, discourses, and technologies of rule that attach to the state are not just sites in which hegemonic power is exercised, but also nodes of contention where subaltern resistance can be articulated and pursued; second, given that these are modalities that are intended to maintain "unstable equilibria" of compromise "in which the interest of the dominant social groups prevail" (Gramsci 1998: 182), they will also tend to constrain the forms of subaltern collective action in ways that can be reconciled with the imperative of reproducing a given hegemonic formation.

Gramsci's comments on law are, as Alan Hunt (1990: 316) has put it, "suggestive but tantalizingly underdeveloped" (see also Litowitz 2000; Cutler 2005). What we intend to do here is to offer some tentative reflections on how lawmaking "participates in the organization of consent" (Cutler 2005: 536) in the context of land acquisitions in neoliberal India, and what implications this has for law struggles as a form of subaltern politics. We are particularly interested in the implications of the fact that the law is integral to the hegemonic imperative of "securing an equilibrium" (Hunt 1990: 316) between dominant and subaltern groups by incorporating "some aspects of the aspirations, interests, and ideology of subordinate groups" (Hunt 1990: 311). We thus propose that when the Indian state responds to the land struggles and law struggles of grassroots social movements and their advocacy networks by initiating lawmaking processes regulating land transfers, this cannot be reduced to an unequivocal expression of democratic accountability, or to an outright stratagem of co-optation on the part of the state. Rather, what we are confronted with is a complex and at times contradictory practice that is aimed at negotiating the oppositional claims of subaltern groups and the economic interests of dominant groups, which are intimately linked to the exploitation of the spaces of accumulation that are currently being pried open by market-oriented reforms. The resultant compromise equilibrium, we argue, is intended to facilitate India's process of neoliberalization; but it remains unstable and frangible due to the fragile public legitimacy of the current regime of dispossession, as the proliferation of anti-dispossession movements in India in recent years testifies. Below we turn our attention to one such movement, in Singur.

5.3 Land Struggle and Law Struggle in Singur

The controversy over the transfer of 997 acres of agricultural land from farmers in West Bengal's Singur block to Indian multinational Tata Motors in 2006 is surely one of the most talked-about land struggles in India over the past decade.[4] The planned Tata Motors factory would produce the company's new flagship car, an affordable Indian version of the "people's car" known as Tata Nano and priced at only Rs. 100,000 (approximately USD 2,500 in 2006). The project itself was widely promoted as a concerted effort at rejuvenating industrial growth in West Bengal, a state that has over the past many decades witnessed an accelerating relative decline in industrial production. Yet, when local farmers learned that their land had been targeted for expropriation, a prolonged resistance erupted that raged throughout the latter half of 2006 as villagers—under the aegis of its organizing committee, the Singur Krishi Jomi Raksha Committee (Committee to Save the Farmland of Singur)—staged rallies and demonstrations, and repeatedly blocked roads and railway tracks. The protesters also clashed with the police on several occasions and hundreds were badly injured or taken into custody (Nielsen 2008, 2010; Roy 2014: 160–7). In the process, the Singur movement attracted the support of a large number of political parties, most notably the Trinamul Congress, the leading opposition party in the state at the time, and non-governmental organizations (NGOs), activist groups, and organizations such as the Association for the Protection of Democratic Rights (APDR), the Institute for

Motivating Self-Employment, the Food First Information and Action Network (FIAN), and the Medha Patkar-led National Alliance of People's Movements (NAPM). But the fierce resistance notwithstanding, the land in Singur was acquired under heavy police cover and amidst much violence in December 2006, and by early 2007 construction work was already well under way.[5]

The Singur movement rose to regional and national fame for several reasons. Occurring in the immediate wake of the violent standoffs over land acquisitions in Kalinganagar and Jagatsinghpur in Odisha, Singur was one of the earliest of India's new "land wars." As such, it was, as one perceptive observer (Bidwai 2007) put it, widely believed to be "paradigmatic" of the ways in which Indian state governments would go about attracting investments and facilitating industrial development in the future. In a context in which the ability of state governments to furnish land was rapidly becoming perhaps "the most important factor in inter-state competition for investment" (Levien 2012: 944; see also Sud 2014: 51), there was a very real fear among social activists and concerned citizens that the *modus operandi* of future land acquisitions would imply little or no consultation with key stakeholders; involve force, violence and coercion, if necessary; and be based on inadequate compensation and no rehabilitation,[6] but rather considerable state subsidies, cheap loans, and favorable tax regimes for investors. Indeed, in the case of Singur, estimates suggest direct and indirect state support for Tata Motors to the tune of a whopping INR 8.5 billion (Mitra 2007). In addition, the fact that a land acquisition was carried out in this manner in a state that was governed by the Communist Party of India (Marxist), a party with a long history of agrarian struggle and a proven track record of pro-poor redistributive agrarian reform, made the events in Singur appear even more dramatic.

As part of their struggle to retain—and later, to reclaim—their land, Singur's unwilling farmers have raised a series of counterclaims. These concerned principally the need for:

> Deliberation, careful planning, necessity to ensure the consent of the peasantry and hence, the need for discussion with various peasant associations, the necessity of a wider dialogue on the roadmap to industrialisation, modernisation and prosperity and finally, the need to balance the industrial drive with agricultural stability and growth. (Samaddar 2009: 159)

Yet when the state government, given its firm commitment to bringing about a quick march towards industrialization (Nielsen 2014: 85–90), showed no interest in renegotiating the terms under which the land acquisition in Singur stood to be carried out, the Singur movement increasingly judicialized as it shifted its politics of resistance from the streets to the court in a strategic attempt at "reading the law back to the state" (Chandra 2013: 55). In late 2006 and early 2007, a total of eleven petitions were brought before the

Calcutta High Court challenging the legality of the land acquisition. Some of these were filed by the unwilling farmers themselves, while several others were forwarded by some of the supportive organizations and activist networks, including the APDR and FIAN. The ensuing law struggle would center crucially on the legitimacy and meaning of the 1894 Land Acquisition Act (Nielsen 2015). It is therefore instructive to cast a quick look at some of the central tenets of this act.

5.4 The Land Acquisition Act of 1894

The Land Acquisition Act of 1894—essentially a colonial measure intended to ensure a smooth acquisition of land for roads, railways, bridges, forts, cantonments, or irrigation works—empowered the state government to acquire land in any locality where such land is likely to be needed for any "public purpose" or for a "company" (Sarkar 2007: 5). The act suggests what may qualify as a "public purpose" but, perhaps unsurprisingly, does not offer any exhaustive definition. With regards to the notion of "company" there are some restrictions attached to acquiring land for a government company, while the acquisition of land for a private company is very restricted.

The acquisition begins when the District Collector issues a public notice to that effect. Those affected by the acquisition have the right to object to the District Collector, who is obliged to consider any such objections before forwarding his final recommendations to the government. Once a final decision on the acquisition has been arrived at, the District Collector begins marking out and measuring the land, and announces that claims for compensation (known as the award) may be made to him (Sarkar 2007: 6–11). The act specifies that the market value of the land at the time of the publication of the notification "shall be taken into account" when the award is fixed. Those affected by the acquisition can challenge it on several accounts. They can, for example, dispute the size of their award and the District Collector's measurement of their land by going to court (Sarkar 2007: 17–27). They can also refuse to accept the award altogether. In such cases, the award is deposited with the court. Singur's unwilling farmers had chosen this option as a means through which to register their opposition to the acquisition of their land. But in reality there is no provision in the Land Acquisition Act of 1894 that allows the affected population to challenge the acquisition per se since their consent is not mandatory under law (Samaddar 2009: 154).

In the wake of the land acquisition in Singur, the act was criticized by social activists, commentators, and politicians on many counts: for example, for being designed for subjecthood and not for citizenship (Kannabiran as cited in Sundar 2011: 177); for facilitating land transfers at a nominal rate of compensation and without rehabilitation; for not recognizing community rights over land; and for not compensating those who tilled the land without having legal title (Guha 2007: 49–52). However, it was the notion of "public purpose" that emerged as the focal point of legal contestation when the petitions of the unwilling farmers were heard in the Calcutta High Court. According to the "notifications for acquisition of land" issued by the Office of the Land Acquisition Collector and District Magistrate and Collector in Hooghly, the land acquired in Singur was indeed acquired for just such a purpose. The key question that animated the debate in the High Court was therefore: Does a car factory constitute a public purpose or not?

When the state's Industries Minister, Nirupam Sen, had earlier been confronted with this question by the media, he had replied: "Of course it is a public purpose, industrialization means employment generation, it means development of society. The entire people of the state will be benefited" (quoted in Chattopadhyay 2006).[7] In the High Court, the state government's legal team echoed the minister's point of view: industrial projects like the one in Singur would generate socioeconomic development for the benefit of society at large. This was true, they maintained, even if one (wrongly) assumed that the acquisition was carried out for the benefit of a private company, in this case Tata Motors.

Understandably, the unwilling farmers and their legal team saw things differently. To them, acquiring agricultural land by force only to lease it out with considerable state subsidies to a private multinational could not be justified as a public purpose. Thousands of farmers would lose their livelihood, but only a couple of hundred highly educated and skilled engineers could hope to find employment in Tata Motors' high-tech factory.[8] And the profits would go to Tata Motors. This, they argued, could not qualify as a public purpose.

When the court delivered its verdict in January 2008 it ruled against the unwilling farmers. The court found that the state government could not successfully be challenged "on the ground that the object of the acquisition is not to carry out a public purpose" (Calcutta High Court 2008: 11). And it further maintained that "public purpose does not cease to be so merely because the acquisition facilitates the setting up of industry by a private enterprise and benefits it to that extent" (Calcutta High Court 2008: 10). In this respect, the court's ruling was far from unique. As several scholars have pointed out, the dominant recent trend in the Indian judiciary has been to virtually adopt a "hands-off attitude" in land acquisition cases, thereby more or less directly giving a green light to "all kinds of land acquisitions," including for private companies (Desai 2011: 96–7). In doing so, the courts have concomitantly come to apply a very wide, or diluted, definition of "public purpose," in effect supporting the neoliberal policies of state governments over the rights of vulnerable sections of society (Desai 2011; Deva 2009; Sundar 2011).

Yet Singur's law struggle had not been entirely in vain. The fact that the nature and functioning of the 1894 Land Acquisition Act had been subject to intense scrutiny and a long public debate that had involved dispossessed populations, activist groups, NGOs, advocacy networks, and political parties soon led to a broad call to change the legal framework under which land acquisitions are carried out in India.

There were, however, other voices that *also* clamored for modification in the legal framework underpinning the exercise of eminent domain, albeit in a very different direction. As Sundar (2011: 175) writes, both "the dominant capitalist class and 'people's movements'... have an interest in working on the state to change existing laws, but often to quite different ends." Under the old legal regime, and unlike what had been the case in Singur, industrialists and business houses had often found land acquisitions to be a cumbersome and slow affair. They had routinely been bogged down in court cases that lasted for years if not decades, and which prevented the timely progress of cost-intensive projects. In what Levien (2011: 67) aptly describes as a moment of "political ju-jitsu," when the "opponent's momentum is used to your own advantage," investors and capitalist developers sought to use the momentum generated by the many popular land struggles that occurred in the wake of Singur to press for legal and policy changes that, while appearing to respond to the critics of land dispossession, would in fact make it easier to acquire land for private accumulation. And their voices carry weight. According to Chatterjee (2008), the corporate capitalist class has emerged as *the* dominant group within the state apparatus after 1991, increasingly displacing other formerly influential groups, particularly the rich farmers and the salariat (see Bardhan 1998). There were, in other words, several competing and not readily reconcilable aspirations, interests, and ideologies that had to be balanced in order to construct a compromise equilibrium.

5.5 Toward a New Law

The contours of this compromise first became visible when the incumbent United Progressive Alliance government began the process of amending the Land Acquisition Act of 1894 (Palit 2012: 3). The Union Ministry of Rural Development had in fact embarked on this process already back in 1998, but work only gathered pace in 2007, that is, in the immediate wake of Singur (Ramesh and Khan 2015: 10–11). In December 2007, the Land Acquisition (Amendment) Bill, 2007, was introduced in the Lok Sabha; it was later amended and rechristened the Land Acquisition (Amendment) Bill, 2009, and was tabled in the Lok Sabha along with the Resettlement and Rehabilitation Bill in that year. In combination, the two bills sought to modify the exercise of eminent domain; redefine the notion of public purpose; and improve the resettlement and rehabilitation of the displaced.

In redefining public purpose, the Land Acquisition (Amendment) Bill introduced the category of "useful to the general public," under which state governments retained the right to acquire any land it saw fit.[9] Subsumed under this category was, most importantly, "the provision of land for infrastructure projects of the appropriate government, where the benefits accrue to the general public." "Infrastructure projects" in turn included power generation and transmission; construction of roads, highways, bridges, airports, ports, or rail systems; mining activities, educational, sports, healthcare, tourism, transportation, space program, housing, water supply projects, irrigation projects, sanitation and sewerage systems. For all intents and purposes, the inclusion of all these activities in the category of "useful to the general public" constituted a considerable broadening of the notion of public purpose as articulated in the act of 1894.

In contrast, the draft bill eliminated the ability of state governments to acquire 100 percent of the land required for a *private* company for a "public purpose." Instead, it introduced a clause that specified that only when a private company had *already* directly acquired 70 percent of the land necessary for a project deemed "useful to the general public" could the government intervene to acquire the remaining 30 percent through eminent domain. This so-called 70:30 clause seemingly reduced the state's controversial role in land acquisitions by making it mandatory that a private company ensure the consent of the majority of the project-affected population. But at the same time it diluted the crucial public–private distinction that critics of the draft argued should be central to *any* use of eminent domain.

While the amended land acquisition draft bill could thus in key respects be seen to expand the scope for state intervention in land transfers, the draft Resettlement and Rehabilitation Bill contained progressive measures for the dispossessed and displaced. The bill recognized the rights of the landless and artisans to compensation, and not just the landowners. And it mandated that those who had their land expropriated be given a house plot and money for construction, compensation for cattle sheds, moving costs, a monthly subsistence allowance at minimum agricultural wages, and a pension of INR 500 per month. Given the absence of any mention of rehabilitation in the act of 1894, this constituted a clear improvement.

The fact that the draft bills were thus strong on rehabilitation but weak on limiting the exercise of eminent domain points to the contours of an emerging compromise equilibrium on land acquisitions in India under which state-led land transfers to private investors could continue and potentially even accelerate, albeit at a considerably higher cost. However, the two bills lapsed with the expiry of the term of the fourteenth Lok Sabha and the subsequent constitution of the fifteenth Lok Sabha in 2009. After that, the idea of simply amending the existing act was abandoned and a complete rewriting of the law initiated (Ramesh and Khan 2015: 11), eventually resulting in the Land Acquisition, Rehabilitation and Resettlement Bill, 2011. This bill was, in turn, redrafted between 2011 and 2013 until it was, in its final avatar, the Right to Fair Compensation and Transparency in Land Acquisition, Rehabilitation and Resettlement Act, 2013, passed by the Lok Sabha in August 2013. It received the assent of the president of India the following month and came into effect on January 1, 2014.

5.6 Mediating the Law Struggle: Treading the Middle Path?

The journey toward a passable new law on land acquisitions had apparently been a bumpy one from the point of view of those entrusted with the responsibility of drafting it. The Right to Fair Compensation and Transparency in Land Acquisition, Rehabilitation and Resettlement Act would be based on its draft predecessors, the drafting of which had over time come to involve a multitude of stakeholders espousing radically different goals. These included different government ministries; the National Advisory Council chaired by Sonia Gandhi; various national and regional political parties; state governments; NGOs; the Federation of Indian Chamber of Commerce and Industry (FICCI); the Confederation of Indian Industry (CII); the Confederation of Real Estate Developers Associations of India; and some multinational corporations, to name just a few (Sud 2014: 45). The difficulties involved in treading a middle path between these many different agendas was highlighted by Union Minister for Rural Development, Jairam Ramesh, in two successive interviews with the Indian magazine *Frontline* in 2013 (Rajalakshmi 2013b; Srinivasan 2013). Both interviews focused on a publicly circulated penultimate draft of the Right to Fair Compensation and Transparency in Land Acquisition, Rehabilitation and Resettlement, and Ramesh spoke vividly of the tightrope that his ministry had had to walk while drafting the bill:

> Industry associations are very apprehensive that it will jack up [the] costs [of acquiring land for private industries]. Vocal sections of civil society feel that I have not gone far enough [to protect land owners]. I have to find a middle path. I have struck a balance... I feel satisfied that civil society is unhappy with the bill for not being progressive enough; industry is unhappy with the bill for being too progressive. If everyone is unhappy, then I think it is a good bill (Rajalakshmi 2013b: 34)... If I have succeeded in displeasing Medha Patkar on the one side and CII / FICCI on the other, it means I have done something right. (Srinivasan 2013: 105)

Chalking out a compromise equilibrium that reconciled, in particular, the demands of industry with the demands of social movements for greater protection of the rights of the individual landowner within a single legal framework had evidently not been easy. The process was ostensibly made all the more challenging because different ministries apparently entertained different views on what the new act should first and foremost work to achieve. While the Ministry of Commerce and Industry ostensibly championed ideas that would have made acquiring land for industries easier, Ramesh's Ministry of Rural Development as well as the Standing Committee on Rural Development appear to have argued for a considerably narrower definition of "public purpose" that would have limited the exercise of eminent domain (Rajalakshmi 2013a: 33–4). Simultaneously, the Ministry of Urban Development wanted lower compensation rates in urban areas, ostensibly to not hamper efforts at carrying out urban development projects (Ramesh and Khan 2015: 51). Multiple sets of voices had thus found their way into the draft bill, but to very different effect.

Both the draft Land Acquisition, Rehabilitation and Resettlement Bill, 2011 and the final Right to Fair Compensation and Transparency in Land Acquisition, Rehabilitation and Resettlement Act, 2013, broke new ground by explicitly linking land acquisition to rehabilitation and resettlement (rather than simply "compensation"). According to Chakravorty (2011: 29), the four main positives in the 2011 draft Land Acquisition, Rehabilitation and Resettlement Bill thus all pointed in the same, important direction: more benefits and rights for the dispossessed. First, landowners of record would be paid much more than they used to; the compensation due would be no less than four times the market price in rural areas and twice the market price in urban areas; second, the approval of 80 percent of landowners would be required when the private sector was involved in projects larger than 100 acres; third, the affected landless population would be compensated with income for twenty years, plus other immediate benefits, and would be included in the rehabilitation and resettlement package; and fourth, any acquisition of 100 acres or more (50 acres in urban areas) would automatically require a rehabilitation and resettlement package.

Although thus increasing both the amount of compensation payable as well as the total pool of beneficiaries, the notion of "public purpose" as defined in the 2011 draft bill gave critics less cause for optimism. As Chakravorty (2011: 30) writes:

> The definition of "public purpose" in the Bill begins with state projects: Two clauses on defence, transportation, irrigation, power, etc., and three clauses on project-affected people, weaker sections, displaced persons, and persons affected by natural calamities. Clause (vi) brings in "Public Private Partnership projects".

> And finally in clause (vii) comes this: "The provision of land in the public interest for private companies for the production of goods for public or provision of public services". This could mean almost anything. Certainly the entire manufacturing sector can be included in this, possibly much of the service sector too. This is business as usual.

To staunch critics, however, it was business *worse* than usual. Rather than truly safeguarding the rights and interests of vulnerable sections of society, they saw the bill in effect making the process of acquiring land *easier* by further diluting the notion of public purpose; and *facilitating* rather than checking the transfer of resources to the private sector (Desai 2011). This, they argued, reflected the persistence of a misleading belief that the solution to the conflict over land acquisitions could be found within the logic of the market (Sarkar 2011: 36).

Between 2011 and 2013, the 2011 draft bill was subject to further revisions and proposed amendments until it was finally passed in August 2013. While the rehabilitation and resettlement clauses were, as we show later, carried over into the final version of the act in only slightly modified form, *Frontline* claimed that attempts had been made between 2011 and 2013 to:

> Water down the present draft... under pressure from industry. Comments received from the Ministry of Commerce and Industry and the Ministry of Urban Development... indicate that there is considerable pressure to dilute the consent, compensation and rehabilitation features. Two rounds of meetings of the GoM [Group of Ministers] have taken place, and by all standards, it appears that the consensus is on making the final version of the Bill investor-friendly.
>
> (Rajalakshmi 2013a: 32)

Frontline claimed that the tendency had been to amend the bill:

> More in response to protests from those representing the interests of the corporate sector rather than taking into account the serious criticisms that have come from organisations representing the interests of farmers and others whose livelihood is linked to land. (Rajalakshmi 2013a: 35)
>
> This consolidates the view that the definition of public purpose has been widened to further the government's objectives relating to foreign direct investment (FDI) in multi-brand retail and other policies. Public purpose also includes acquisition of land for use by public–private partnership (PPP) projects where the ownership of the land will continue to be vested in the government.
>
> (Rajalakshmi 2013c: 32)

Fears that the bill would once again prove to be strong on rehabilitation and resettlement, but weak in terms of limiting the potential for the overt misuse of eminent domain were not unfounded. In the final version of the act, "public purpose" included *inter alia* infrastructure projects relating to agriculture, agro-processing, cold-storage facilities; industrial corridors and mining activities, national investment, and manufacturing zones; and projects for sport, health care, and tourism. A state government could acquire land "for its own use" for any of these very broadly defined "public purposes." Yet it would be wrong to conclude that the 2013 act therefore ignored landowners' interests. Land for public–private partnerships could only be acquired under a 70:30 clause; and land "for private companies for public purpose" only under an 80:20 clause. Both clauses provide considerable protection to landowners in cases where private enterprises are involved. In addition, the act specifies that irrigated multi-crop land—akin to that which was acquired in Singur—could only be acquired as a last resort, or under exceptional circumstances. And it makes it compulsory to conduct a broad social impact assessment to determine whether a proposed acquisition in fact serves a public purpose; the social impact assessment is then to be evaluated by an expert group to determine whether the social costs and adverse social impacts outweigh the benefits; if so, the project is to be abandoned.

The clauses on rehabilitation and resettlement are also extensive. The compensation due to the land losers in rural areas is based on the market value of the land, multiplied by a factor between one and two, depending on the distance to an urban area. To this is added a solatium of 100 percent, in addition to the value of buildings and attached assets similarly acquired. The rehabilitation scheme includes landowners, agricultural laborers, tenants, sharecroppers, and artisans working in the affected area for three years prior to the acquisition and whose primary sources of livelihood stand to be affected. Rehabilitation includes provision of housing in case of displacement; provisions for land for land rehabilitation; a choice between an annuity, a one-time payment, or employment in any industry coming up on the acquired land; subsistence grants for displaced families; one-time grants to artisans and small traders; and a one-time resettlement allowance. Included in the Third Schedule of the act is also a very long list of infrastructure facilities and basic minimum amenities to be provided to the displaced to ensure that the resettled population can "secure for themselves a reasonable standard community life."

In spite of these safeguards, the act's Fourth Schedule constituted a grave cause for concern. This schedule lists and exempts from purview thirteen pieces of legislation under which land can *also* be acquired, including the Land Acquisition (Mines) Act, 1885 and the Coal Bearing Areas Acquisition and Development Act, 1957, which have inflicted huge misery on tribal populations in mineral-rich states.[10] When land is acquired under any of these thirteen acts, the extensive rehabilitation and resettlement clauses in the act are not activated. Since the bulk of land acquisitions in India have in fact historically taken place under one or the other of these thirteen acts, the new act would in practice apply to only a fraction of the land acquisitions occurring in India, even if clause 105(3) in the act specified that the exempted thirteen acts should be brought within its full ambit within one year. As we suggest in our conclusion, it is precisely this combination of seemingly generous provisions for resettlement and rehabilitation with a widening of the definition of public purpose that constitutes the locus of the attempt to negotiate a compromise between dominant and subaltern groups that will enable land acquisitions—and more generally, the project of neoliberal restructuring—to proceed in the long run.

5.7 The Modi Regime and Attempted Amendments

Barely six months after the Right to Fair Compensation and Transparency in Land Acquisition, Rehabilitation and Resettlement Act came into effect, India had a new BJP-led government that was committed to further liberalizing and economically "enabling" reforms as perhaps no other Indian government in the past had ever been. When the current prime minister, Narendra Modi, campaigned in 2014, he framed his campaign around the trope of *acche din*, or "good times," promising that he would do for India what he had done for his home state of Gujarat over more than a decade, namely generate high economic growth rates that would, or so it was projected, translate into "good times" for most Indians, and not least the "neo-middle class" (Jaffrelot 2015) who aspire for upward mobility and an improved quality of life. The means to this end was to "expand modern infrastructure amenities and mass employment opportunities through rapid industrial growth across the country" (Ruparelia 2015: 755). One of the first signs of this policy emphasis was the government's attempt to amend the Right to Fair Compensation and Transparency in Land Acquisition, Rehabilitation and Resettlement Act, a move that followed intense lobbying and pressure by industry groups in the wake of Modi's electoral win (Verma 2015). This was done by introducing an ordinance in late December 2014 that modified several key provisions in the 2013 act. Among the provisions that were modified were the elimination of the earlier mentioned 80:20 and 70:30 clauses (for private and public–private projects respectively) for five categories of projects: defense, rural infrastructure, affordable housing, industrial corridors, and infrastructure projects. Moreover, while the 2013 act had made social impact assessments mandatory for all projects except irrigation, the ordinance did away with this provision for projects falling into the five abovementioned categories. Several other protective clauses were also done away with; for example, the definition of public purpose was expanded to allow governments to acquire land for private hospitals and educational institutions; the definition of the kind of private entities for whom land can be acquired was widened; and the stipulation that acquired land lying unused for five years must be returned was removed (Ramesh and Khan 2015: 124–30). At the same time, the government brought the thirteen hitherto exempted acts in Schedule Four under the purview of the act (and issued a limited job guarantee to affected families), as indeed it was compelled to do as per "safeguards" built into the 2013 act (Ramesh and Khan 2015: 125).

As may be expected, industry was delighted with this turn of events. Both CII and the Associated Chambers of Commerce and Industry of India (ASSOCHAM) announced that they "wholeheartedly" welcomed the ordinance, and the CII—in a statement that testifies to the importance of business lobbying for changes to the 2013 act—declared that it was particularly pleased that the government had "incorporated our suggestion to exempt important sectors like defense, rural electrification, rural housing, and industrial corridors from the mandatory 80 percent consent from affected families" (cited in Rajalakshmi 2015).

Because ordinances lapse after some time and cannot be repromulgated endlessly, the government sought to turn it into law by passing the Right to Fair Compensation and Transparency in Land Acquisition, Rehabilitation and Resettlement (Amendment) Bill, 2015, in the Lok Sabha (where it holds a majority) in March 2015. However, the bill met with strong opposition both inside and outside parliament. Social movements and farmers' organizations—including the BJP's own farmers' front, the Kisan Sangathan—protested the bill as an infringement on their rights and hard-won victories in the making of the 2013 act. The proposed new bill ultimately foundered in the Rajya Sabha where the National Democratic Alliance (NDA) does not have a majority. For that reason, the ordinance was repromulgated three times between December 2014 and May 2015 until it was finally allowed to lapse in August 2015. At the time, Modi blamed "so many rumors" for derailing the passing of the bill, while also reassuring concerned farmers that "every voice in the nation is important and the voice of farmers has a special place" (cited in *The Assam Tribune*, August 30, 2015). At this point, a Joint Parliamentary Committee (JPC) was set up to deliberate on the proposed changes to the 2013 act, but at the time of writing this it is still unclear what the outcome of the JPC will be.

5.8 Conclusion

The trajectory of the Right to Fair Compensation and Transparency in Land Acquisition, Rehabilitation and Resettlement Act, 2013, is an unfinished one, and there is much to suggest that we are likely to see renewed struggles over land and law in the not too distant future. Given this prospect, it is interesting to note that, as Chandra (2015: 50) has recently pointed out, whereas "business" has increasingly acquired the ideological dominance to define what "development" means in India today, effectively equating it to "a package of vaguely defined terms including 'urbanisation,' 'industrialisation,' and 'infrastructure creation,' in which it is assumed that the private sector will take the lead," business has not yet fully acquired a matching political dominance that allows it to pursue this agenda with full consent or compliance—not even with a Modi-led government in place in Delhi. This runs contrary to Chatterjee's reading of the unbridled power of capital in contemporary India. When we consider, further, that the failure, so far, at least, to water down the provisions of the 2013 act was paralleled by the BJP's electoral debacle in Bihar and clear setbacks on Modi's home ground in Gujarat, it is evident that the struggles yet to come will be subject to political equations and social forces that make their outcome less of a foregone conclusion than what much of the post-election hype of 2014 would have us believe.

Whatever the eventual outcome of future amendment processes, what we have tried to do in this chapter is to sketch out a perspective that might enable us to approach more generally the study of law struggles and lawmaking on the deeply conflictual terrain of land acquisitions in a way that is both nuanced and critical, and—most importantly—oriented toward deciphering the complex equations of enablement and constraint that social movements are confronted with as they mobilize in and through the legal domain to contest dispossession.

Viewed through such a perspective, the Right to Fair Compensation and Transparency in Land Acquisition, Rehabilitation and Resettlement Act, 2013, seeks to carve out a "middle path" that, in our view, seeks to construct a compromise equilibrium between the interests and claims of dominant and subaltern groups along two axes. First, as a concession to the long-standing demands of social movements that have challenged forced displacement, the act puts forward a series of progressive propositions on compensation and rehabilitation. It proposes an inclusive definition of project-affected people; seeks to entrench legally enforceable rights to rehabilitation; and raises the rates of compensation for land and labor lost. Arguably, this amounts to what Gramsci (1998: 161) labeled a sacrifice "of an economic-corporate kind" that is made by dominant groups in the interest of eliciting the consent of subaltern groups. In this context it may thus make it easier for the state to turn "barricaders" into "bargainers" (Levien 2013: 358) when land is acquired.

Second, the act also illustrates how the construction of a compromise equilibrium through the making of such concessions operates within certain limits, and ultimately has to bolster "the decisive function exercised by the leading group in the decisive nucleus of economic activity" (Gramsci 1998: 161). This is manifest, above all, in the remarkably wide definition of public purpose contained in the act, which, in our reading, serves two crucial purposes. On the one hand, it makes it possible for the Indian state to consolidate the restructuring of India's political economy in a direction that creates spaces of accumulation for private and globalized capital, and thus to emerge as a more fully-fledged "neoliberal state" (Harvey 2006). On the other hand, the fact that public purpose has been defined in such wide terms might work to constrain the political space in which social movements can articulate political projects that challenge the form and direction of development in contemporary India; and in particular in terms of the role of the state and its obligation to be accountable to those groups who have lost the most and gained the least from neoliberalization.

Notes

1. This chapter draws extensively on arguments we have made elsewhere (Nielsen and Nilsen 2015). We thank Achin Chakraborty and Anthony D'Costa for their constructive comments.
2. Neither land dispossession and displacement, nor the popular resistance to it are phenomena exclusively associated with economic liberalization and the ascent of market forces, even if popular resistance has certainly proliferated over the past ten years when compared to, for instance, the early Nehruvian period (Pedersen 2017). Indeed, Chakravorty (2013: xxiii) estimates that more than 90 percent of all land "converted" since 1947 was taken by the state for state infrastructure projects.
3. See Pal (Chapter 6, this volume) for an overview of economic arguments around eminent domain.
4. While no comprehensive survey of the social composition of Singur's unwilling farmers exists, an average land holding in Hooghly district where Singur is located is 1.63 acres (Government of West Bengal 2007: 36). The average landholding in Hooghly is thus "marginal" (i.e., less than 2.5 acres) in the official system of classification. More generally, in Hooghly district, a full 97.18 percent of all holdings are either "marginal" or "small"' (2.5–5.0 acres). A survey conducted by Nielsen in Singur in 2009 covering more than 140 affected households revealed an average landholding of less than one acre per household prior to the land acquisition.
5. In late 2008, however, Tata Motors decided to abandon their plant in Singur. But the land remains in their possession and it has been impossible to carry out a legally valid process of land restitution.
6. In accordance with the provisions of the 1894 Land Acquisition Act, each land loser in Singur was offered a cash compensation based on the estimated market value of the land at the time of the acquisition (approximately INR 600,000 per acre for mono-crop land; INR 900,000 per acre for multi-crop land) in addition to a 30 percent solatium (i.e., compensation for inured feelings as distinct from financial loss or physical suffering), a 12 percent per annum interest (calculated for the period between the issuing of the notification and the actual acquisition and

payable as a one-time amount), and a 10 percent incentive payable to those landowners who voluntarily handed over their land to the state and agreed not to contest the acquisition through the legal system (Nielsen 2008: 236).

7. Syntax in the original.
8. To D'Costa (2014), this form of relatively "jobless," or "job-poor" growth is a more general feature of what he calls "compressed capitalism," that is, the coexistence of different phases of capitalism, such as primitive accumulation *and* advanced forms of accumulation embedded in innovation-led economic expansion in a global setting.
9. This section is based on Levien's (2011) analysis of the two drafts.
10. It bears mention that the SEZ Act which was included in the Fourth Schedule in an earlier draft of the bill was removed in the final version. This may be an indication of how SEZs have come to be more widely regarded as first and foremost "enclaves for transnational capital to flourish" (Sampat 2010: 172) that are not easily justified as serving a public purpose.

References

Banerjee-Guha, S. 2013. "Accumulation and Dispossession: Contradictions of Growth and Development in Contemporary India." *South Asia*, 36(2): 165–79.

Bardhan, P. 1998 (1984). *The Political Economy of Development in India.* New Delhi: Oxford University Press.

Bedi, H. P. 2013. "Special Economic Zones: National Land Challenges, Localized Protest." *Contemporary South Asia*, 21(1): 38–51.

Bidwai, P. 2007. "West Bengal on the Wrong Track? The Singur Syndrome." In *Singur and Nandigram and . . . —The Untold Story of Capitalised Marxism*, edited by D. Sen and D. Bhattacharya, Kolkata: Kanoria Jute and Industries Limited Sangrami Shramik Union, pp. 36–7.

Calcutta High Court. 2008. "Ordering Portion of the Judgment Dated 18th January, 2008," Calcutta High Court, Kolkata.

Chakravorty, S. 2011. "A Lot of Scepticism and Some Hope." *Economic and Political Weekly*, 46(41): 29–31.

Chakravorty, S. 2013. *The Price of Land: Acquisition, Conflict, Consequence.* New Delhi: Oxford University Press.

Chandra, K. 2015. "The New Indian State." *Economic and Political Weekly*, 50(41): 46–58.

Chandra, U. 2013. "Beyond Subalternity: Land, Community, and the State in Contemporary Jharkhand." *Contemporary South Asia*, 21(1): 52–61.

Chatterjee, P. 2008. "Democracy and Economic Transformation in India." *Economic and Political Weekly*, 42(16): 53–62.

Chattopadhyay, S. S. 2006. "Land Reform Not an End in Itself." *Frontline*, 23(25). Available at: <http://www.frontline.in/static/html/fl2325/stories/20061229001903600.htm> (accessed February 1, 2016).

Cutler, A. C. 2005. "Gramsci, Law, and the Culture of Global Capitalism." *Critical Review of International and Political Philosophy*, 8(4): 527–42.

D'Costa, A. P. 2014. "Compressed Capitalism and Development: Primitive Accumulation, Petty Commodity Production, and Capitalist Maturity in India and China." *Critical Asian Studies*, 46(2): 317–44.

Desai, M. 2011. "Land Acquisition Law and the Proposed Changes." *Economic and Political Weekly*, 46(26–7): 95–100.

Deva, S. 2009. "Public Interest Litigation in India: A Critical Review." *Civil Justice Quarterly*, 28: 19–40.

Government of India. 2013. "The Right to Fair Compensation and Transparency in Land Acquisition, Rehabilitation and Resettlement Bill, 2013." New Delhi: Government of India.

Government of West Bengal. 2007. *District Statistical Handbook Hooghly, 2006*. Kolkata: Bureau of Applied Economics and Statistics, Government of West Bengal.

Gramsci, A. 1998. *Selections from the Prison Notebooks of Antonio Gramsci*. London: Lawrence and Wishart.

Green, M. 2002. Gramsci Cannot Speak: Presentations and Interpretations of Gramsci's Concept of the Subaltern. *Rethinking Marxism*, 14(3): 1–24.

Guha, A. 2007. *Land, Law and the Left: The Saga of Disempowerment of the Peasantry in the Era of Globalization*. New Delhi: Concept Publishing.

Gupta, A. and Sivaramakrishnan, K. 2011. "Introduction: The State in India after Liberalization." In *The State in India after Liberalization: Interdisciplinary Perspectives*, edited by A. Gupta and K. Sivaramakrishnan, London: Routledge, pp. 1–27.

Harvey, D. 2006. *A Brief History of Neoliberalism*. London: Oxford University Press.

Hunt, A. 1990. "Rights and Social Movements: Counter-Hegemonic Strategies." *Journal of Law and Society*, 17(3): 309–28.

Jaffrelot, C. 2015. "What 'Gujarat Model'?—Growth without Development—and with Socio-Political Polarisation." *South Asia*, 38(4): 820–38.

Jenkins, R., Kennedy, L., and Mukhopadhyay, P. (eds). 2014. *Power, Policy and Protest: The Politics of India's Special Economic Zones*. New Delhi: Oxford University Press.

Kapoor, D. 2011. "Subaltern Social Movement (SSM) Post-Mortems of Development in India: Locating Trans-Local Activism and Radicalism." *Journal of Asian and African Studies*, 46(2): 130–48.

Levien, M. 2011. "Rationalising Dispossession: The Land Acquisition and Resettlement Bills." *Economic and Political Weekly*, 46(11): 66–71.

Levien, M. 2012. "The Land Question: Special Economic Zones and the Political Economy of Dispossession in India." *Journal of Peasant Studies*, 39(3–4): 933–69.

Levien, M. 2013. "Regimes of Dispossession: From Steel Towns to Special Economic Zones." *Development and Change*, 43: 381–407.

Litowitz, D. 2000. "Gramsci, Hegemony, and the Law." *Brigham Young University Law Review*, 515: 515–51.

Mallon, F. 1995. *Peasant and Nation: The Making of Postcolonial Mexico and Peru*. Berkeley, CA: University of California Press.

Mitra, A. 2007. "Santa Claus visits the Tatas—Freebies from a Debt-ridden Government." *The Telegraph*, March 30. Available at: <http://www.telegraphindia.com/1070330/asp/opinion/story_7580979.asp> (accessed February 2, 2016).

Nielsen, K. B. 2008. "'Not on Our Land!' Peasants' Struggle against Forced Land Acquisition in India's West Bengal." In *Rights and Legal Empowerment in Eradicating Poverty*, edited by D. Banik, Farnham: Ashgate, 217–46.

Nielsen, K. B. 2010. "Contesting India's Development? Industrialisation, Land Acquisition and Protest in West Bengal." *Forum for Development Studies*, 37(2): 145–70.

Nielsen, K. B. 2014. "Saving the Farmland: The Making of Popular Anti-Land Acquisition Politics in Singur, West Bengal." PhD Dissertation, University of Oslo.

Nielsen, K. B. 2015. "Law and *Larai:* The (De)Judicialisation of Subaltern Resistance in West Bengal." *Journal of Contemporary Asia*, 45(4): 618–39.

Nielsen, K. B. and Nilsen, A. G. 2015. "Law-Struggles and Hegemonic Processes in Neoliberal India: Gramscian Reflections on Land Acquisition Legislation." *Globalizations*, 12(2): 203–16.

Palit, A. 2012. "The Land Acquisition, Resettlement and Rehabilitation (LARR) Bill 2011: Providing Solutions or Raising Questions?" *Journal of Emerging Knowledge on Emerging Markets*, 4: 1–14.

Pedersen, J. D. 2017. "'The Dog That Didn't Bark' (Very Loudly)—Large-scale Development Projects with Little Protest in Nehru's India." In *Industrialising Rural India: Land, Policy and Resistance*, edited by K. B. Nielsen and P. Oskarsson, London: Routledge, pp. 21–39.

Rajalakshmi, T. K. 2013a. "A Law and Its Losers." *Frontline*, 29(26): 31–5.

Rajalakshmi, T. K. 2013b. "We Have to Strike a Balance." *Frontline*, 29(26): 34–5.

Rajalakshmi, T. K. 2013c. "No Meeting Ground." *Frontline*, 30(7): 32–4.

Rajalakshmi, T. K. 2015. "Ordinance Raj." *Frontline*. Available at: <http://www.frontline.in/the-nation/ordinance-raj/article6852240.ece> (accessed March 28, 2016).

Ramesh, J. and Khan, M. A. 2015. *Legislating for Justice: The Making of the 2013 Land Acquisition Act*. New Delhi: Oxford University Press.

Roseberry, W. 1995. "Hegemony and the Language of Contention." In *Everyday Forms of State Formation: Revolution and the Negotiation of Rule in Modern Mexico*, edited by G. Joseph and D. Nugent, Durham: Duke University Press, pp. 356–66.

Roy, D. 2014. *Rural Politics in India*. New Delhi: Cambridge University Press.

Ruparelia, S. 2015. "Minimum Government, Maximum Governance: The Restructuring of Power in Modi's India." *South Asia*, 38(4): 755–75.

Samaddar, R. 2009. "Prescribed, Tolerated and Forbidden Forms of Claim Making." In *Social Justice and Enlightenment: West Bengal*, edited by P. K. Bose and S. K. Das, New Delhi: Sage, pp. 153–79.

Sampat, P. 2010. "Special Economic Zones in India: Reconfiguring Displacement in a Neoliberal Order?" *City & Society*, 22(2): 166–82.

Sampat, P. 2013. "The 'Goan Impasse': Land Rights and Resistance to SEZs in Goa, India." LDPI Working Paper no. 53, The Land Deal Politics Initiative, The Hague.

Sarkar, P. (ed) 2007. *The Land Acquisition Act, 1894 with Short Notes along with West Bengal Amendments*. Kolkata: Books-n-books.

Sarkar, S. 2011. "The Impossibility of Just Land Acquisition." *Economic and Political Weekly*, 46(41): 35–8.

Srinivasan, G. 2013. "Seeking Consensus." *Frontline*, 30(8): 105.

Sud, N. 2014. "Governing India's Land." *World Development*, 60: 43–56.

Sundar, N. 2011. "The Rule of Law and the Rule of Property: Law-Struggles and the Neo-Liberal State in India." In *The State in India after Liberalization: Interdisciplinary Perspectives*, edited by A. Gupta and K. Sivaramakrishnan, London: Routledge, pp. 175–93.

The Assam Tribune. 2015. "Won't Renew Ordinance on Land Acquisition: Modi." August 30. Available at: <http://www.assamtribune.com/scripts/detailsnew.asp?id=aug3015/at043> (accessed March 10, 2016).

Verma, S. 2015. "Subverting the Land Acquisition Act, 2013." *Economic and Political Weekly*, 50(37): 18–21.

6

Land Acquisition and "Fair Compensation" of the "Project Affected"

Scrutiny of the Law and Its Interpretation

Malabika Pal

6.1 Introduction

It is important to be fair but more important to "appear" fair. This principle seems to have influenced the decision to change the title of the Land Acquisition, Rehabilitation and Resettlement Act, 2013 of the Government of India to the Right to Fair Compensation and Transparency in Land Acquisition, Rehabilitation and Resettlement Act, 2013 (hereafter the New Land Act, also known by its acronym LARR).[1] It is clear that there is an effort to take into account public perceptions of the acquisition and rehabilitation process carried out by the state by the use of its power of eminent domain.[2] This act came in the wake of widespread protests over several years against the Land Acquisition Act, 1894.[3] For almost a decade already, the government had been trying to work out a national rehabilitation policy that would be legally enforceable. This legislation subsumes both the acquisition and rehabilitation aspects, giving a signal to people that the government has finally worked out the rehabilitation package and hence is in a position to provide an assurance of "fair" compensation.[4]

The land question in India has been fraught with controversies since independent India sought to embark on a path of industrialization. For example, the colonial government used eminent domain to acquire land for many government projects, in particular, railways and irrigation. The continued use of these powers even in independent India was justified on the grounds that an enormous amount of land was needed for public sector projects.

Constitutional amendments repealed the provisions relating to property rights, namely Articles 19(1)(f) and 31(2), so that they no longer acted as hindrances to the acquisition process. In the 1962 and 1984 amendments to the Land Acquisition Act of 1894, a procedure for land acquisition for private companies was specified. Over the years, the power of eminent domain has been used to acquire land on behalf of the private sector, suggesting the growing influence of Indian big business and a perceptible shift in the state–capital relationship. Consequently, the resort to eminent domain on behalf of capital has intensified in the post-1991 liberalization period. With millions of people awaiting proper resettlement and rehabilitation, resentment against the process of acquisition and the use of the justification of "public purpose" has been growing in India. In recent years, the "social consent to operate [which] means securing the free, prior and informed consent of affected communities to part with their resources and also the consent to do business in their community" (Chakravorty 2013: 8; Kakani, Raghu Ram, and Tigga 2009: 135) has become imperative for any project to be successful.

The change of title in the New Land Act can be seen as an attempt to shift the focus from the event of land acquisition to the compensation process for those dispossessed and displaced. By doing that, the government has in effect legitimized the inevitability of acquisition of large tracts of land by the state for industrialization. The real question that has concerned analysts studying India's struggle with this issue is whether the law has provisions that would ensure greater fairness than the earlier legislation or this is just window dressing in an attempt to give a semblance of fairness before the May 2014 general elections. Has the democratic polity been able to exert pressure to achieve "real" fairness or has there been mere tokenism to appease the vociferous voices?

To answer this question, an attempt must be made to reach a clearer understanding of the meaning of the term "fair" in this context. To do this, in Section 6.2 I examine the compensation to the "project affected," which includes those whose land has been acquired as well as those who did not own land but were dependent on the land that was acquired.[5] Background is provided to identify the circumstances when the use of eminent domain should be considered justified. We study the path followed by the interpretational debate on the takings law in the United States in order to draw upon the principles arrived at regarding what constitutes "fair compensation." We find that the ethical perspective given by Frank Michelman has been dominant. In fact, Michelman has been described as one of the most influential constitutional scholars of the twentieth and twenty-first centuries (Underkuffler 2006; see also Wenar 1997). We also discuss his utilitarian framework in which he showed that incorporation of demoralization costs into the cost–benefit analysis helps in choosing only those projects that are socially beneficial. Section 6.3 examines the New Land Act or LARR to find out whether the claim of "fairness" is a legitimate one and briefly outlines the legislative changes in the New Land Act that the central government sought to introduce. Section 6.4 examines the role of the judiciary in interpreting the Land Acquisition Act, 1894. This is done because judicial interpretation affects morale in a significant way. Also, one needs to see whether the judiciary is taking public perceptions into account while arriving at the verdicts. The concluding section examines the relevance of Michelman's framework in the context of the land question in India today. It highlights the even greater need to take account of social discontent for policy formulation because the land acquisition and rehabilitation issues have resulted in widespread protests and have also become an area of intense research and public scrutiny.

6.2 The Compensation Question

Compulsory acquisition of land through the use of the power of eminent domain has long been a subject of debate, particularly in the two largest democracies of the world, the United States and India.[6] The reason for this is the inherent contradiction between the democratic process and the state's power to take land without consent. As the needs of economic and social development intensify in India and the demand for land grows, the tensions seem to magnify and threaten the very basis of the democratic polity. In this section I cover the rationale behind eminent domain, the takings law in the USA, and fair compensation to subsequently analyze India's New Land Act.

6.2.1 *Rationale for the Power of Eminent Domain*

The origin of the term "eminent domain" can be traced to the jurisprudential writings of seventeenth- and eighteenth-century scholars. Hugo Grotius,[7] the earliest and most important of the writers, wrote of this power (quoted in Chakravorty 2013: 112):

> The property of subjects is under the eminent domain of the state, so that the state or he who acts for it may use or even alienate and destroy such property, not only in the case of extreme necessity, in which even private persons have a right over the property of others, but for ends of public utility, to which ends those who founded civil society must be supposed to have intended that private ends should give way. But it is to be added that when this is done the State is bound to make good the loss to those who lose their property.

The rationale for the power of eminent domain has been traced to the reasons for the formation of the state. Epstein (1985), in his analysis of the takings question in the United States, provides the philosophical background to the political development that forms the basis of an expanding economy. Epstein argued that according to the natural rights tradition, each individual is endowed with certain rights to life, liberty, and property, which must be protected by the state. Since the resources required for the formation and operation of the state would not come voluntarily from the people, the state must be vested with powers to take private property. Specifically, two major restraints—that eminent domain be permitted only for public use and just compensation be paid—were meant to prevent the creation of the Hobbesian sovereign. Epstein goes on to elaborate that Thomas Hobbes had emphasized the dangers of unconstrained self-interest and had argued for a comprehensive contract by which all individuals surrender their liberty and property to a sovereign. However, he points out a problem in this Hobbesian idea by citing Max Weber's argument regarding the state's monopoly of the use of physical force to argue that a legal monopolist would profit like any other monopolist (Epstein 1985: 8). He further brings in John Locke's counterview that although all men were not driven by self-interest, there would be many such people and therefore a representative government was needed (Epstein 1985: 9). Epstein argues that although Locke categorically prohibited the taking of private property by the "supreme power" of the state, the Lockean conception has been made workable by the provision that property may be taken upon payment of just compensation (Epstein 1985: 12). The eminent domain clause can be seen as a specific protection for individuals against the state. When applied in the context of the land question today in developing countries such as India, it can be seen as a protection against capital in the interest of the larger good.[8]

In examining the economic rationale for endowing the state with powers to acquire property without the owners' consent, the question arises when a property needs to be acquired for public use, why should the state not purchase it. Ronald Coase identified three conditions under which bargaining among private parties will lead to the efficient outcome irrespective of the initial assignment of rights: the existence of well-defined property rights, the absence of transactions costs, and symmetrical wealth effects.[9] However, Coase argued that to suppose that transacting is costless is "a very unrealistic assumption" and in its absence governmental regulation may "lead to an improvement in economic efficiency" (Coase 1960: 18). State intervention is often upheld when contiguous parcels of land held by several owners need to be assembled. This entails high transactions costs and the problem of speculative hold-outs arises, particularly in cases of strategically located properties. Some may delay in entering the bargain because buyers would be willing to pay more for the subsequent plots rather than abandon the project altogether. In an assembly problem in which plots owned by different people must be combined for a single project, the purchase of one lot depends on the others, that is, the transactions are interconnected. Holding out in the form of pure and mixed strategy equilibria always occurs when complementarity is sufficiently high. The existence of complementarity means that the value of the combined plot is greater than the stand-alone value of the individual plots. The owner of a plot or owners of two or more separate plots find it in their interest to hold out and thereby delay the negotiations to capture a large share of the residual surplus (Menezes and Pitchford 2004; see also Roychowdhury and Sengupta 2012).

Another possibility is that the existing owners could wait and not enter the bargaining process in order to capture the external benefits such as the rise in price of the adjoining land resulting from the completion of the project (Grossman and Hart 1980). Further, a situation could arise where the owners may not want to sell for any price that the state is willing to offer (Alpern and Durst 1984). It has been suggested, therefore, that eminent domain does not have many close substitutes for resolving multiparty strategic bargaining. As justification for the use of eminent domain, the means–ends distinction has been made (Merrill 1986). "Means" refers to the need for land assembly and "ends" refers to the objective of providing public goods. Four possible cases can be identified. The one that involves a public good with an assembly problem is the one in which the use of eminent domain is justified by both approaches. Since the government can finance public goods through taxation, there is no need to use the eminent domain power when an assembly problem is not there. In the case where a private good requires land assembly, as occurs when a car manufacturer wants to set up a plant,[10] the means approach justifies the use of eminent domain but the ends approach does not.[11]

Munch (1976) examined the prices paid for land acquired for three large projects by the Chicago Department of Urban Renewal and concluded that prices paid under eminent domain differ systematically from the "fair market value" standard. Higher-valued properties tended to get proportionately larger settlements (about one-third more than market value) because they hired better lawyers, and low-valued properties received less than market value. She could not obtain a comparable set of data for free-market assemblies and concluded that her results "leave unproved" the superiority of eminent domain to market acquisition (Munch 1976: 495).

We can say that the eminent domain clause with the two limitations of public purpose and just compensation was meant to be a protection for individuals against the state. The economic justification lies in huge costs of assembling plots of land since individual owners could delay the bargaining process to obtain higher gain for themselves. In case of a private good the use of the eminent domain route has often been justified on the grounds of ends. However, there is no unequivocal evidence that eminent domain is superior to the free market for land assemblies.

6.2.2 *Takings Scrutiny in the United States*

Takings have been defined as "constitutional law's expression for any sort of publicly inflicted private injury for which the Constitution requires payment of compensation" (Michelman 1967: 1165). The United States Constitution provides an eminent domain (or takings) clause: "nor shall private property be taken for public use, without just compensation." The debate regarding takings of property is centered primarily on the following questions: what constitutes a taking requiring compensation, whether it is legitimate, and what constitutes just compensation. In this, two kinds of inquiry have contributed to what has been called a clearing of the confusion that was typical of this field.[12] One is the interpretational debate regarding the meanings of four terms in the eminent domain clause: "private property," "taken," "public use," and "just compensation." In the second type of inquiry, some court cases are considered landmarks in the sense of establishing certain important principles regarding the takings question. In India, the issue of what constitutes taking has been largely neglected. Although the question of legitimacy of acquisition is an oft-litigated issue, it is the compensation question that has become the focus in recent times. In what follows, we summarize the terminological debate in the United States to get a better understanding of the term "fair."

In deciding whether a particular government activity requires compensation, courts and scholars have used a number of tests. The dominant nineteenth-century physical invasion test, which received its classic conceptualization from Justice Harlan,[13] was based on the layman's idea of *property as things*—property is a thing over which the owner has rights of free use, enjoyment, and disposal. A taking occurs when the government takes physical possession of some privately owned thing. However, the government can use various means to try to reduce the value or inhibit the development of the property through regulation so that it has a much reduced market value when the government buys it. Often the government can "regulate" certain activities out of business without physical invasion and thus get away without paying compensation.[14]

The concept of property underlying the physical invasion test came under strong scrutiny during the twentieth century in Wesley Hohfeld's writings (Hohfeld 1919). According to him, property cannot be things; property can only be property rights. If one regards property as "things" then intellectual property and divided legal control over things would remain unexplained. Bruce Ackerman (1977) argued that each resource owner can be viewed as holding a bundle of rights vis-à-vis other potential users. Thus the takings clause would be viewed as follows:

> Whenever the state takes any user right out of Jones's bundle and puts it in any other bundle, private property should be understood to have been taken ... Even if

> Jones's bundle contains but a single user right, it is nonetheless protected against a taking by the clause. (Ackerman 1977: 28)

Another important approach to the taking problem is the diminution of value test, which arose from Justice Holmes' decision in *Pennsylvania Coal Co. v Mahon.*[15] He laid emphasis on economic value, thus shifting the focus of the interpretation to the second term of the clause: "taken." According to this test, the government might invade or property rights might be interfered with, but a taking occurs only if economic values are diminished by a significant percentage because of the government's action. There is the problem of setting the percentage that draws the line between a taking and a regulation. It was argued that the provision was more a guard against unfairness rather than against mere value diminution (Sax 1964). The focus of the discourse shifted to the "public use" term with the decision in *Berman v Parker.*[16] A privately owned department store was taken and transferred using the eminent domain power as part of a neighborhood redevelopment scheme that would add to the orderliness and beauty of the district. The "public use" term was interpreted to mean "public purpose" (order and beautification) even though the state did not intend to manage or occupy the building. This aspect has often come up in Indian court judgments while deciding whether a particular land acquisition serves public purpose.[17] For example, in the Singur case where the West Bengal state government claimed public purpose by acquiring land on behalf of a large, profit-making, private auto manufacturer, the Tatas, the farmers, and their supporters rejected this notion. To this the government responded that if land acquisition led to industrial investments and thus jobs then the outcome serves a public purpose (see also Chapter 5 of this volume).

The aspect of the takings debate that is most relevant to India is that elaborated in the influential work of Frank Michelman and Joseph Sax. They took the interpretational attempt forward by emphasizing the justice aspect of the term "just compensation." Sax (1964) argued that once property is thought of as economic value the question becomes to what kind of competition should existing values be exposed. Michelman (1967) argued that diversion of resources would alter the personal welfare situation of a person and will have the impact of redistribution of welfare. What constitutes a taking would depend on whether the correct theory of justice requires that the act be compensated. The central concept of the takings clause was not property but "economic values" or "welfare." His utilitarian standard required the balancing of the welfare gains of a proposed redistribution and the welfare costs borne by the "losers" and the cost incurred by the government as compensation. Epstein (1985) argued that a state action triggers takings scrutiny if it alters any incident of any private party's property bundle. Such an act must be for "public use" if it is to be legitimate at all: it must protect existing property entitlements or create a net social gain that must be distributed in proportion to pre-existing property entitlements. This would reduce "rent seeking," which arises when groups devote resources to manipulate property rights in their favor. This aspect is particularly relevant for India, where there is a rampant practice of procuring agricultural land at throw-away prices and then changing the land use to nonagricultural. The subsequent sale and resale is often done at exorbitant prices, many times the original price. The gains from this process are not shared with the original owners.[18]

One American case that has had an influence on the Supreme Court of India in recent times is the landmark case of *Kelo v City of New London*.[19] In this case, the United States Supreme Court ruled in favor of the pharmaceutical giant Pfizer and against the middle-class people who were living in the 90-acre Fort Trumball neighborhood of New London, Connecticut, which was acquired and significantly expanded the scope of public purpose. Upholding the taking, Justice John Paul Stevens wrote for the majority that the city's economic development plan, which will provide appreciable benefits to the community including increased tax revenue and new jobs, "unquestionably serves a public purpose." Justice Sandra Day O'Connor wrote for the dissenters that the decision means that "the sovereign may take private property currently in ordinary private use, and give it over for new ordinary private use." The residents had rejected the city's compensation package saying no amount of money could compensate them. The decision made every home, church, corner store vulnerable to being replaced by commercial development since the latter increases tax revenue and every region needs more tax revenue. Somin (2015), in a book-length analysis of the case, argued that the close 5–4 verdict in this case was an error, which the Supreme Court should reverse.

In the United States, there has been a scrutiny of each of the terms of the takings clause and landmark judgments have had an impact on how the interpretational debate has progressed. The aspect of the takings debate that is most relevant to any analysis of India's land acquisition process is the insight provided by Michelman (1967), who argued that diversion of resources changes the personal welfare of individuals and thus affects morale. This is magnified when the justification of economic development widens the scope of eminent domain so that every plot of land becomes vulnerable to acquisition.

6.2.3 *The Constraint of "Fair" Compensation*

Michelman (1967) argues that the only "test" for the requirement of compensation is the test of fairness: is it fair to put into effect a particular social measure without compensating for the private loss inflicted in the process?

He drew upon the utilitarian property theory of Bentham and argued that collective allocational decisions that are accompanied by particular distributions are compensable or not depending on whether they are "critically demoralizing." According to Bentham, property is a collection of rules that govern the exploitation of resources and therefore form the basis of expectations. Unpredictable redistributions would therefore destroy the security of future enjoyment of the product and would affect the morale. However, not all redistributions are totally demoralizing and a line can be drawn between compensable and non-compensable collective impositions.

According to Michelman's utilitarian standard, in deciding whether a government action was a taking that requires compensation, three factors need to be considered: efficiency gains, demoralizing costs, and settlement costs. Efficiency gains are "the excess of benefits produced by a measure over losses inflicted by it, where benefits are measured by the total number of dollars which prospective gainers would be willing to pay to secure adoption, and losses are measured by the total number of dollars which prospective losers would insist on as the price of agreeing to adoption" (Michelman 1967: 1214). This can be denoted by B (benefits) – C (costs). Demoralization costs (D) consist of "the total of (1) the dollar value necessary to offset disutilities which accrue to losers and their sympathizers specifically from the realization that no compensation is offered,[20] and (2) the present capitalized value of lost future production (reflecting either impaired incentives or social unrest) caused by demoralization of uncompensated losers, their sympathizers, and other observers disturbed by the thought that they themselves may be subjected to similar treatment on some other occasion." (Michelman 1967: 1214). Settlement Costs (S) are "measured by the dollar value of time, effort, and resources which would be required in order to reach compensation settlements adequate to avoid demoralization costs" (Michelman 1967: 1214). In other words, demoralization costs are those that arise when there is no compensation, settlement costs arise when compensation is paid. According to Michelman, a government measure whose dollar benefits exceed costs should nevertheless not be adopted if the net benefit is exceeded by both demoralization costs and settlement costs. Symbolically, a utilitarian does not undertake the project if $(B - C) < \min(D, S)$. If net benefits are positive and greater than either settlement or demoralization costs (or both), the lower S or D should be incurred. That is, the government should pay if $(B - C) > S$ and $S < D$. On the other hand, the government need not pay if $(B - C) > D$ and $D < S$.

We note that per the framework proposed by Michelman, an acquisition should not occur even if the monetary benefits exceed costs, when the demoralization and settlement costs exceed the benefits. He identified certain "sources" of these costs; one such source is in the distinction between strategically determined losses and natural hazards such as earthquakes, cyclones, floods, and so on. Michelman elaborates that demoralization costs would also arise in case of uncompensated government actions, where the efficiency gains are doubtful, where there is uncertainty about reciprocal benefits, where there is a lack of political will to extract mitigating concessions for the losers, and other situations where there is a feeling that similar situations will occur in the future. Michelman's framework has been subjected to intense scrutiny.[21] In this chapter, the focus is on the significance of public perceptions because there was an effort to hold consultations with the public before the enactment of the New Land Act. That this was indeed the case is evident from the fact that a book on the making of the New Land Act has been titled *Legislating for Justice* (see Ramesh and Khan 2015).

Fisher (1988) argues that there are two types of "demoralization costs" that would be substantial in many contexts. First, what Michelman calls the "dollar value [of] outrage" (Michelman 1967: 1215),[22] in which people unconvinced by government actions may view them as immoral.[23] Fisher argues that this would result in "a kind of discomfort that a judge who wishes to maximize efficiency must take into account in deciding whether to classify the actions in question as 'takings'." The second type is "search" costs, which result when a judicial decision denies compensation in defiance of popular perceptions that compensation must be paid and thereby undermines people's faith that by and large the law is in conformity with their sense of justice. People would be unwilling to make decisions without "looking up" the rules and this would entail deadweight losses that must be taken into account by economists. Also, there is the possibility that judicial decisions may in turn have an impact on popular views regarding the circumstances in which compensation is appropriate.

In the context of land acquisition in India today, incorporation of demoralization costs is extremely important. This is because since the process of economic liberalization began, the need by industry for land has increased by leaps and bounds. The government's role in the process has changed to becoming an agent for the private sector to facilitate land availability. The purpose of the acquisition is stated to be public purpose but the land losers, as alluded to earlier in the Singur case, know the dubious nature of such claims. Despite the backlog of millions of people yet to be rehabilitated, the pace of acquisition has continued unabated. All these have meant that demoralization costs have increased significantly in the last three decades. India needs to choose only those projects that will be socially beneficial. Use of Michelman's standard gives a framework to filter out economically efficient projects that have very high demoralization costs. In a democracy, one must find a way of recognizing these costs and factoring them into the analysis to achieve true fairness.

6.3 The Right to Fair Compensation and Transparency in the Land Acquisition, Rehabilitation and Resettlement Act, 2013

The Indian government was forced to take into account the discontent over the land acquisition process carried out over the decades since independence. This is evident from the various attempts made to formulate a legally enforceable national rehabilitation policy, since the Land Acquisition Act, 1894 only dealt with the process of acquisition and determination of monetary compensation. The latest notified version was the National Rehabilitation and Resettlement Policy (NRRP), 2007, which was abandoned after it came in for sharp criticism.[24] After taking suggestions from the Standing Committee on Rural Development (2007–8/2012) and the National Advisory Council, a new bill was proposed in 2011 titled the Land Acquisition and Rehabilitation and Resettlement Bill. The final version is the latest Right to Fair Compensation and Transparency in LARR, 2013. This section analyzes the background to the enactment of this New Land Act and the pressures that compelled the government to attempt a semblance of fairness. It then examines the new legislation to find out whether the claim of fairness is a legitimate one. This is done by a scrutiny of the provisions to see if an attempt is made to actually take into account the demoralization costs discussed earlier. The provisions of the land Ordinance will also be discussed briefly.

6.3.1 *Discourse on the Compensation Question prior to the New Land Act*

Land acquisitions had been going on both prior to and after independence without any uniform and enforceable compensation framework. Millions of people have been displaced and protests have taken place in various parts of the country.[25] There had been a steady demand for amendment of the colonial Land Acquisition Act, 1894 and its replacement by new legislation. Prior to the recent enactment of the New Land Act, there have been seventeen amendments to the prevailing Land Acquisition Act, 1894, with two important ones being in 1962 and 1984. When plans to amend the archaic act were initiated and a bill circulated for public response in 1998, it came in for sharp criticism (Asif 1999; Pinto 1998). Asif (1999) noted that the government has acknowledged that land acquisition and displacement have resulted in protest movements marked by growing militancy and argued that the absence of transparency makes the act one of the most misused civil laws in the country.[26] Since the district collector's role is decisive in the entire process of acquiring land and awarding compensation, he criticized this collector-centric procedure and called for a paradigm shift to democratize the acquisition process. He argued that land should be made available for genuine public purposes and that the land losers should not be impoverished in the process. Pinto (1998) echoed the same viewpoint calling for a national debate on the issue.

A national debate started in 2007 with the violence in Nandigram and Singur in West Bengal and the relocation of the Tatas to Gujarat. The poll debacle of the left-front government placed the land and rehabilitation issues in the forefront of discussion and debate. In fact, Guha (2007; also Chapter 12 in this volume) while examining the case of the pig-iron manufacturing plant, Tata Metaliks, which was set up in Kharagpur in 1992, argues that the acquisition case did not see any of the popular opposition witnessed in Singur and Nandigram.[27] Apart from Singur, many cases have been documented and scrutinized by researchers. For instance, specific case studies have been analyzed for land acquisition in Assam (Gohain 2006), Hyderabad (Reddy and Reddy 2007), Andhra Pradesh (Balagopal 2007), Bhatta-Parsaul in Uttar Pradesh (Kumar 2011), Rajasthan (Levien 2011), Jharkhand (Lahiri-Dutt, Krishnan, and Ahmad 2012), and in Pune, Maharashtra (Sathe 2014), among others.

We categorize the discourse on the land question around the three issues: whether acquisition should take place at all, when compensation should be paid, and what constitutes "fair" compensation. Regarding the first issue, Sampat (2013) examines the eminent domain power in the Indian context and like Iyer (2007) argues: "acquisition needs to be made contestable not merely in regard to compensation, but also in relation to the public purpose" (Sampat 2013: 43). She further argues: "Given the expanded scope and the conflation of private interest with public purpose, the power of eminent domain now seems set to engender large scale 'accumulation by dispossession' (Harvey 2005); its only counter being popular resistance to dispossession" (Sampat 2013: 46). It has been emphasized that it is not imperfect land markets that have resulted in resistance to acquisition but the nonexistence of land markets: there are often no "willing sellers" for a particular land, although there are buyers. Chakravorty (2013) calls these lands "priceless" and advocates that they should be kept out of the purview of acquisition.

The second issue relates to the question of when compensation should be paid. In India, when consent has not been forthcoming, the legal owners have been forced out of their lands and compensation denied. For instance, when mining projects face resistance, residents have been forced to leave the area through blasting operations in adjacent plots. There are instances when ponds, which are the mainstay of villagers, have been polluted or work on dams started so that the people residing in adjacent areas may be moved out on grounds of threat of subsidence. It is only when land is acquired using the Land Acquisition Act that one can make a claim for compensation. Unlike in the United States, where the question of what constitutes a "taking" has been subject to intense scrutiny, this has not been done in India. Only in the case of those people who do not directly own land but are dependent for their livelihoods on the land being acquired that the question arises whether compensation should be paid at all.

The question of what constitutes "fair" compensation has been the focus of public discourse in India. Numerous suggestions have been given with the view to making the people partners in the process of development. The Land Acquisition Act, 1894 took the market value as the bench mark and a solatium was awarded over and above that to factor in the compulsory nature of the transaction. However, this method came in for severe criticism (see, for instance, Morris and Pandey 2007). Various suggestions have been made, including an option pricing method (Marjit 2010) and land auctions (Ghatak and Ghosh 2011), among others. The crux of the arguments is that in cases of compulsory acquisition there is no willing seller–willing buyer market scenario, so market prices cannot be used as an indicator of value for the parties involved.[28] It is against the backdrop of this kind of discourse that the New Land Act came into force.

6.3.2 *The "Fairness" Claim of the New Land Act*

The New Land Act seems to shout out loud its intention of being "fair." This emphasis is quite understandable given the context in which it is being enacted; it is clear that public perceptions are of utmost importance. The question, however, is: when the very foundation of eminent domain as understood in the theoretical literature is that it must be accompanied by just compensation, is it at all required that a thing so obvious must be separately emphasized? Is the title making a statement that an unfair practice is being rectified?[29] That India had for long deviated from the views as put forth by the natural rights tradition and it is only now that continued pressure from the voting public has forced the state to reconsider its position? The preamble is full of words that attempt to emphasize justice: "a humane, participative, informed and transparent process of land acquisition," "provide just and fair compensation," and "that affected persons become partners in development."

An examination of the provisions of the act show that regarding the issue of when acquisition should take place, consent is needed only in the cases of private companies [at least 80 percent of the affected families; 2(2)(b)(i)] and of public–private partnerships [at least 70 percent of the affected families; 2(2)(b)(ii)].[30] This has been done to circumvent the hold-out problem. Further, it is specified that:

> The appropriate Government shall ensure that (a) there is a legitimate and bona fide public purpose for the proposed acquisition which necessitates the acquisition

> of the land identified; (b) the potential benefits and the public purpose referred to in clause (a) shall outweigh the social costs and adverse social impact as determined by the Social Impact Assessment that has been carried out; .... [8(1)]

Using Michelman's framework, the phrase "adverse social impact" would include the demoralization costs. The challenge lies in factoring in what has also been referred to as the "psychic externalities" (Kennedy 1981). This refers to a kind of discomfort that people experience when they regard certain government acts as immoral. It has been argued that a judge while deciding whether a particular action is a taking requiring compensation must take these kinds of externalities into account. In recent times, these have threatened to outweigh all benefits. The increase in compensation mandated by the New Land Act could help reduce demoralization; but it still continues to be based on market value. We note that where land is acquired using the urgency clause [17(1) in the old act and 40(1) in the New Land Act],[31] no social impact assessment is required. In the New Land Act, the specific reasons cited are land acquisition for national defense and emergencies arising out of natural calamities or any other emergency with the approval of Parliament. This implies that the costs and benefits associated with the acquisition will not be calculated but the compensation amount will be determined by the collector.[32]

The collector-centric approach of the old act has been retained. While determining the compensation amount, the seventh point specified in Section 28 states that the collector must take into account: "any other ground which may be in the interest of equity, justice and beneficial to the affected families." This is an open-ended provision subject to different interpretations. A novel aspect of the act is the provision for setting up a National Monitoring Committee for Rehabilitation and Resettlement, but it is to be set up only "whenever necessary." There is a provision for establishment of a Land Acquisition, Rehabilitation and Resettlement Authority, but it would consist of just one person, called the "presiding officer." We further note that the settlement costs associated with compensation, if calculated, would be quite high in India since land records are not up to the mark. Section 3(i) gives the costs associated with acquisition. These consist of the amount of compensation including solatium, costs of damages caused to the land and standing crops in the process of acquisition, cost of acquisition of land and buildings, infrastructure cost for resettlement, and administrative cost. According to Michelman's framework, the settlement costs are vital to any ex-ante decision making. In the Indian context, in an attempt to get on with "industrialisation, development of essential infrastructural facilities and urbanization," the settlement costs cannot be ignored in the cost–benefit calculation. In fact, the urgency clause permits the collector to take possession of land even before the award of compensation has been made. The widespread misuse of this clause was regarded as one of the major flaws of the Land Acquisition Act, 1894. Also, the delay in getting compensation could turn out to be disastrous for the families who subsist on the produce of the land. An interest rate of 12 percent [30(3)] would be nominal considering the usurious rates charged in rural areas. This difference itself could lead land losers into a debt trap. On the whole, we find that there are no provisions in the New Land Act that make the claim of "fairness," "humane," or "transparent" justified. The collector is the all-powerful entity and this colonial construct would continue to "rule"; in fact, by bringing in public–private partnerships, this power has only expanded.

An Ordinance was promulgated on December 31, 2014, which made such draconian changes to the provisions of the New Land Act that the latter appeared humane by comparison. The most significant change was the removal of the three important restraints introduced by the New Land Act: social impact assessment, consent requirements, and restriction on acquisition of multicrop land for four additional categories of projects apart from national defense. These were rural infrastructure including electrification, affordable housing and housing for the poor, industrial corridors, and infrastructure including projects under public–private partnership where the ownership of land continues to vest with the government. Although this provision specified four categories, they covered most of the cases for which land is required. Moreover, in both the old and new land acts, it was the urgency clause that exempted projects from cost–benefit analysis and courts had often struck down attempts to use it. The Ordinance sought to legitimize the urgency provision for a whole range of projects. The other change that significantly expanded the scope of eminent domain was the substitution through clause (yy) in Section 3 of the New Land Act: the term "private entity" in place of "private company," where the former would include "any entity other than a Government entity or undertaking and includes a proprietorship, partnership, company, corporation, non-profit organisation or other entity under any law for the time being in force." Further, in Section 101 the period for return of unutilized land was changed from five years to "a period specified for setting up of any project or for five years, whichever is later." The time required for setting up the project could be increased so much that the provision would be nullified. The New Land Act had increased the accountability of officers who were entrusted to implement the law. The Ordinance diluted this by mandating that prior consent was needed from the government before the officers could be penalized.

The Ordinance met with severe criticism from across the board. The central government, however, only retreated from its position after several months when the Ordinance, which was re-promulgated for the third time, was allowed to lapse on August 31, 2015. A bill incorporating the changes was passed in the Lok Sabha in March 2015 but failed to garner the requisite numbers in the Rajya Sabha of the Indian Parliament. The resolve of the ruling party (BJP) to continue to push through the draconian changes shows that they had the confidence that these sweeping changes would be accepted in the name of economic development. However, most of the other political parties were not willing to face the wrath of the farmers and their concerted resistance indicates that they did not want to risk an electoral debacle on this highly contentious issue.

6.4 Role of the Judiciary in Interpretation of the Land Acquisition Law in India

Judicial interpretation has a crucial role in a democracy not only in actual implementation of the law but also in influencing the perceptions of the people and therefore their morale. For an aggrieved land loser, the judiciary is the only recourse to remedy. It has often been argued that courts have failed to act as guardians of the rights of people (see, for instance, Gonsalves 2010) and to ensure that the two restraints placed on the power of eminent domain, namely that it be used for public purpose and that just compensation be paid, are adhered to. The lack of clear directions regarding many of the contentious issues and the continual expansion in the scope of public purpose has meant that public discontent has risen. In this section, a preliminary attempt is made to broadly examine the stance taken by the judiciary in India in interpreting the land acquisition legislation.

There are conflicting opinions in India regarding the important question of whether an acquisition serves a public purpose and whether it should be undertaken at all. In the United States, the exercise of eminent domain can be challenged on the grounds that it is not for a legitimate public purpose. Section 6(3) of the Land Acquisition Act, 1894 stated that a declaration of the government that the land is being acquired for a public purpose was conclusive evidence that the land is needed for public purpose and unless it is shown that there has been a colorable exercise of power, the courts cannot go on to examine whether it is a public purpose or not. The Supreme Court observed in the case of *Somawanti and Others v State of Punjab and Others (And connected Petitions)*:

> Now whether in a particular case the purpose for which land is needed is a public purpose or not is for the State Government to be satisfied about. If the purpose for which the land is being acquired by the State is within the legislative competence of the State the declaration of the Government will be final subject, however, to one exception. That exception is that if there is a colourable exercise of power the declaration will be open to challenge at the instance of the aggrieved party.[33]

The exception pertains to situations in which it appears that the custodian of power is influenced in its exercise by considerations or ends outside of those for which the power is sanctioned. In such cases, one argues that there is a pretention of achieving a legitimate goal.

In another landmark judgment, *State of Bihar v Kameshwar Singh*,[34] it was observed by Justice Mahajan that:

> The right point to be determined in each case is whether the acquisition is in general interest of the community as distinguished from the private interest of an individual ...The sovereign power to acquire property compulsorily is a power to acquire it only for a public purpose. There is no power in the sovereign to acquire private property in order to give it to private persons. Public purpose is the content of the power itself.

Justice Mahajan quoted from Willoghby's Constitutional Law:

> As between individuals, no necessity however great, no exigency however imminent, no improvement, however valuable, no refusal, however unneighbourly, no obstinacy, however unreasonable, no offers of compensation, however extravagant, can compel or require any man to part with an inch of his estate.

In *R L Arora v The State of Uttar Pradesh*,[35] which related to a case in which land was acquired for an industrialist in Kanpur for a textile machinery parts factory, the Supreme Court set aside the acquisition and concluded that the intention of the legislature was that land should be acquired only when the work constructed was directly useful to the public and the public would be entitled to use the work for its own benefit. However, this stance was overturned in the case of *Somawanti* mentioned earlier.

The courts have also upheld the interest of the individual in some cases. In the case of *Gadigeppa Mahadevpa Chikkumbi v State of Karnataka*,[36] where the acquisition deprived the petitioner of his sole means of livelihood as an agriculturist, his fundamental right to pursue his occupation as assured under Article 19(1)(g) of the Constitution of India was taken away; it was held that the acquisition of land in the garb of public interest was unsustainable.[37] However, there have been instances in which the judiciary has upheld acquisitions on behalf of the private companies as public purpose acquisition as long as a part of the compensation cost was paid out of the state exchequer.[38] In a more recent case,[39] the Supreme Court extended the scope of public purpose stating: "If the project taken as a whole is an attempt in the direction of bringing foreign exchange, generating employment opportunities and securing economic benefits to the State and the public at large, it will serve public purpose." However, within the first six months of 2011, the Supreme Court quashed attempts by state governments to use emergency provisions in the law to deny farmers their right to object to acquisition.

It set aside three land acquisition proceedings—one by the government of Uttar Pradesh for a jail in Sahranpur, another for acquiring land in Greater Noida in Uttar Pradesh for planned industrial development, and the acquisition of the Greater Noida Extension Area (GNEA) in Gautam Buddha Nagar district of Uttar Pradesh. Widespread protests against the acquisitions may have influenced the verdicts.

Another contentious issue relates to whether compensation should be paid at all in case of compulsory acquisition. By the Constitution (forty-fourth Amendment) Act, 1978, the property clauses of Articles 19(1)(f) and 31(2) were omitted and the right to property ceased to be a fundamental right. Article 300-A, which states that no person shall be deprived of his property save by authority of law, was included in Part XII of the Constitution. The obligation of the state to pay compensation for property acquired was obviated by this amendment. In the case of *Jilubhai Nanbhai Khachar v State of Gujarat,*[40] the Supreme Court explicitly stated that "Payment of market value in lieu of acquired property is not sine qua non for acquisition." Another extreme view was put forth in *State of Bihar v Kameshwar Singh,*[41] by Mahajan, that ordinarily the government has no right to acquire private property to give it to private persons even on payment of compensation. However, it is mostly agreed that acquisition must be accompanied by compensation.

The most contentious aspect has been the amount of compensation. In *Jilubhai Nanbhai Khachar v State of Gujarat* it was stated that the adequacy of the amount of compensation cannot be questioned in a court of law. However, the validity of irrelevant principles used in the determination of the amount is amenable to judicial scrutiny. The general principles determining compensation in India were laid down in Sections 23 and 24 of the Land Acquisition Act, 1894. These have been elaborated and expanded by the judiciary[42] following a case-by-case adjudication practice.[43] Some of the interpretations of the act with respect to the principles regarding estimation of market value and the final award have been discussed in Pal (2013). Innumerable cases have been filed in the context of the earlier act. Singh (2012) argues that the old act was inherently prone to litigation over compensation and that the use of eminent domain "can guarantee neither efficiency nor fairness of the outcome."[44]

We find that although for an aggrieved land loser the judiciary is the only recourse to remedy, courts have not put forth a clear and consistent stand on whether an acquisition serves a public purpose, and regarding payment and the amount of compensation. This has led to confusion and discontent rather than contributing to building a basis for an understanding of the key issues of public purpose and fair compensation like that in the United States, where, as was mentioned earlier in the chapter, detailed judicial reasoning has contributed immensely to the interpretational debate.

6.5 Relevance of Michelman's Framework in the Indian Context

We find that the discourse on land acquisition in India is mainly around the two restraints: public purpose and just compensation. In recent times, acquisition of land by the government on behalf of companies has become highly contentious. While some argue for voluntary transactions, others call for some regulation of these voluntary transactions; still others argue that in the case of large-scale land acquisition for industry, the state's participation is essential (Banerjee et al. 2007; Ghatak and Ghosh 2011). In fact, Ghatak and Ghosh (2011: 71) argue:

> The market often works well in arranging bilateral transactions, but its effectiveness drops exponentially as the number of parties to the transaction grows large, especially in a country like India where property rights are poorly defined, land records are fuzzy and courts work at a glacial pace.

In Singur itself, the 1,000 acres of land acquired belonged to 12,000 owners, indicating that transaction costs are high due to negotiations with innumerable owners for a particular plot of land.[45] As the last decade has shown, land acquisitions under the old act had taken place for private companies purportedly for public purpose but almost all the surplus created went to the new owners.[46]

In the Indian context, it would be useful to calculate the settlement costs involved in compensation in addition to the usual cost–benefit analysis. It is only right that the gains from the land transfer should be shared with the owners of the asset to make them partners in development.[47] Using Michelman's framework, one can say that incorporating both settlement and demoralization costs in the utilitarian standard on an ex-ante basis would mean that not every project that serves a public purpose and has positive efficiency gains should be undertaken, that is, many projects would not be cost effective. This would reduce the scope of using eminent domain and also make the land transfer less prone to social unrest. The more doubtful the "public purpose," the higher the demoralization costs associated with the project. It has been documented how the stated public purpose has not materialized in numerous projects and those affected have often faced multiple displacements without any compensation (see Hardikar 2013). The solatium of 100 percent that the New Land Act stipulates would not be able to capture these costs since many may not be willing to part with land at any price.

India is at a crucial juncture in its development path and the need for land for purposes of industrialization will only grow. The New Land Act has been called pro-peasant and anti-corporate and it has been argued that the costs of projects would escalate, the latter used as justification for the Ordinance. Our analysis of the provisions show that using Michelman's framework to understand the takings conundrum, neither efficiency nor equity considerations would be met if settlement and demoralization costs were not considered explicitly before a project is undertaken. The failure of the state apparatus to explicitly take account of demoralization costs resulted in the democratic process taking over and voting the three-decade-old left-front government in West Bengal out of one of its strongest bastions following the incidents of Nandigram and Singur. This can be considered a watershed in India's struggle with the issues of land acquisition and rehabilitation, forcing some state governments to review their own rehabilitation policies. Any claim of "fairness" in the national legislation must be backed by provisions that substantially alter the process that has been adopted so far. Only when people are secure that their property and livelihood would be protected would they have faith and become true "partners in development." Thus, even if some projects which have positive benefits (B) – costs (C) are rejected due to high settlement costs (S) and demoralization costs (D), the economy will not be worse off since ultimately development is meant to be inclusive and for the benefit of all.

To conclude, this chapter has attempted to place the Indian land acquisition issue in the context of the conception of eminent domain and the two restraints of public purpose and just compensation. It is argued in the chapter that once we analyze the issues using the utilitarian standard of Michelman, all three issues of whether acquisition should take place at all, whether compensation should be paid, and the amount of compensation can be brought in so that the scope of the state's power to acquire land is curtailed. This will benefit the economy by channeling scarce resources to their most socially beneficial purposes. Also, by making the landowners partners in a development process that encompasses both agriculture and industry, the goal of equity can be better served.

Acknowledgments

I thank Anthony D'Costa for his incisive comments at various stages in the revision process.

Notes

1. This was done on the insistence of Rahul Gandhi, the vice-president of the All India Congress Party and a contender for the post of prime minister in the Parliamentary elections in May 2014. An alternative title suggested by the National Alliance of People's Movements (NAPM) was "Development Planning, Resettlement and

Rehabilitation Bill." There is a clear indication that the ruling government wanted to emphasize the "fairness" part, rather than "development."

2. These protests have intensified into militant opposition in some parts of India.
3. Fisher (1988) highlights the significance of public perceptions of the takings doctrine.
4. The Bharatiya Janata Party (BJP), which came to power in May 2014, attempted to alter the legislative framework. However, it failed to garner support from other political parties and had to retreat from its position. The changes that were sought to be introduced were drastic and will be discussed subsequently.
5. Section 3 of Chapter I of the New Land Act provides the categories of people who would be considered "project affected."
6. Singh (2012) notes that the use of eminent domain for private projects has been more controversial in land-abundant United States than in relatively land-scarce England. This is because the power to grant permission rests with the office of the deputy prime minister, which ensures that there is no competition by local authorities to give away land to the private companies at cheap rates. Chandra (2008) shows in his analysis of the Singur case in West Bengal how the states were competing with each other in offering the "best package" of subsidies to industry.
7. De Jure Belli Et Pacis (1625).
8. We note that this protection is effective only when the acquisition is accompanied by just compensation.
9. In two articles, Coase (1959, 1960) propounded what has become the celebrated Coase Theorem.
10. Two instances are used of eminent domain for acquisition of land for General Motors in Poletown, Detroit, USA, and for Tata Motors in Singur, West Bengal, India.
11. The recent land acquisition for the private sector shows that often land of a specific nature is not required, although there is a land assembly problem. Justification in terms of the ends approach gives rise to strained reasoning triggering protests all over the country.
12. In India, as elaborated later, this area is fraught with confusing rules and judgments.
13. In *Mugler v Kansas* 123 U.S. 623 (1887).
14. See Fischel (1995) for a detailed study of what is called regulatory takings in the USA.
15. 260 US 393, 412–16 (1922).
16. 348 US 26, 32–4 (1954).
17. For instance, in *R L Arora v The State of Uttar Pradesh* [AIR 1962 SC 764], it was stated that land should be acquired only when the work constructed would be used directly by the public for its own benefit. See Section 6.4 of this chapter.
18. For instance, Pune Divisional Commissioner Prabhakar Deshmukh was accused in this connection. Available at: <http://www.moneylife.in/article/how-rti-activist-exposed-pune-div-commissioner-prabhakar-deshmukhs-landscam/36456.html?utm_source=PoweRelayEDM&utm_medium=Email&utm_content=Subscriber%2361608&utm_campaign=Today%27s%20Exclusives%2020%20feb%202014> (accessed February 20, 2014).

19. 545 U.S. 469 (2005).
20. This means that these costs do not overlap with the cost term C in the calculation of efficiency gains. They arise due to the lack of compensation and disappear if compensation is paid.
21. The economic critique of Michelman's approach has been given in Blume and Rubinfeld (1984), Kaplow (1986), Fischel and Shapiro (1988). A summary of the critique is given in Pal (2013). Fisher (1988) argues that although economists' well-taken point mandates a substantial modification of the second and third terms of Michelman's formula, that does not deprive it of its significance.
22. Fisher (1988: fn. 26, 1779) quoting Ackerman (1977: 46–7), referred to this as "the outrage and demoralization of good citizens who believe themselves [or others] victims of unprincipled behaviour" and "the long-run disutility of citizen disaffection with the state's decision making processes."
23. (Michelman 1967: 1217–18) considers some reasons for such perceptions of immorality.
24. Pal (2013) argues, using Michelman's framework, that the predetermined act of using eminent domain was taken at par with random natural disasters. None of the provisions of the NRRP (2007) tried to address the demoralization costs.
25. Estimates of displacement vary, for example, from 60 million (Fernandes 2008) to 100 million (NAPM). A White Paper on the number of displaced would help in proper assessment.
26. One wonders whether it is to combat this kind of criticism that the word "transparency" has also been added to the title of the New Land Act.
27. Guha (2004), in his detailed research of Paschim Medinipur district in West Bengal, had revealed the policy failure of the left-front government.
28. Chakravorty (2013) focuses on the price of land in his book on land acquisition, conflict, and consequence.
29. Ramesh and Khan (2015: 3) also allude to this when they argue: "Few countries employ the same unbridled use of acquiring authority that Indian authorities have had."
30. We note that acquisition for public-private partnerships (PPPs) have been inserted into the Act.
31. This route has been used far too often to transfer land to companies.
32. The Supreme Court of India strongly warned against the use of this power arbitrarily against poor farmers, emphasizing that depriving him of property without a hearing is "constitutional anathema." See *State of Punjab and another v Gurdial Singh* AIR 1980 SC 319.
33. AIR 1963 SC 151.
34. 1952 SCR 1056: AIR 1952 SC 252.
35. AIR 1962 SC 764.
36. AIR 1990 Kant 2.
37. See also *Bamandas Mukherjee v State of West Bengal* AIR 1985 Cal 159.
38. Singh (2012) cites the case of *Indrajit C Parekh v State of Gujarat* [AIR 1975 SC 1182] in which the Supreme Court upheld the contention of the Gujarat government that a contribution of even one rupee from the exchequer is sufficient to validate

the acquisition for the companies under Part II of the Land Acquisition Act, 1894. See also *Pratibha Nema v State of MP*, AIR 2003 SC 3140.

39. [1] *Sooraram Pratap Reddy and Others;* [2] *Suraram Krishna Reddy And Another;* [3] *V. Krishna Prasad;* [4] *A.L. Sadanand;* [5] *Malla Reddy And Others;* [2] *District Collector, Land Acquisition And Others;* [3] *Government of Andhra Pradesh and Others;* [4] *District Collector and Others* 2008 (9) SCC 552.
40. AIR 1995 SC 142; Supp (1) SCC 596; (1994) Supp 1 SCR 807.
41. AIR 1952 SC 252.
42. In *State of West Bengal v Arun Kumar* (1984) CLJ 220 (DB), it was held that Section 23 is not exhaustive.
43. *Union of India v Pramod Gupta,* 2005 AIR SCW 4645; JT 2005 (8) SC 203; 2005 (7) scale 187; AIR 2005 SC 3708; (2005) 12 SCC 1.
44. Singh's analysis (2012: 48) of 325 High Court Cases in Punjab and Haryana revealed that litigation benefits high value properties more because the burden to prove the market value is on the owner although all the relevant information regarding land records is possessed by the government and richer farmers have more resources for litigation.
45. Levien (2011) has argued: "The law only requires that a majority of land be purchased, not a majority of land-holders agree to surrender their land."
46. Chandra (2008) shows that the claim of fulfilling the public purpose of "generating employment" does not stand the test of scrutiny.
47. Levien's (2012) study of Mahindra World City (MWC) on the outskirts of Rajasthan, on whose behalf the Rajasthan State Industrial and Investment Corporation (RIICO) acquired land, revealed that even if four times the price had been paid, MWC would still have been left with a large profit.

References

Ackerman, B. A. 1977. *Private Property and the Constitution.* New Haven, CT: Yale University Press.

Alpern, A. and Durst, S. 1984. *Holdouts*! New York: McGraw-Hill.

Asif, M. 1999. "Land Acquisition Act—Need for an Alternative Paradigm." *Economic and Political Weekly*, 34(25): 1564–6.

Balagopal, K. 2007. "Illegal Acquisition in Tribal Areas." *Economic and Political Weekly*, 42(40): 4029–34.

Banerjee, A., Bardhan, P., Basu, K., Chaudhury, M. D., Ghatak, M., Guha, A. S., Majumdar, M., Mookherjee, D., and Ray, D. 2007. "Beyond Nandigram: Industrialisation in West Bengal." *Economic and Political Weekly*, 42(17): 1487–9.

Blume, L. and Rubinfeld, D. L. 1984. "Compensation for Takings: An Economic Analysis." *California Law Review*, 72: 569–628.

Chakravorty, S. 2013. *The Price of Land: Acquisition, Conflict and Consequence.* New Delhi: Oxford University Press.

Chandra, N. K. 2008. "Tata Motors in Singur: A Step towards Industrialisation or Pauperisation." *Economic and Political Weekly*, 43(50): 36–51.

Coase, R. H. 1959. "The Federal Communications Commission." *Journal of Law and Economics*, 2: 1–40.

Coase, R. H. 1960. "The Problem of Social Cost." *Journal of Law and Economics*, 3: 1–44.

Epstein, R. A. 1985. *Takings: Private Property and the Power of Eminent Domain*. Cambridge, MA: Harvard University Press.

Fernandes, W. 2008. "Paying the Price for Someone Else's Displacement." Available at: Council for Social Justice and Peace—Goa, <http://www.csjpgoa.org/ecology-and-environment/forests-and-tribal-issues/paying-the-price-for-someone-else%E2%80%99s-displacement/> (accessed February 20, 2014).

Fischel, W. A. 1995. *Regulatory Takings: Law, Economics and Politics*. Cambridge, MA.: Harvard University Press.

Fischel, W. A. and Shapiro, P. 1988. "Takings, Insurance and Michelman: Comments on Economic Interpretations of 'Just Compensation' Law." *Journal of Legal Studies*, 17(2): 269–93.

Fisher, W. W., III. 1988. "The Significance of Public Perceptions of the Takings Doctrine." *Columbia Law Review*, 88(8): 1774–94.

Ghatak, M. and Ghosh, P. 2011. "The Land Acquisition Bill: A Critique and a Proposal." *Economic and Political Weekly*, 46(41): 65–72.

Gohain, H. 2006. "Land Question in Assam—A Short Note." *Economic and Political Weekly*, 41(32): 3459–60.

Gonsalves, C. 2010. "Judicial Failure and Land Acquisition for Corporates." *Economic and Political Weekly*, 45(52): 37–42.

Grossman, S. J. and Hart, O. D. 1980. "Takeover Bids, the Free Rider Problem and the Theory of the Corporation." *Bell Journal of Economics*, 11: 42–64.

Guha, A. 2004. "Land Acquisition in a West Bengal District." *Economic and Political Weekly*, 39(42): 4620.

Guha, A. 2007. "Peasant Resistance in West Bengal a Decade before Singur and Nandigram." *Economic and Political Weekly*, 42(37): 3706–11.

Hardikar, J. 2013. *A Village Awaits Doomsday*. New Delhi: Penguin Books.

Harvey, D. 2005. *A Brief History of Neoliberalism*. New York: Oxford University Press.

Hohfeld, W. N. 1978 [1919]. *Fundamental Legal Conceptions as Applied in Judicial Reasoning*. Westport: Greenwood Press.

Iyer, R. 2007. "Towards a Just Displacement and Rehabilitation Policy." *Economic and Political Weekly*, 42(30): 3103–7.

Kakani, R. K., Raghu Ram, T. L., and Tigga, N. S. 2009. "Insights into Land Acquisition Experiences of Private Businesses in India." In *Land—A Critical Resource for Infrastructure*, edited by R. Lall, India Infrastructure Report series of the 3iNetwork and the Infrastructure Development Finance Company (IDFC). New Delhi: Oxford University Press, pp. 135–42.

Kaplow, L. 1986. "An Economic Analysis of Legal Transitions." *Harvard Law Review*, 99(3): 509–617.

Kennedy, D. 1981. "Cost-Benefit Analysis of Entitlement Problems: A Critique." *Stanford Law Review*, 33: 387.

Kumar, A. 2011. "The Battle for Land: Unaddressed Issues." *Economic and Political Weekly*, 46(25): 20–3.

Lahiri-Dutt, K., Krishnan, R., and Ahmad, N. 2012. "Land Acquisition and Dispossession—Private Coal Companies in Jharkhand." *Economic and Political Weekly*, 46(6): 39–45.

Levien, M. 2011. "Rationalising Dispossession: The Land Acquisition and Resettlement Bills." *Economic and Political Weekly*, 46(11): 66–71.

Levien, M. 2012. "The Land Question: Special Economic Zones and the Political Economy of Dispossession in India." *The Journal of Peasant Studies*, 39(3–4): 933–69.

Marjit, S. 2010. "Guaranteeing Future Claims of Farmers in Land Acquisition: An Option Pricing Approach." *Economic and Political Weekly*, (45)52: 80–2.

Menezes, F. and Pitchford, R. 2004. "A Model of Seller Holdout." *Economic Theory*, 24: 231–53.

Merrill, T. W. 1986. "The Economics of Public Use." *Cornell Law Review*, 72: 61–116.

Michelman, F. I. 1967. "Property, Utility, and Fairness: Comments on the Ethical Foundations of 'Just Compensation' Law." *Harvard Law Review*, 80(6): 1165–258.

Morris, S. and Pandey, A. 2007. "Towards Reform of Land Acquisition Framework in India." *Economic and Political Weekly*, 42(22): 2083–90.

Munch, P. 1976. "An Economic Analysis of Eminent Domain." *Journal of Political Economy*, 84(3): 473–97.

Pal, M. 2013. "Rehabilitation of the 'Project Affected': Eminent Domain and Just Compensation." In *Law and Economics*, edited by S. Gangopadhyay and V. Santhakumar, vol. 2: Practice. New Delhi: Sage, pp. 1–18.

Pinto, A. 1998. "Fillip to Land Transfers—Land Acquisition Bill, 1998." *Economic and Political Weekly*, 33(49): 3107–9.

Ramesh, J. and Khan, M. A. 2015. *Legislating for Justice—The Making of the 2013 Land Acquisition Law*. New Delhi: Oxford University Press.

Reddy, R. V. and Reddy, B. S. 2007. "Land Alienation and Local Communities. Case Studies in Hyderabad-Secunderabad." *Economic and Political Weekly*, 42(31): 3233.

Roychowdhury, P. and Sengupta, K. 2012. "Transparency, Complementarity and Holdout." *Games and Economic Behaviour*, 75: 598–612.

Sampat, P. 2013. "Limits to Absolute Power: Eminent Domain and the Right to Land in India." *Economic and Political Weekly*, 48(19): 40–2.

Sathe, D. 2014. "Vicissitudes in the Acquisition of Land: A Case Study." *Economic and Political Weekly*, 49(7): 74–7.

Sax, J. L. 1964. "Takings and the Police Power." *The Yale Law Journal*, 74: 36–76.

Singh, R. 2012. Inefficiency and Abuse of Compulsory Land Acquisition: An Enquiry into the Way Forward." *Economic and Political Weekly*, 47(19): 46–53.

Somin, I. 2015. *The Grasping Hand: Kelo v City of New London and the Limits of Eminent Domain*. Chicago, IL: Chicago University Press.

Underkuffler, L. S. 2006. "Frank I. Michelman: Kelo's Moral Failure." *William and Mary Bill of Rights Journal*, 15: 377.

Wenar, L. 1997. "The Concept of Property and the Takings Clause." *Columbia Law Review*, 97: 1923–46.

7

The *Adivasi* Land Question in the Neoliberal Era

Rajesh Bhattacharya, Snehashish Bhattacharya, and Kaveri Gill

7.1 Introduction

The high rate of economic growth in India since the mid-1980s has led to a sharp increase in the demand for land and other natural resources by capital. This has resulted in widespread dispossession of peasants and other communities from their traditional habitats, loss of control over their means of production and subsistence and access to common property resources. Consequently, in recent times there has been a huge upsurge in conflicts between the state and private capital on one side and the dispossessed communities on the other, with various political and civil society groups often joining the latter in their resistance to the former.

This is particularly salient in the context of tribal communities (*adivasis*) due to their multiple and intimate connections with land and natural resources, particularly forest resources in the case of forest-dependent communities.[1] The significance of the "land question" in *adivasi* areas becomes evident from the fact that as per the census of India, 2011, the Scheduled Tribes, with more than 100 million people constituting over 8 percent of the total population, are heavily concentrated in states that are rich in forest and mineral resources.[2] The dispossession of *adivasis* of their customary and traditional rights to land and forest resources looms large across modern Indian history, both colonial and postcolonial. The *adivasis* have been brought to the heart of political economy in India in the current conjuncture by, first, the political resistance of the *adivasis* to their forcible dispossession by the state and private capital (Dandekar and Choudhury 2010), and second, largely in response to these movements, a series of radical legislation actions such as the

Panchayats (Extension to Scheduled Areas) Act, 1996 (PESA) and the Scheduled Tribes and Other Traditional Forest Dwellers (Recognition of Forest Rights) Act, 2006 (FRA).[3] These acts seek to significantly decentralize the governance structure of the areas that are densely populated by the *adivasi* communities (i.e., the Fifth Schedule areas as defined by the Constitution) by enabling self-governance mechanisms, as well as to provide constitutional recognition of a wide range of traditional rights of the *adivasis* over forest lands and resources. Together, PESA and FRA could alter the relation between the *adivasis* and the state in a more fundamental way than any other legislation in independent India.[4] These legislative acts not only seek to grant autonomy to *adivasi* communities defined at the local level, but also make it mandatory to obtain their consent before any acquisition of the resources can take place.

However, despite this legislation, dispossession of the *adivasis* has continued—and even intensified—in recent times. Further, since their promulgation, PESA and FRA have faced continuous opposition to their implementation, including attempts to dilute those provisions in the acts that protect *adivasi* autonomy and rights over land and forest resources. The dual processes of recognition and violation of *adivasi* rights present a specific conundrum. In legislating PESA and FRA, the Indian state comes through as an empowering agency that recognizes the rights of the *adivasis* and enables them to exercise these rights (even if pressured into doing so by popular movements); in forcible acquisition of *adivasi* land, through violation and/or dilution of the same legislative acts, the state comes across as a predatory force robbing people of their rights and dislocating them at will (often coerced by corporate capital into abrogating its responsibilities to people). It appears that the state gives with one hand what it takes away with the other.

One resolution of the conundrum that has been advanced in the existing literature privileges the role of political movements and civil society organizations in pressurizing the reluctant state—captive to powerful economic interests within and outside—to pass progressive legislation. In such circumstances, it is not surprising that the state often reneges on its own commitments to the people by violating or not implementing the legislation in their true spirit, unless otherwise forced to by popular pressure.[5] While being sympathetic to this position, we, however, argue that insofar as the state is seen to operate within the neoliberal discourse of governance, there is a need to problematize the latter itself in order to develop a more nuanced understanding of the nature and implications of state interventions.

While the neoliberal economic discourse insists on well-defined property rights and free markets as instruments of good governance, the use of extra-market force, in the form of forcible acquisition of land, constitutes a violation of the same principles. In this chapter, we argue that the framing of laws such as PESA and FRA need not necessarily represent the forced hand of the state;

rather, it might inaugurate a space of rights-based negotiation/consultation as part of a neoliberal governance structure that seeks to replace discretionary violence by mandatory "construction of consent" (Harvey 2005) *as supplements to the market.* Thus PESA and FRA could also be used to universalize neoliberal principles and deepen their hegemonic reach.

The rest of the chapter is organized as follows. In Section 7.2, we point to the problematic character of land as a commodity, in general, and in the presence of an "outside" of capital, in particular. Specifically, we emphasize the constitutive role of violence in sustaining land as private property and commodity through an erasure of community. Section 7.3 traces the long history of dispossession of *adivasis* from their land in colonial as well as in independent India. In Section 7.4, we discuss the new mode of governance of *adivasi* land inaugurated by PESA and FRA as a potentially sharp break from the past. We discuss how failures in implementation of PESA and FRA, and the continued salience of violence in acquiring *adivasi* land, present a particularly rigid face of neoliberalism bent on securing the free flow of land for capital. In Section 7.5, however, we argue that by adopting an expansive notion of property rights and by instituting mandatory consent-seeking from the *adivasis* for any acquisition of their resources, the new governance structure inscribed in PESA and FRA points to a more protean face of neoliberal order, which can instrumentally deploy *adivasi* "rights" and "autonomy" to facilitate access to resources for expansion of capital. Section 7.6 concludes.

7.2 Land, Property, and Community

The "land question" has traditionally been central to the narrative of transition from precapitalist to capitalist economies. At the heart of the "land question" lies the processes of dispossession of direct producers from independent access to means of production (and subsistence), most importantly, land and natural resources. Marx (1977) used the term "primitive accumulation" to refer to the destruction of noncapitalist social spaces for commoditization of labor power, means of production and means of subsistence, that is, for the creation of the conditions of existence of capitalist production. Central to Marx's discussion of primitive accumulation is the emphasis on violence, most prominently, in instituting private property in land as an instrument of dispossession. The term "primitive accumulation" has gained new currency in contemporary India in the context of land acquisition for industrial, infrastructural, and developmental projects, which often see a violent encounter between corporate capital and the state on one hand, and resistant people on the other.[6]

In this chapter, we depart from the traditional use of the concept of primitive accumulation within the narrative of transition from precapitalism to capitalism. Our chapter is theoretically informed by a range of recent works that argue that primitive accumulation does not constitute the prehistory of capitalism, but is rather one of the *conditions of existence* of capitalist production (see Bhattacharya 2010; Harvey 2003; Sanyal 2007, among others). The central idea that informs these works is that capital must continuously engage with its (noncapitalist) "outside" for access to resources for its expanded reproduction (Chatterjee 2008; Sanyal 2007).

Land is a peculiar "commodity." It is bought and sold in the market just like any other commodity. Yet, as Polanyi (2001) reminds us, land is simply a gift of nature. It is not socially produced, and hence, it cannot be a "commodity" in the true sense, that is, an object socially produced for sale. Polanyi refers to land as a "fictitious" commodity; land becomes a commodity by virtue of the legal system that allows it to be monopolizable and alienable. Under the laws of private property, the globe can be parceled among individuals with exclusive property rights over their parcels.

In the liberal theory of property—for example, in Locke's (1988) labor theory of property—rights to life, liberty, and property are natural rights. While all land belonged originally to humanity in common, what Locke refers to as the "state of nature," right to life necessitates some individual possession of nature's resources for securing food, shelter, clothing, and so on. An individual has a right to apply her labor to land to secure her necessities, and thereby create private property in land. In Locke, private property in land follows from private property in labor power, and therefore, from *individuated* human beings. The reality or possibility of communal reproduction of life based on communal labor on common land is elided. The transition from common land to private property in land via application of private labor is justified by potential private benefits of enclosures, which, in turn, act as incentives for individuals to increase productivity of enclosed land. Thus, land as private property came to be justified by deservingness grounded in a social discourse on productivity.

It is a short step from the productive use of land to the productive character of the people using the land. In colonial discourse, the Indian subjects were categorized in a racially hierarchical order into "superior" castes, that is, semi-civilized population groups engaged in settled agriculture, and "inferior" tribes, that is, uncivilized people who were engaged in hunting-gathering, slash-and-burn agriculture, or pastoralism. This characterization justified right by conquest on productivist grounds, for example, taking away of tribal land, as was stated in the official debates during the period of emergence of colonial Indian Forest Acts (Guha 2000).[7]

Private or public property (as opposed to commons) is predicated on the violent erasure of the community. Property, in turn, creates commodities and markets. In a community that relies on nature for useful objects for their collective survival, nature is a bundle of use-values. On the contrary, in a society of private individuals, that is, private property-owning individuals, reproduction of life requires consumption of products of private labor exchanged in the market as commodities (which have exchange-values, in addition to use-values). In Locke's "state of nature," collective human existence is structured around use-values, where the communities collectively appropriate use-values. Use-values belong to, and acquire their significance, in the space of social interaction and not private consumption. In contrast, in a commodity economy, social relations between individuals are mediated by commodities such that use-values of commodities become a matter of private consumption, and exchange-values become the medium of social interaction (Basu 2012).[8]

In a commodity economy, the meanings of commodities—including land as the fictitious commodity—are derived from the socially constructed notion of the "self" rather than the community. In a community economy, on the other hand, the meaning of land—as the "repository of a seemingly infinite variety of potential use-values 'spontaneously provided by Nature'" (Harvey 1982: 334)—is derived from connection of community to commons, which privileges the "unique," non-homogenizable and non-substitutable character of land. When such a community economy encounters the commodity economy, land fails to circulate between them as a commodity since use-values in a community resist homogenization in terms of (exchange-) values in the commodity space.

In such a context, homogenization of land in the commodity space is premised on the destruction of the community. The replacement of community by a society of private-property-owning individuals entails violence, because two very different (wo)man–nature relations are counterposed to each other. In the context of contemporary acquisition of *adivasi* land in India by the government for industrial or mining projects, Basu (2012: 224) argues, "violence signifies the absence of communication, the presence of difference in language, ethics, values or law." The fiction of land as property arising out of the application of individual labor is sustained by suppressing the history of violence involved in destruction of the community.

The flip side of commodification of land is the commodification of labor power—the labor power of those who do not deserve property, and are hence dispossessed of land through primitive accumulation. With both labor power and land emerging as commodities, we have not just an economy characterized by private property and commodity production, but specifically a capitalist economy, where the circulation of land as commodity is driven by the logic of capital. The owner of land has a claim to rent, that is, a cut of the surplus value produced on land. Any stream of revenue (such as an annual rent) can be considered interest on some imaginary, fictitious capital. Land becomes a pure financial asset, which is bought and sold according to the rent it yields. Capital circulates through land markets promoting activities on the land that conform to the production of the highest (surplus) value. Thus land, which is an infinite bundle of use-values in the "state of nature," takes its most "civilized" form in capitalism, that is, a pure financial asset indistinguishable from other speculative forms of capital.

It must be noted here that even though private property is the quintessential form of property in capitalism, it often masquerades as "public" property, for example, when nationalized resources like forests, minerals, and so on, are leased to private capitalists for profit-making activities. This is particularly true in the *adivasi* context, where resources from which *adivasis* are dispossessed (forest lands, farmland, etc.) are not converted into private property per se, but are nationalized before being made available for use by capitalists.

In contemporary India, the encounter between the state and corporate capital on one hand, and the *adivasis* on the other, exemplifies the clash between the commodity and the noncommodity character of land. When the tribal people of Niyamgiri hills in Odisha rejected the multinational company Vedanta's proposal to mine bauxite in those hills, they asserted the "uniqueness" of their land and rejected the value-calculus that sought to homogenize lands in the commodity space.[9] Incidents like this, and they are far from rare in contemporary India, signal the limits of commodification of land and the circulation of capital in land markets. Given that neoliberal policies focus on well-defined property rights and well-functioning markets, our intervention in this chapter thus constitutes a problematization of the contemporary neoliberal governance of land in India.

7.3 The Dispossession of *Adivasis* in India

The long history of systematic and continuous dispossession of tribal populations from their natural habitats began in the colonial period. The Indian Forest Act of 1878, by one stroke of the pen, erased customary rights of the rural population to forest products and declared the forests, covering one-fourth of the land area, to be the property of the state. The act made a clear distinction between "rights" and "privileges" over forest resources, stating that "customary usage" of forest resources by villagers was not their right, but "merely [a] privilege" accorded to them by the state (Sivaramakrishnan 1995: 14).

The independent Indian state carried this endeavor further by replacing "privileges" with "concessions." National interests trumped all other claims of forest-dependent communities (Kulkarni 1987). The National Forest Policy of 1952 stated that while communities residing in the neighborhood of forests may naturally make use of forest products, "[s]uch use, however, should in no event be permitted at the cost of national interests. The accident of village being situated close to a forest does not prejudice the right of the country as a whole to receive the benefits of a national asset" (Government of India 1952). Since national interests were identified with industrialization, the dispossession of tribal population groups due to mining, construction of dams, roads, factories, and other development projects accelerated with the big push for industrialization after independence.

At present, 80 percent of the tribal population of India is concentrated in an almost contiguous belt that spans the states of Andhra Pradesh, Odisha, Jharkhand, Chhattisgarh, Madhya Pradesh, Maharashtra, and Gujarat (Fernandes 2008). A significant part of this belt is forestland on which the *adivasi* population is dependent for resources, while other lands are common property resources. Such areas with a preponderance of *adivasi* populations are identified by the government as the Fifth Schedule Areas and are accorded protection under the constitution (Government of India 2002), while tribal lands in the Northeast are accorded protection under the Sixth Schedule.

Despite the constitutional protection, the tribal population in these regions has been regularly displaced for development projects through "legalized" land alienation by using colonial-era acts such as the Land Acquisition Act of 1894 (Fernandes 2008). While, in the absence of proper national-level data, it is extremely difficult to determine the exact figure for development-induced displacement, reliable yet conservative estimates put the number between 21 million and 60 million (Fernandes 2007; Mahapatra 1999: 5). Estimates recognized by the Planning Commission say that 55 percent of all displaced persons up to the year 2001 were tribal people (Government of India 2002), while among the total number of people displaced due to development projects, 47 percent are from the tribal population (Government of India 2014). As Roy (1999: 19) puts it, "[t]he ethnic 'otherness' of their victims [the tribal people] takes some of the pressure off the nation-builders. It's like having an expense account. Someone else pays the bills."

Interestingly, it is not only the "productivist" paradigm (valorizing the "commodity" form of land) that has provided the artillery against the *adivasis*. There is an entire discourse that has taken shape over a long time focusing on the uniqueness of land, and hence, as argued earlier, its noncommodity character. This is the discourse on wilderness, wildlife, biodiversity, climate change and so on, which can be broadly categorized as the "conservationist" argument. In this discourse, land is unique, nonreproducible, and nonsubstitutable and thus must be protected from market transactions in property through legislation cordoning off land from the reach of the market.

The Indian Forest Conservation Act, 1980, by subordinating the tribal communities to conservation objectives and by viewing them as the "encroachers," dealt a blow to tribal rights over forest resources (Kumar and Kerr 2012). The forest-dependent communities came to be seen as exploiting nature for narrow material interests of subsistence, and thus being incapable of greater ecological concerns. In 2002, the Ministry of Environment and Forests, following a Supreme Court ruling, ordered the eviction of all forest encroachers within a period of six months. Almost 150,000 hectares of land were "cleared," leading to eviction of an estimated 300,000 tribal and nontribal forest-dwelling families (Kumar and Kerr 2012). The public outcry following this, coupled with intense pressure from social movements (Vaidya 2016), led to the enactment of the Forest Rights Act, 2006 (FRA). Yet, according to a Ministry of Rural Development report, forest degradation rather than encroachment remains the main problem (Government of India 2009: 111–12).

For accelerated economic growth, land needs to be drawn into the commodity space so that it can freely flow from low-value to high-value uses. But *adivasis* often look at their natural habitat as "unique" because of their specific material and cultural connection to land premised on community-conditioned use-values procured from nature. This creates a barrier to commoditization of *adivasi* land. Given these limits to circulation of land as commodity in *adivasi* areas, forcible acquisition by the state is often called forth to supplement the market. However, in the neoliberal era, the state itself is subjected to powerful critiques for its interventions in the market. Thus, the neoliberal order needs a new governance structure to deal with limits of the land market in *adivasi* areas.

7.4 State Recognition of Tribal Rights and the Political Economy of Capitalism

In postindependence India, the governance of the tribal areas (or the Fifth Schedule Areas) deliberately privileged the Union executive (the governor of the state) over individual state legislatures, ostensibly to protect the *adivasis* from onslaughts on their economic, cultural, and political life by the majority communities (Kurup 2008). However, in the actual governance of such areas, the objective of nation building often trumped the goal of tribal welfare or autonomy, as we have noted in the previous section. Given the general trend toward decentralization in the last three decades (premised on the neoliberal critique of the centralized state) and the emphasis on market-led as opposed to state-led development, governance of tribal areas had to move away from the "trusteeship" model around the figure of the governor in Fifth Schedule Areas to a "self-government" model based on decentralized institutions.

This was accomplished by the seventy-third amendment to the constitution in 1992, which created a decentralized governance structure across rural India, including the Scheduled Areas (Pal 2000). In the tribal areas, it replaced the traditional authority structures of the *adivasis* by uniform, administratively defined, and decentralized institutions called the *Panchayati Raj* institutions (PRIs). Unlike the rest of India, where these structures filled a vacuum of democratic local governance, in the Scheduled Areas they actually set up dual authorities of the *adivasis* and the state. After the *adivasis* mobilized to protest this backdoor attempt to dilute their autonomy—invested in their own traditional bodies of authority and governance—and the courts deemed the extension of PRIs to Scheduled Areas unconstitutional (namely, the ruling of the Andhra Pradesh High Court in 1995), the government enacted the PESA in 1996 (Pal 2000).

At its core, PESA recognizes "customary law, social and religious practices and traditional management practices of community resources" (Government of India 1996: 1) prevalent in Fifth Schedule Areas, and aims to devolve power to local *adivasi* populations to allow them to genuinely self-govern in certain critical spheres. In the act, provisions are made for a definition of a *Gram Sabha* (village assembly), the basic village-based unit of governance in rural India that is appropriate to *adivasi* areas, and is not just an extension of the PRIs as it operates in the rest of the country.

> [T]he *Gram Sabha*-centric PESA system cannot function unless the *Gram Sabha* is located in a habitation, the natural living unit of the community... It is only thus that the Gram Sabha, comprising all adult members of the habitation, can play the pre-eminent role in all consultative stages and in final decision-making, as provided for in PESA. (Government of India 2013: 289)

The Forest Rights Act of 2006 (FRA) further complemented the fundamental focus of the PESA on autonomy of *adivasis*. The FRA, in letter and spirit, is perhaps even more pathbreaking in many ways. In its preamble, it acknowledged that

> the forest rights on ancestral lands and their [Scheduled Tribes and other traditional forest dwellers'] habitat were not adequately recognized in the consolidation of State forests during the colonial period as well as in independent India resulting in historical injustice to the forest dwelling Scheduled Tribes and other traditional forest dwellers who are integral to the very survival and sustainability of the forest system. (Government of India 2006: 1)

Moreover, in justifying why this law was being enacted, it argued "it has become necessary to address the long-standing insecurity of tenure and access rights of forest dwelling Scheduled Tribes and other traditional forest dwellers including those who were forced to relocate their dwelling due to State development interventions" (Government of India 2006: 1).

Under the FRA, *adivasi* and other communities were given no less than thirteen rights, including right of ownership over minor forest produce, community rights of entitlements to resources, rights for conversion of *pattas* or leases on forest lands to titles, *adivasi* rights which are accepted as rights of tribals under any traditional or customary law of the concerned tribes, and right to in situ rehabilitation in cases where traditional forest dwellers have been illegally displaced from forestland without rehabilitation prior to 2006 (Government of India 2006: 3–4). In terms of institutional procedures for determining forest rights, FRA vests the key authority and power with the *Gram Sabhas* (Government of India 2006: 7).

Thus both PESA and FRA ostensibly recognize the community mode of existence and the centrality of common land in *adivasi* communities. The emphasis on custom and tradition in these acts is also recognition of *adivasi* beings and doings that may be different from those in societies where social reproduction of life is market-based to a significant extent. This accentuates the problematic character of land, discussed in the previous section. While the neoliberal order seeks to "'rigidly oppose' any centralization of market-inhibiting policy competencies" (Harmes 2006: 741) by weakening the central state, decentralization itself may support market-inhibiting forces in certain areas, for example, in *adivasi* areas due to the specificities of these communities discussed before. Further, in *adivasi* areas, the problem of compensation for the dispossessed people based on any monetary evaluation of costs and benefits faces a "language" barrier, so to speak:

> Market valuation of rights, which are trampled, to bring water and forest resources within the ambit of market transactions is impossible. Community life, nature, and culture—which the *adivasis* lose—are not purchased in the market. Nor does there exist any alternative against which their opportunity cost can be calculated. All these are *unique* to the forest dwellers, so how can one calculate their market value? (Basu 2012: 224; emphasis added)

Given these vexing problems, it is no wonder that implementation of PESA and FRA, in spirit, has been a highly contested process. In the neoliberal model of governance, the emphasis is on "market-preserving federalism" to constrain the central government in its "market-inhibiting" interventions by "locking-in inter-jurisdictional competition" (Harmes 2006: 727). In the context of India, federalism has been instrumentally used by investors to involve the states in a race to the bottom. It is only to be expected that state governments would be reluctant to implement central legislative acts like PESA and FRA for fear of losing out on investment because of the land problem.

Hence, it is no wonder that numerous issues plague the operationalization and implementation of PESA and FRA across all nine states where *adivasis* are concentrated.[10] It has been recommended that the *Gram Sabhas* should be reconfigured on the basis of *adivasi* habitations, so that "the reconstituted village reflects tribal homogeneity and conforms to the territorial-cum-habitation boundaries of traditional tribal *Sabhas*" (Government of India 2013: 291). However, state governments often disregarded or bypassed this definition of *Gram Sabha* in the PESA regulations in defining local administrative units.

Similarly, the FRA is found to be poorly implemented across the states, as the forest bureaucracy has not ceded control in most cases (Government of India 2010). Moreover, individual rights to forest resources continue to be given priority over community rights by the implementing government authorities. The distribution of power between the *adivasi* communities and frontline officials from the state government is so asymmetrical that it works against the letter and spirit of the acts, and there are reports of continued violence by Forest Department officials against *adivasis* and forest dwellers (Kumar and Kerr 2012).[11]

As should be evident from this discussion, PESA and FRA, at least in letter and spirit, constitute a radical transformation of the state's approach to governing the *adivasis*—from a mode of centralized trusteeship to devolution of power to the communities and recognition of their community rights. The resistance to effective implementation of these legislative acts is evidence of a rigid face of neoliberalism, insensitive to differences and bent on battering down all barriers to the free flow of land and natural resources to capital. However, in Section 7.5, we argue that neoliberalism is also capable of a more flexible stance: more protean and more capable of accommodating differences in terms of social, economic, and legal structures.

7.5 Of Rights and Consultation: A New Mode of Neoliberal Governance?

The response of the state in the form of PESA and FRA can be understood as a reaction to the spread, depth, and scale of *adivasi* resistance to the corporate takeover of such resources aided by the government over the last couple of decades. This, to our mind, is indeed a valid—albeit partial—explanation of the seemingly paradoxical enactment of PESA and FRA in the neoliberal era. A regime of capitalist accumulation requires a stable social order, more so when such accumulation is driven primarily by private capital as in the neoliberal era. When an intensification of people's movements threatens to disrupt such stability, the state may be forced to yield some concessions to maintain the regime and to minimize costs involved in applying force to contain the unrest.

However, a deeper probing of the neoliberal mode of governance reveals another, more fundamental, reason for the recent expansion of a rights-based approach to the *adivasi* land question, which is often missed by the existing critiques of neoliberalism. The neoliberal goal of creating a self-regulating market economy is predicated on a forced separation between economic and political spheres (Polanyi 2001), where the processes of social reproduction are subjugated to the logic of the market. Such a separation, which Polanyi referred to as the *dis-embedding* of the economy from the society is, in turn, based on the existence of a generalized commodity relation in all aspects of production, including land. However, as we have argued before, commodification of land is often contested on the ground by communities, which relate to it in terms of community-conditioned use-values. This signals the limits of the market and the mobility of capital in land markets. Therefore, market forces have to be supplemented by force—applied by the state and backed by the legal system—in breaking down barriers to land markets. In the neoliberal model of a self-regulating market economy, violence is a nonmarket and, hence, political intervention. Violence shatters the myth of the strict separation between the economic and the political spheres that grounds the neoliberal order. Violence thus appears to be both essential to and disruptive of the neoliberal order. The problem with violence is that it resuscitates the prehistory of property and confirms the presence of community.

Dispossession by force, when it becomes a political issue, has two unsavory consequences. First, resistance to such dispossession often puts into question the sanctity of the existing property rights by invoking the history of conquest behind the existing property rights and by foregrounding the ethics of alternative relations to land. Second, resistance to forcible dispossession often leads to insurgency and rebellion, which presents a law and order problem that negatively affects the security of property. The sanctity and security of property is best preserved when the separation between the political and the economic is sustained, that is, property is not politicized, which is of course what the utopian neoliberal project is all about.

In our understanding, the rule of property and markets, which together constitute the core of the very sparse and elegant neoliberal policy prescription, requires a mode of governance that directly addresses the question of dispossession, that is, negotiates the limits of the market in land beyond the realm of violence. In fact, it should be obvious that a generalized rule of property rights requires rules to address "dispossession" of property rights. If the rule of law has to have a nonarbitrary and universal premise (including the specified domain of exceptions), then that rule of law must be sophisticated enough to recognize diversity of property rights on the ground and the diverse modes of exchange of property rights in order to fill up a legal void where dispossession thrives. We do not seek to de-emphasize the persistence of violence in supplementing market forces when it comes to the acquisition of *adivasi* land in the neoliberal era. However, we understand PESA and FRA, despite the many problems of their implementation we have noted before, as constituting a new modality of negotiation with the *adivasis* over land. This new modality is premised on the valorization of community, captured through the recognition of nonstandard rights, that is, traditional and community rights, as well as nonmarket communication, that is, consultation and consent.

The intimate connection of *adivasis* to land—based as it is to a significant extent on "commons" and non-market exchanges—resists integration of the *adivasis* by the state on the plane of individual property rights and social interactions primarily through market exchange. Both property rights and markets need to be supplemented if the *adivasis* need to be integrated under a uniform order of governance. It may be instructive to take an example from outside the Indian context. At the height of the "land grab" in African countries following the global commodity price spike in 2008, and facing/fearing popular resistance, a World Bank report made the following point:

> Failure to map and record land rights, even if only at the community level, makes it difficult to identify boundaries and legitimate owners as a basis for engaging in mutually agreed to land transfers. Recording rights provides outside investors with "somebody to talk to," *a legitimate and authorized partner to negotiate on the nature of investments and on compensation.*
>
> (Deininger and Byerlee 2011: 98, emphasis added)

It is interesting to note the emphasis on "talking," on face-to-face interaction. In ideal markets, people do not talk to each other, commodities do. If individuals do not want to sell their land, the market ends right there. The "rule" of markets also recognizes the boundary of the market, that is, where individuals refuse to participate in the market. The space of negotiation/consultation/conversation appears as an anomaly in an ideal market. In our understanding, however, negotiation/consultation/conversation is a supplement that must address not the absence of the market per se, but the presence of a limit to the market ordained by its own liberal logic, a limit that must, however, be crossed somehow, to gain access to resources that cannot be secured within the domain of capital. Consultation, or "construction of consent" (Harvey 2005), emerges as a necessary mode of engagement of capital with its "outside," particularly when violence (as the alternative mode of engagement) works against the hegemony of the neoliberal order.

The importance of consent in the context of the *adivasi* land question is easily seen if we look at the implications of FRA for transfer of tribal land. We have already seen that FRA recognizes a variety of rights vested in the individual or the community, making forceful taking of *adivasi* land more difficult. For example, no transfer of tribal land is allowed unless FRA has been implemented in the concerned area, that is, the rights of forest dwellers have to be settled before any land transfer can be made. Moreover, under FRA, the

Gram Sabha has to give consent to any transfer of land. At the same time, through delineation and registration of individual or community rights, FRA defines the parties who must give consent and the extent of resources over which their consent is required.

With the institutionalization of such a legal–administrative order, dispossession can now be addressed within a new paradigm of resettlement/rehabilitation. However, resettlement/rehabilitation, being based on a cost–benefit analysis, is ultimately based on the calculation of exchange-values. In place of impersonal market prices, we now have negotiated rehabilitation/resettlement packages. The purpose of consultation is to communicate the costs and benefits of giving up land for industrial or mining projects. The loss of use-values to the *adivasis* giving up land is reduced to a monetary measure of cost (evaluated at market prices), while the gain in use-values or values (if those use-values are to be secured through the market as commodities) is aggregated in a monetary measure of benefit. Thus, "community" and "autonomy" can be instrumentally used, in and through the consultative process governed by PESA and FRA, to extend the logic of market valuation into the space of community use-values.

The possibility of expansion of the neoliberal order through PESA and FRA has been noted before (Naidu 2011; Savyasaachi 2011). On the unique connection between human beings and nature that is not premised on ownership, as seen in significant degree among the *adivasis*, Savyasaachi (2011: 59) argues that

> it is not possible to belong to a place and own it as well, because belonging comes with mutuality, which is premised on the acceptance, that there is no legitimate ground for ownership of the "right to disposal" over that which has not been created with human work.

In general, individual titles have been preferred over community rights in cases of implementation of FRA in India, as we have noted before. This has been an instrument for drawing *adivasi* communities into the ambit of private property rights commensurate with the culture of commodities and the logic of capital. A World Bank (2006) report argues that secure and efficient tenure over natural resources, such as forests, provides communities with the incentives to invest as ownership offers opportunities for capitalizing of forest resources (Ramdas 2009). There is evidence that some state governments may actively promote export/urban market-oriented agriculture on individual titles granted under FRA, tying the FRA in to employment programs such as the National Rural Employment Guarantee Scheme (NREGS):

> Now tribals can cultivate their lands with dignity without any fear. Tribals can plant rubber plants, mango, cashewnut, orange, lime or palm oil as per the local

> conditions. The State Government would also develop lands in tribal areas and the tribals will be paid daily wages under NREGS programme though they are working in their own land. (News report quoted in Ramdas 2009: 69)

On the other hand, the delineation and recording of rights can be used for "invisible" dispossession. At one extreme, the provisions of PESA and FRA can be manipulated to cheat *adivasis* of their resources. There is ample evidence that communities can be manipulated into supporting a corporate takeover of land; sometimes fake documents proving consent of the *Gram Sabhas* are produced to get clearances from the government (Yadav 2014).[12] It has been argued that, for corporate projects in areas of indigenous habitation across the world, consultative processes are often manipulated.

> [A] frequent indigenous criticism is that World Bank-sponsored consultation meetings are used primarily by project officials and consultants to give legitimacy to the project, gain access to communities, divide opinion and pressure local indigenous communities to accept the external agendas of governments, development agencies, big business and NGOs. (Griffiths 2005: 15)

But, in a subtler sense, the recording of rights has often meant a contraction, rather than an expansion of access to resources for the *adivasis*, as the following account by a Savara woman from Srikakulam district of Andhra Pradesh suggests:

> I don't know whether I was (more) free before the act or after the act. Earlier I was a "thief" in the eyes of the law, but learnt to survive. Now I am "legal" and have legally lost my land as the government took all and gave me nothing. We have "legally" been granted "two acres" of community land, whereas all this is ours (pointing to the hills beyond). We reject these titles. (Quoted in Ramdas 2009: 72)

It has been widely noted that the settlement of rights under FRA has predominantly been restricted to individual rights. The settlement of community rights has not only faced resistance from the forest bureaucracy but has received a lukewarm response from the communities themselves. Often, this is because "[p]eople did not feel the need to claim community rights because they had never faced any problem in accessing community resources in forest areas" (Samarthan 2012: 32). Recording and delineation of rights (de jure) is most commonly seen as an expansion of rights, but from the point of *adivasis* with traditional access to nature (de facto), this often means just the opposite: exclusion from a world to which (s)he always belonged.

The current focus on dispossession and rights is as much a prerogative of the neoliberal capitalist order as it is the outcome of consciousness born out of historic and ongoing struggles against dispossession. This is not to deny the importance of struggles against dispossession, but rather to highlight the specific response in the neoliberal era to such struggles. A generalized regime for property rights can harbor different modes of social exchanges, market as well as nonmarket. The neoliberal universalism is not based on homogeneity, but ordered differences. While not denying the importance of FRA, PESA, and so on, to the tribal population and protection of their concerns, one must also be attentive to the possibility of differentiated integration of resistant groups, of assimilation of the "outside" as "difference."

7.6 Conclusion

In this chapter we have sought to deflect the existing debate on policies related to *adivasis* away from violence, even when recognizing that violence is everywhere, and unimaginably destructive in its consequences for *adivasis,* toward consultation/consent as the new mode in which the state seeks to address the *adivasis*. We trace this transformation to the emergence of a protean regime of property and market, willing to accept its limits even as it seeks to extend inexorably beyond it.

We argue that legislation such as PESA and FRA represent a radical departure in the mode of governance of tribal population in India in terms of the institutionalization of autonomy of *adivasi* communities and recognition of traditional, customary rights, particularly in the context of forest lands and other natural resources. The land question in the *adivasi* areas is more complex than in other places as the *adivasis* have historically remained outside the private property relations and their community ethics often resist inclusion within the generalized commodity space. Thus a new mode of governance recognizing nonstandard rights (community, traditional, customary) and institutionalizing consultation in lieu of or as a supplement to market exchange was called for. The significance of the emergence of PESA and FRA should be understood in this context.

This is not to argue that under the neoliberal regime, the *adivasi* land question can be satisfactorily resolved within the PESA–FRA governance structure. The widespread failure of implementation of the laws and the continued—even increased—dispossession of the tribal population over the last two decades paint a completely different picture. To our mind, such failures are symptomatic of the friction between the rigid face of the neoliberal order, which seeks unfettered flow of land and resources for impatient capital, and the emergent governance structure, which supposedly seeks to "empower" the *adivasi* communities. We argue that the neoliberal order can be flexible enough to make the new governance structure an instrument in advancing its hegemonic reach. The quest for a completely unregulated market and supreme freedom of capital—a quest that attempts to undermine the rights-based legal framework—on one hand, and the thrust for an idealized institutional governance mechanism—based on legal recognition of heterogeneity and the limits of property and markets—on the other, mark the contradictory utopian pursuits of capitalist development in India in the current conjuncture.

Acknowledgments

We would like to thank Amit Basole, Pranab Basu, Suraj Jacob, Priya Sangameswaran, both the editors of this volume, and an anonymous reviewer for helpful comments on earlier drafts of this chapter. We would also like to thank the participants in the conference on "The Return of the Land Question: Dispossession, Livelihoods, and Contestation in India's Capitalist Transition" on March 4–6, 2014, sponsored by University of Melbourne, IDSK Kolkata, and IIM Calcutta.

Notes

1. The word *adivasis* means original inhabitants or indigenous people. In India, they are referred to as Scheduled Tribes as defined in the Indian Constitution. In this chapter, we have used the words *adivasis*, tribal people, and Scheduled Tribes interchangeably.
2. More than 90 million *adivasis* live in forested and mineral resource-rich areas. Almost 60 percent of forest cover of the country is found in tribal areas. The states with the largest forest cover in terms of absolute area are also states with substantial tribal populations: Andhra Pradesh, Chhattisgarh, Odisha, Maharashtra, and Madhya Pradesh. With regard to mineral resources, three states with substantial tribal populations—Odisha (22.1 percent), Chhattisgarh (31.8 percent) and Jharkhand (26.3 percent)—account for 70 percent of India's coal reserves, 80 percent of its high-grade iron ore, 60 percent of its bauxite, and almost 100 percent of its chromite reserves.
3. The resistance to the dispossession is often linked to armed insurgencies against the Indian state in many of the *adivasi*-dominated areas, led by the Communist Party of India (Maoist) and some other similar organizations. More than eighty districts in the heartland of India, most of them with a concentrated *adivasi* population, have been declared Left Wing Extremism (LWE)-affected areas by the government of India. See the website of the Ministry of Home Affairs, Government of India (available at: <http://mha.nic.in/naxal_new> (accessed March 27, 2016)) for updated information on policies with respect to LWE-affected areas. This has prompted the government to undertake special "developmental" policies in these areas, including prioritization of the implementation of PESA and FRA. Panchayats represent a decentralized form of governance by a committee at the village level.

4. For a fascinating account of the state's maneuvering in drafting the FRA in its most expansive ideal, see Vaidya (2016).
5. For example, the architects of the FRA foresaw that a mere promulgation of legal rights has no bearing on their actual implementation, and the latter would only happen "through struggle, it will not be through the state" (Vaidya 2016).
6. Acquisition of land by the state is not a new phenomenon in India. Under British rule, the colonial state acquired land for "public" purposes using the clause of "eminent domain" under the Land Acquisition Act of 1894. Post-independence, this framework remained the basis for land acquisition by the Indian state, till the enactment of a new act in 2013, which continues to be a contested legislation. We briefly trace the evolution of land acquisition in the *adivasi* areas by the state, during the colonial era as well as in independent India (both before and after the onset of economic reforms in 1991), in Section 7.3 of this chapter.
7. Through the equation between states of nature, common lands, nonsettled cultivation and "wilde wastelands," the doctrine of enclosure trumped the rights of pre-existing inhabitants (Whitehead 2012: 2).
8. Sweezy (1942) argued that Marx dropped use-value as the object of analysis of political economy because he was looking at social relations, whereas use-values belong to private relations between individuals and goods. Our understanding of use-value is informed rather by Basu (2012), who has argued that in the absence of private property rights and individuation in society, that is, in the context of communities and commons, use-values do not refer to private consumption and subjective relations between individuals and goods, but rather to the community-given meaning to relations between human beings and want-satisfying characteristics of nature held in common. Thus our use of use-value, following Basu, is a departure from its dominant use in the traditional Marxian literature.
9. See Sethi (2016b) on the legal battles surrounding this issue, and the Odisha state's arguments asking the Supreme Court to squash or to revisit the twelve *Gram Sabhas'* unanimous resolutions against mining in the area.
10. It lies beyond the purview of this chapter to discuss in detail the vast empirical literature documenting poor implementation of both PESA and FRA across the states. *Inter alia,* see chapter 8 of Government of India (2013) for a succinct discussion of implementation of PESA and the minutes of the review meeting conducted by the Ministry of Tribal Affairs with the Secretaries of various states on the status of implementation of FRA (Government of India 2015).
11. For example, the Union Ministry of Tribal Affairs recently succumbed to pressure to re-interpret the FRA and allow the Maharashtra forest department to regain control over the lucrative trade in forest produce such as *tendu* leaves and bamboo (see Sethi 2016a).
12. See Choudhury (2016) for an example of a more blatant illustration of such instances, in the context of iron mining in the tribal forest lands of Keonjhar district in Odisha.

References

Basu, P. K. 2012. "Rethinking the values of the Left." *Rethinking Marxism*, 24(2): 221–39.

Bhattacharya, R. 2010. "Capitalism in Post-Colonial India: Primitive Accumulation Under Dirigiste and Laissez Faire Regimes." Doctoral Dissertation, University of Massachusetts, Amherst. Available at: <http://scholarworks.umass.edu/cgi/viewcontent.cgi?article=1204&context=open_access_dissertations> (accessed March 31, 2016).

Chatterjee, P. 2008. "Democracy and Economic Transformation in India." *Economic and Political Weekly*, 43(16): 53–62.

Choudhury, C. 2016. "The Keonjhar Takeover." *Outlook*, February 8. Available at: <http://www.outlookindia.com/magazine/story/the-keonjhar-take-over/296512> (accessed March 26, 2016).

Dandekar, A. and Choudhury, C. 2010. "PESA, Left-Wing Extremism and Governance: Concerns and Challenges in India's Tribal Districts." Report of a Study Commissioned by Ministry of Panchayati Raj, Government of India, Institute of Rural Management, Anand.

Deininger, K. W. and Byerlee, D. 2011. *Rising Global Interest in Farmland: Can it Yield Sustainable and Equitable Benefits?* Washington, D.C.: World Bank Publications.

Fernandes, W. 2007. "Mines, Mining and Displacement in India." In *Proceedings of 1st International Conference on Managing the Social and Environmental Consequences of Coal Mining in India*, edited by G. Singh, D. Laurence, and K. Lahiri-Dutt, New Delhi and Dhanbad: Indian School of Mines University, pp. 333–44.

Fernandes, W. 2008. "Indian Tribes after Sixty Years—A Study."*Sanhati*. Available at: <http://sanhati.com/articles/1094/> (accessed March 31, 2016).

Government of India. 1952. "National Forest Policy." Government of India, New Delhi.

Government of India. 1996. "The Provisions of the Panchayats (Extension to the Scheduled Areas) Act." Government of India, New Delhi.

Government of India. 2002. "Tenth Five Year Plan (2002–2007)." Planning Commission, Government of India, New Delhi.

Government of India. 2006. "The Scheduled Tribes and Other Traditional Forest Dwellers (Recognition of Forest Rights) Act." Government of India, New Delhi.

Government of India. 2009. "Draft report of the Committee on State Agrarian Relations and Unfinished Task of Land Reforms." Ministry of Rural Development, Government of India, New Delhi.

Government of India. 2010. "Report of the National Committee on Forest Rights." Ministry of Environment and Forest, Government of India, New Delhi.

Government of India. 2013. "Towards Holistic Panchayat Raj—Twentieth Anniversary Report of the Expert Committee on Leveraging Panchayats for Efficient Delivery of Public Goods and Services." Ministry of Panchayati Raj, Government of India, New Delhi.

Government of India. 2014. "Report of the High Level Committee on Socio-Economic, Health and Educational Status of Tribal Communities in India." Ministry of Tribal Affairs, Government of India, New Delhi.

Government of India. 2015. *Minutes of the Review Meeting on the Implementation of Forest Rights Act, 2006*. Ministry of Tribal Affairs. Available at: <http://www.tribal.nic.in/

WriteReadData/CMS/Documents/201507301119105175071merged_document.pdf>, accessed September 21, 2016.

Griffiths, T. 2005. "Indigenous Peoples and the Global Environment Facility: Indigenous Peoples Experiences of GEF-Funded Biodiversity Conservation – A Critical Study." Forest Peoples Programme, Moreton-in-Marsh. Available at: <http://www.forestpeoples.org/documents/ifi_igo/gef/gef_study_jan05_eng.pdf>, accessed March 31, 2016.

Guha, R. 2000. *The Unquiet Woods: Ecological Change and Peasant Resistance in the Himalaya*. Berkeley, CA: University of California Press.

Harmes, A. 2006. "Neoliberalism and Multilevel Governance." *Review of International Political Economy*, 13(5): 725–49.

Harvey, D. 1982. *The Limits to Capital*. Oxford: Basil Blackwell.

Harvey, D. 2003. *The New Imperialism*. Oxford and New York: Oxford University Press.

Harvey, D. 2005. *A Brief History of Neoliberalism*. Oxford: Oxford University Press.

Kulkarni, S. 1987. "Forest Legislation and Tribals: Comments on Forest Policy Resolution." *Economic and Political Weekly*, 22(12): 2143–8.

Kumar, K. and Kerr, K. 2012. "Democratic Assertions: The Making of India's Recognition of Forest Rights Act." *Development and Change*, 43(3): 751–71.

Kurup, A. 2008. "Tribal Law in India: How Decentralized Administration Is Extinguishing Tribal Rights and Why Autonomous Tribal Governments Are Better." *Indigenous Law Journal*, 7(1): 87–126.

Locke, J. 1988 (1689). *Two Treatises of Government*. Cambridge: Cambridge University Press.

Mahapatra, L. K. 1999. *Resettlement, Impoverishment, and Reconstruction in India: Development for the Deprived*. New Delhi: Vikas Publishing House.

Marx, K. 1977 (1867). *Capital: A Critique of Political Economy*, vol. 1. New York: Vintage Books.

Naidu, S. 2011. "The Potential of Forest Rights Act." *Sanhati*, January 29. Available at: <http://sanhati.com/excerpted/3197/> (accessed March 31, 2016).

Pal, M. 2000. "Panchayats in Fifth Scheduled Areas." *Economic and Political Weekly*, 35(19): 1602–6.

Polanyi, K. [1944] 2001. *The Great Transformation: The Political and Economic Origins of Our Time*. Boston, MA: Beacon Press.

Ramdas, S. R. 2009. "Women, Forest Spaces and the Law: Transgressing the Boundaries." *Economic and Political Weekly*, 44(44): 65–73.

Roy, A. 1999. *The Cost of Living*. London: Flamingo.

Samarthan. 2012. "Recognition of Community Rights under Forest Rights Act in Madhya Pradesh and Chhattisgarh: Challenges and Way Forward." UNDP 2012. Available at: <http://www.undp.org/content/dam/india/docs/DG/recognition-of-community-rights-under-forest-rights-act-in-madhya-pradesh-and-chhattisgarh-challenges-and-way-forward.pdf> (accessed March 31, 2016).

Sanyal, K. 2007. *Rethinking Capitalist Development: Primitive Accumulation, Governmentality and Post-Colonial Capitalism*. New Delhi and Abingdon: Routledge.

Savyasaachi. 2011. "Forest Rights Act 2006: Undermining the Foundational Position of the Forest." *Economic and Political Weekly*, 46(15): 55–62.

Sethi, N. 2016a. "Tribal Ministry Relents Over Forest Rights Act." *Business Standard,* January 2016. Available at: <http://www.business-standard.com/article/economy-policy/tribal-ministry-relents-over-forest-rights-act-116010700071_1.html> (accessed March 26, 2016).

Sethi, N. 2016b. "Odisha Says Tribals Had No Power to Reject Vendanta Bauxite Mining." *Business Standard,* March 2016. Available at: <http://www.business-standard.com/article/current-affairs/odisha-says-tribals-had-no-power-to-reject-vedanta-bauxite-mining-116030101027_1.html> (accessed March 26, 2016).

Sivaramakrishnan, K. 1995. "Colonialism and Forestry in India: Imagining the Past in Present Politics." *Comparative Studies in Society and History,* 37(1): 3–40.

Sweezy, P. 1942. *The Theory of Capitalist Development.* New York: Monthly Review Press.

Vaidya, A. 2016. " 'Word Traps' and the Drafting of India's Forest Rights Act." In *Staking Claims: The Politics of Social Movements in Contemporary Rural India,* edited by U. Chandra and D. Taghioff, New Delhi: Oxford University Press (forthcoming).

Whitehead, J. 2012. "John Locke, Accumulation by Dispossession, and the Governance of Colonial India." *Journal of Contemporary Asia,* 42(1): 1–21.

World Bank. 2006. "Unlocking Opportunities for Forest-Dependent People in India." Main Report, vol. 1. Report Number 34481-IN. Agriculture and Rural Development Sector Unit, South Asian Region.

Yadav, A. 2014. "As Coal Blocks Face De-Allocation, a Rush for Environmental Clearance." *The Hindu,* March 2014. Available at: <http://www.thehindu.com/business/Industry/as-coalblocks-face-deallocation-a-rush-for-environment-clearance/article5756721.ece> (accessed March 31, 2016).

Part III

Regional Perspectives

8

Noncultivating Households Owning Land in an Agrarian Economy

Some Observations from Andhra Pradesh

R. Vijay

8.1 Introduction

India's rural economic structure has been undergoing changes, largely due to various public policies such as land reform measures, a green revolution package, and recently, land acquisition for nonagricultural use. Consequently, there are changes in the decision-making process of the individual economic agents. One aspect of the structural change is the growing importance of households owning land but not cultivating the land themselves (Vijay 2012a). Vijay (2012a) shows that the share of these households is increasing and the share of land owned by them is also increasing; they are the "new landlords" in the rural areas. An increase in the share of households who own land but do not cultivate it can have major implications for the growth of the agrarian sector as these households do not have any incentive to invest in agriculture. In addition, if these households lease out the land, extraction of ground rent from the peasantry is a major form of surplus extraction in the economy, creating constraints to agrarian transformation. In the postindependence period, there were various attempts to reduce the importance of these noncultivating, landowning households (NCHs) in the rural areas. One of the early attempts in public policy was to transform the agrarian structure through a series of land reform measures, to eliminate the influence of agents who were not cultivators and who were perceived to have adverse effects on the cultivating agents. These measures were able to reduce the role of NCHs who were identified as intermediaries. However, in the new liberal phase there has been a revival of households who own land but do not self-cultivate it, in other words, a revival of a new class of landlords in the economy. A question that arises in this context is, why do these households want to hold on to land in the general context of unviability of agriculture?

An increase in the share of NCHs implies that, over time, households who were previously cultivators are becoming NCHs but continuing their interest in their land. If one assumes that a decrease in the share of cultivators implies diversification of households from cultivation to noncultivation occupations, why do these households continue to hold on to land? Does land represent "cultural" value? Or is land seen as a source of "political power" for the upper class, like the traditional landlords in rural areas? Or is ownership of land seen as a store of value as well as a productive asset? In other words, does the idea of downwardly rigid land prices and lack of alternative sources of income for agricultural labor households induce them to enter the tenancy market and because of rental income induce NCHs to hold on to land?

The analysis in this chapter is conducted on two levels: using secondary data sets and substantiating the claims by presenting information from village surveys. Three different rounds of the National Sample Survey Organization (NSSO) (37th, 48th and 59th) on "Assets and Liabilities" have been used to identify the relative importance of noncultivating households. This source can provide broad generalizations but the actual process generating these households might not be captured in these surveys. To substantiate, the present analysis uses a detailed village survey conducted in nine villages in erstwhile Andhra Pradesh conducted by Rao and Bharathi as part of a project "Land and Rural Poverty" in 2003–4.[1] There is no claim that these villages are representative, but what is happening in them is likely to provide information on trends that may be occurring in other villages.

The rest of this chapter is divided into five sections. In Section 8.2, I analyze the different rounds of "Assets and Liabilities" of NSSO data to bring out the importance of NCHs owning land. In Section 8.3, I present information of the location of the nine villages and the main sources of irrigation. In Section 8.4, I group the household data into classes based on labor and land relations. This class-based analysis is used to describe the structure of the village economy and the importance of NCHs in the rural economy. In this section, farm sector households are classified and a basic structure of the village economy obtained. In addition, land market transactions are analyzed to see whether the market is transferring land to NCHs. In Section 8.5, two conditions for the sustenance of NCHs owning land is presented. In this section, the role of increasing land prices and tenancy as potential income sources for NCHs owning land is discussed. A combination of these two streams of income may explain the sustenance of the NCHs. The final section concludes.

8.2 Noncultivating Households in Rural Areas: A View from NSSO Rounds

The NSS (National Sample Survey) reports on "Households Assets and Liabilities in India" (1986, 1998, 2005) give some indicators of the changing rural structure. The survey was conducted for rural as well as urban areas but this chapter concentrates only on the rural sector. In all these data collection rounds, rural households are classified into two major groups. One is the cultivators who are identified as households operating at least 0.002 hectares of land,[2] and the second, the noncultivators who operate no land or operate less than 0.002 hectares. The noncultivators are further classified into "agricultural laborers," "artisans," and "other" households. In this classification, the cultivators have a relation with the agricultural labor households as well as with the artisans, and in turn, the artisans have a relation with noncultivators and also the cultivators. This relation facilitates current or future production. But the "other" is a nondescriptive category that could include different types of agents. They could be traditional landlords, input and output traders, moneylenders, commission agents, or urban workers staying in villages. The farm sector consists of individuals who take part in the agricultural production process and the nonfarm sector consists of individuals who might facilitate production (artisanal community or input traders) or facilitate transfer of goods from rural to urban or urban to rural or within the rural sector (output traders), or might be rentiers (landlords).

The structural changes witnessed in India show a decline in the farm sector, predominantly due to a decrease in the share of cultivator households and an increase in the share of "other" households in the rural sector (Table 8.1). The "other" households are increasing their share to total households and their impact on the production structure is also growing. During the period 1971–81 there was an increase in households reporting cultivation. A declining proportion of agricultural labor households accompany this. For the rest of the groups, there are marginal changes. In other words, the farm–non-farm composition remained more or less the same between 1971 and 1981. However, there are drastic changes between 1981 and 2002. For example, there is a sharp decline in households depending on the farm sector. This decline is mostly due to the decline in cultivating households. The cultivating households declined from 76.3 percent to 59.7 percent, while agricultural labor households remained more or less the same. In nonfarm activities, the increase is sharp: from 10.8 percent to 20.7 percent (Vijay 2012a).

At the structural level, it would be interesting to see whether an increase in households in the nonfarm sector is also accompanied by an increase in land owned by them but which is not cultivated by them. At the national level, the

Table 8.1. Percentage distribution of rural households by occupational category at the national level (1971, 1981, 1991, and 2002)

Household occupational group	Percentage of households			
	1971 (26th)	1981 (37th)	1991 (48th)	2002 (59th)
Cultivators	72.4	76.3	66.1	59.7
Agricultural labor	14.6	11.3	14.2	14.4
Total farm sector	**87.0**	**87.6**	**80.3**	**74.1**
Artisans	2.4	1.6	3.8	5.2
Others	10.6	10.8	15.8	20.7
All noncultivators	27.6	23.7	33.9	40.3
All households	100.0	100.0	100.0	100.0

Source: National Sample Survey Organization (2005): "Households Assets and Liabilities in India," NSS 59th Round, Report number 500, June 30, 2002

Table 8.2. Relative importance of households owning land but not cultivating land in the rural areas at the national level

	% of NCH	% of NCH owning land	% of NCH owning land to total rural households owning land	Value of land owned by NCHs
1981	23.0	19.1	4.5	3.8
1991	34.0	29.6	10.0	7.5
2002	40.3	36.7	14.7	9.8

Note: NCH refers to noncultivating households.

Source: Calculated by authors using 37th, 48th, and 59th rounds of NSSO reports on "Assets and Liabilities in India"

share of noncultivating households owning land to total noncultivating households is increasing, from 19.1 percent to 36.7 percent from 1981 to 2002. The share in the value of land owned by these households has also increased, from a mere 4 percent to around 10 percent in the corresponding period (Table 8.2).

The NSS report (No. 500) also gives the value of assets and their distribution for cultivators and noncultivators. The cultivating households predominate in asset distribution. On average, they hold Rs. 372,000 worth of assets. The noncultivating households on the average hold Rs. 107,000 worth of assets (Table 8.3). The predominance of agricultural labor households in the noncultivating households, whose asset position is likely to be very poor, must have pulled down the average for the noncultivating households. However, the noncultivators show 38.2 percent of their assets in the form of land (Table 8.3). In asset composition, noncultivating households show a difference from cultivators with a substantial share in financial assets, household durables, and buildings, while the cultivators show a predominance of land

Table 8.3. Value and composition of assets owned by cultivators and noncultivators (2002)

	Cultivators	Noncultivators	All rural households
Average value of asset	Rs. 372,632	Rs. 107,230	Rs. 265,606
Share of asset in total asset	83.71	16.28	100
Composition of assets			
Land	68.1	38.2	63.2
Buildings	20.1	41.4	23.5
Livestock and poultry	2.3	1.3	2.1
Machinery and equipment	3.8	3.4	3.7
Household durables	4.2	10.0	5.1
Financial assets	1.6	5.4	2.2
Dues recoverable	0.1	0.2	0.1
Total	100.0	100.0	100.0

Source: National Sample Survey Organization (2005): "Households Assets and Liabilities in India," NSS 59th Round, Report Number 500, June 30, 2002

in their asset base. The widely different portfolio of the noncultivating households possibly indicates their similar economic behavior as elsewhere in diversifying their assets when compared to cultivating households. Thus the noncultivating households show certain linkages with India's growth, which also indicates the diversification of the rural sector into nonfarm activities. Their interest in land seems to be important. To the extent these households own land and do not self-cultivate it, they continue to exert influence on the land market, lease market, and thus the production structure of agriculture, which has implications for the process of agrarian transformation.

The NCHs, as defined in the NSS "Asset and Liabilities" round, is a heterogeneous group consisting of agricultural laborers, artisans, and "other" households. "Other" households can be either labor-supplying households or households owning land but not cultivating it themselves. It was shown by Vijay and Sreenivasulu (2013) that the percentage value of land owned by NCHs owning land in the highest two class groups is a meager 4.7 percent of households in the NCHs, but they own 45 percent of the value of land owned by NCHs. The three lowest asset class groups, which could be identified as predominantly "too poor to farm" households, make up 51 percent of households in the NCH category, while they own a meager 10 percent of the value of land owned by NCHs. Given this, it follows that in the evolving structure of the rural economy, households that are "too busy to farm" are significant in terms of the share of value of land owned by these groups. Based on the NSS report 500, one can arrive at the share of households who own land but do not cultivate, but not whether they lease it out, keep it fallow, or use other contractual arrangements to access land. This lacuna can be filled when one uses primary data collected from villages.

8.3 Structure of the Village Economy and the Importance of Noncultivating Households

Next only to Tamil Nadu, Andhra Pradesh witnessed the highest shift in occupation from cultivators to noncultivators. Based on different NSS rounds on "Assets and Liabilities," the number of noncultivating households was 57 percent and the proportion of NCH owning land to total noncultivating households was 50.8 percent in 2002. They owned about 16.2 percent of the total value of land (Vijay 2012b). The presence of agricultural labor is high and it is also a state that has a high incidence of pure tenancy. Between 1991 and 2001, the number of cultivators declined from an already low level of 27.74 percent to 25.47 percent (Directorate of Economics and Statistics 2006). Interestingly, districts considered to be the most developed in the state, namely the Godavari districts, with a hundred-year history of canal irrigation and a provision of double crops has the lowest proportion of cultivating households. The East Godavari and West Godavari districts recorded 14.90 and 15.06 percent of households as cultivators; this proportion decreased to 12.92 and 14.20 percent respectively by 2001 (Directorate of Economics and Statistics 2006). At the other extreme, there are districts such as Warangal and Adilabad, which have witnessed an increase in the proportion of cultivators. Given the extreme diversity, an attempt was made to study the agrarian structure by Rao and Bharathi (2010). This chapter draws on that survey.

Erstwhile Andhra Pradesh (which now comprises two states: Andhra Pradesh and Telengana) has basically three regions based on the level of development. This section introduces the villages in terms of their location based on their region and gives general information on the nature of irrigation and the crops cultivated.

Agrarian institutions, including the market institutions, have a tendency to undergo changes and transform themselves, and these changes are associated with the level of development of the region and/or subregion. However, using levels of development, the state is sometimes grouped into five regions (Rao and Subrahmanyam 2002). The top place in the scale of development, with a high index of output per hectare, is South Coastal Andhra comprising the Krishna Godavari delta regions, namely the districts of East Godavari, West Godavari, Krishna, Guntur, Prakasam, and Nellore. Next in importance comes the region of North Telangana, comprising the districts of Nizamabad, Adilabad, Karimnagar, and Khammam. North Coastal Andhra, comprising Srikakulam, Vizianagaram, and Visakhapatnam occupies third place. Rayalaseema is fourth, comprising Cuddapah, Kurnool, Ananthpur, and Chittoor. Fifth is South Telangana, comprising Rangareddy, Hyderabad, Mahaboobnagar, Medak, and Nalgonda. The regional differences are partly historical and partly due to differential public investments and complimentary private investments. As the

Table 8.4. Location, number of households, and sources of irrigation in the surveyed villages

District	Mandal*	Village	Number of households surveyed in village	Percentage of land irrigated by different sources			
				Canal	Tank	Wells	Rainfall
West Godavari	Veravasaram	Mentipudi	90	100	–	–	–
	Ganapavaram	Kothapalli	208	100	–	–	–
	Koyyalagudem	Seethampet	170	–	–	54.77	46.22
Mahaboobnagar	Atmakur	Arepalli	338	53.26	18.86	18.23	9.65
	Bhootpur	Tatiparthy	216	–	12.42	32.34	55.25
Karimnagar	Dharmapuri	Chinnapur	216	61.47	–	16.69	21.85
	Dharmapuri	Nagaram	171	29.26	–	65.7	5.05
Srikakulam	Jalumuru	Jonanki	151	–	92.22	–	7.78
	Elcherla	BantalaKoduru	177	–	–	58.09	41.91

Notes: *A district in a state is divided into mandals, which were earlier identified as talukas.

Source: Vijay and Sreenivasulu (2013)

institutions are likely to undergo changes with levels of investment and consequent levels of development, the attempt in the village survey was to distribute the ten sample villages across the five regions. The attempt to cover at least two villages in each region succeeded in all regions except Rayalaseema, where even one village could not be covered. In total nine villages were covered, three in South Coastal Andhra and two villages each in North Coastal Andhra, North Telengana, and South Telengana.

Table 8.4 provides information on the *mandal* (sub-districts or administrative divisions under districts) from which the villages are selected, the number of households surveyed, and sources of irrigation in the surveyed villages. Three villages were surveyed in the West Godavari district, two of the villages (Mentipudi and Kothapalli) are irrigated from canal water and one village (Seethampet), which is not part of the irrigation command area, is in the dry zone in the district with nearly 54 percent of the land irrigated by wells. Two villages were selected from Mahaboobnagar district, a district known as a drought prone district, which witnesses large-scale labor migration to urban areas. One of the villages is rain fed and the second uses canal water as a source of irrigation. Karimnagar district, which is part of Telengana area has had significant intervention by the Communist Party of India (People's War Group), now called the CPI (Maoist) Party. The party had a presence in the two villages surveyed in this area in the past but not in the present period. One of the villages surveyed was predominantly irrigated by canal water while in the second village wells form the main source of irrigation. The two villages selected in north coastal Andhra are Jonanki and Bantalakoduru. Jonanki is a village with significant presence of tanks as a source of irrigation while the other village uses a well as the principal source of irrigation (Vijay 2012a).

In Godavari district, the two canal-irrigated villages grow mostly paddy, while the third village, Seethampet, has nearly 52 percent of the land under tobacco, followed by 22 percent under paddy, 9 percent under sugarcane, and 4 percent under maize, which is mostly for chicken feed. Seethampet has a diversified cropping pattern and the crops cultivated are inputs to agroindustry. Arepalli, the village in Mahaboobnagar district with a significant proportion of land irrigated by canal, also has a high proportion of land assigned to paddy cultivation (53 percent). In the dry village of Tatiparty in Mahaboobnagar district, the principal crop is the inferior cereal, jowar (40 percent), followed by castor (28 percent). In the two villages in Karimnagar district paddy is the main crop (Chinnapur at 45 percent and Nagaram at 37 percent). In Chinnapur, maize is the second main crop (11 percent), which is partly consumed and partly sold in the market. In Nagaram, in addition to paddy, turmeric is grown for the market and forms a significant proportion of the crops cultivated (17 percent). In the case of the two villages surveyed in Srikakulam district, the tank-irrigated village, Jonanki, has a dominance of paddy crop (72 percent); in the second village, Bantalakoduru, paddy is the important crop (45 percent) followed by pulses and chilies (Vijay and Sreenivasulu 2013).

8.4 Classification of Households in the Farm Sector

Given the extreme diversity in the farm sector in the rural economy, an attempt was made to study the agrarian structure by Rao and Bharathi (2010). This chapter depends heavily on that survey and the methodology they used in classifying households. The surveyed villages represent different regions and the method of classification of households chosen by the authors was a five-class analysis. Presenting agriculture in terms of class analysis was in vogue during the late 1960s and 1970s and recently again used by Ramchandran, Rawal, and Swaminathan (2010). In a sense, the report by Rao and Bharathi (2010), and by implication this chapter, revisits such a framework and presents a different understanding compared to the land-based classification used in identifying structures by Vyas (2003). In our survey a complete enumeration of all households residing in the village was conducted. A detailed questionnaire was canvassed to all the households in the village to capture the characteristics of the households (members of the households, caste, migrations, etc.), the resource position of the households (land owned, operated, type of land, instruments owned, number of working family members, etc.), and the exchanges the households enter into in the land, labor, and credit markets. Those households directly related to agricultural production were classified into five classes based on their interaction in the labor market.

They are: noncultivating households, rich peasantry, middle peasantry, poor peasantry, and agricultural labor households.

8.4.1 *Class Composition of the Surveyed Villages*

The total number of households in the farm sector in the nine villages is 1,737 and they own 3,795.56 acres. The average size owned works out to be 2.18 acres per household; if agricultural labor households are excluded, it works out to be 3.02 acres per household. At the aggregate level, the economy is dominated by the poor peasantry. They form 43.87 percent of the households and own 35.43 percent of the land. In terms of numbers, the agricultural labor households form the second largest group in the villages and constitute 27.63 percent of the households. The poor peasantry and agricultural labor households together are 70.5 percent of the rural households. At the other extreme are the labor-demanding households: the rich peasantry and noncultivating households. The rich peasants form 4.37 percent of the households and operate 16.90 percent of the land. But the noncultivating households are numerically larger and constitute 5.58 percent of rural households, which is larger than the share of rich peasant households. These noncultivating households own 20.10 percent of the land. The middle peasantry constitutes 18.53 percent of households and owns 26.41 percent of the land (Table 8.5).

In the nine villages surveyed it was found that 5.5 percent of households are NCHs and they own approximately 20 percent of the land. There are two issues that arise in this context. One is the distribution of the NCHs over villages: are there some villages with a higher share of NCHs and a higher share of land owned by this class? Second, the surveys were conducted among rural households; but are there households who reside in urban areas that may own land in rural areas? Can one provide a "guesstimate" on the extent of land owned by nonresident households who by implication are also NCHs?

The villages show marked heterogeneity in terms of number and share of land owned by NCHs. In terms of numbers, Chinnapur village has the highest

Table 8.5. Distribution of households and land owned across class groups

	Number of households (%)	Area owned, in acres (%)
Agricultural laborer (AGL)	480 (27.63)	43.62 (1.15)
Poor peasants (PP)	762 (43.86)	1,344.64 (35.42)
Middle peasants (MP)	322 (18.53)	1,002.54 (26.41)
Rich peasants (RP)	76 (4.37)	641.5 (16.90)
Noncultivating households (NCH)	97 (5.58)	763.2 (20.10)
Total	1,737 (100)	3,795.56 (100)

Source: Field survey

Table 8.6. Land-based importance of NCHs in the surveyed villages

Village	No. of NCHs	Share of NCHs to total households (%)	Share of land owned by NCHs (%)	Share of land owned by NCHs and nonresident households (%)
Mentipudi	9	10	38.1	65.4
Kothapalli	6	2.8	15.8	67.8
Seethampet	12	7.1	24.1	24.1
Arepalli	14	4.1	11.5	11.5
Tatiparthy	0	0	0	0
Chinnapur	36	16.7	26.5	26.5
Nagaram	4	2.3	5.8	11.8
Jonanki	13	8.6	49.4	49.4
Bantalakoduru	3	1.6	60.3	60.3
Total	97	5.5	20.1	25.2

Source: Field survey

share of NCHs followed by Mentipudi. The dry village in Mahaboobnager (Tatiparthy) does not have NCH households. But if one analyzes the data on land owned by NCHs, the villages from Srikakulam have the highest share of land owned by NCHs. The village of Bantalakoduru is the only village with a traditional landlord from a landlord family. They owned nearly 175 acres of land in the village. The villages of Chinnapur and Mentipudi have a higher share of land owned by NCHs (Table 8.6).

A particular feature of the survey was that it enumerated all households who had stayed in the village. So if nonresident households own land in the village they would not be captured in the survey. In Mentipudi and Kothapalli villages, there is a major difference between land owned and land operated. This difference is also expressed in the difference between land leased in and land leased out only in these two villages. So one can assume that the difference between land owned and land operated is due to the presence of nonresident households and these households are leasing out. Given that they are not resident households and leasing out the land one can identify them as NCHs. If this line of thinking is correct, then the share of land owned by NCHs and nonresident households in the developed coastal districts is much higher and comparable to land owned by NCHs in the traditional landlord-dominated districts (Table 8.6).[3]

In a static sense, one is witnessing an important role for noncultivating households in the surveyed villages. But in the long run, what also needs to be analyzed is whether the noncultivating households are both purchasing and selling land from and to peasant classes. The survey collected data on land market transactions by all households for a period of five years (1999–2003) preceding the survey. The data is presented in Table 8.7. Assuming that the data is reasonably representative, the area of land sold is more than the area of land purchased. While 175 acres are sold, the purchases total only 106 acres.

Table 8.7. Distribution of land transacted in land market across classes between 1999 and 2003

	Sale		Purchase		Net purchase (area purchased-area sold)
	No. of transactions	Area transacted (in acres)	No of transactions	Area transacted (in acres)	
Agricultural labor households (AGL)	–	–	–	–	–
Poor peasantry (PP)	41 (37.27)	43.64 (24.94)	24 (38.70)	19.79 (18.67)	−23.85
Middle peasantry (MP)	44 (40.00)	62.30 (35.61)	24 (38.70)	27.55 (26.01)	−34.75
Rich peasantry (RP)	14 (12.72)	23.48 (13.42)	3 (4.83)	6.4 (6.03)	−17.08
Noncultivating households (NCH)	11 (11.00)	45.53 (26.02)	11 (17.74)	52.25 (49.29)	+6.72
Total	110 (100)	174.95 (100)	62 (100)	105.99 (100)	−68.96

Notes: Net purchase = Area purchased-area sold; figures in brackets are proportion to totals.
Source: Vijay (2012)

It may not be farfetched if one presumes that the difference between the sales and purchases is accounted for by nonresident households. If one takes the net purchases, all classes record a negative net purchase, while NCHs record a positive net purchase. In other words, during the five-year period, NCHs are accumulating additional land while other classes are experiencing dispossession of their land. Thus, the market is facilitating land transfers to the NCH from the other cultivating households.

8.5 Why NCHs Avoid Selling Land

One of the important tendencies seen in the Indian economy is the reduction in the share of rural households and a decrease in the share of cultivator households. A decrease in the share of cultivators would imply that cultivators in earlier periods are shifting out of agriculture and becoming noncultivators but continue to have an interest in land. A question that arises in this context is why do households who have diversified out of agriculture not sell their land? This section provides an explanation of one potential source of income in the future for NCHs (in the short and long run) due to owning land.

8.5.1 *Land as a Secure Asset*

If one assumes that erstwhile cultivating households are converting to NCHs, these households have three options: to sell land, to lease out land, or keep the land fallow. Let us consider the first option. Land as an asset has two components, as a store of value and a productive asset. In the Indian context, the land market is an inactive market and the extent of land entering the land market is dependent on supply-side factors, that is, distress conditions faced by landholding households in the rural areas. On the demand side there is an excess demand for land, which is leading to consistent increases in price of land over time. According to Vakulabharanam et al. (2011) writing on the crop holiday in West Godavari district, the price of land "has risen tremendously over the last 15 years, with prices going up more than 10–20 times. This has created a rebirth of absentee landlordism and tenancy" (2011: 14). At the national level, Chakravorty (2013: 52) shows that "productivity is not the primary determinant of prices of agricultural land." He maintains that prices of agricultural land in Punjab are twenty times higher or more than what can be explained by productivity alone. This trend is true for large areas of Haryana, and large parts of Kerala and Tamil Nadu, which are proximate to urban areas (Chakravorty 2013).

Chakravorty (2013) and Vakulabharanam et al. (2011) do not consider this price rise a bubble, but a permanent feature. For Chakravorty, increasing land prices are due to the easing of credit, while for Vakulabharanam, the increase in prices may be structural. If one considers a structural explanation, it could be that the excess demand for land is related to its feature as a "secure" asset with low probability of a decrease in prices of land implying low levels of uncertainty in prices. The feature of land prices, that is, downwardly rigid prices and low variability in land prices leading to low uncertainty, makes this a secure asset compared to other assets in the economy. In the context of a changing rural sector, where NCHs are increasing, the NCHs might sell land if there were entry barriers to enter the non-farm sector (NFS), such as the need for cash income to enter the NFS. If one wants to be a fertilizer dealer or commission agent, households would need some income to establish this venture and therefore might want to sell part of their land. So in the long term land is a secure source of income for households but in the short term it could be a "dead investment" if it does not generate any income. One option for NCHs to earn income in the short run is through productive use of land by leasing it out.

8.5.2 *Tenancy as a Source of Yearly Income*

If households want to hold on to land but do not want to cultivate the land themselves, they can keep it fallow or lease it out. If the owner believes that land is also a productive asset, they might want to derive income by leasing out the land.

Tenancy is important in the surveyed villages. The total leased in land is 651 acres (Table 8.8), which is 16.66 percent of the total operated land. If one takes the leased out land, the total is 331.5 acres, or 11.27 percent of the total owned land. The difference between the extent of land leased in

Table 8.8. Distribution of area leased in and leased out across classes

	Leasing in		Leasing out		Net lease in area
	No. of transactions	Area leased in (in acres)	No. of transactions	Area leased out (in acres)	
AGL	–	–	7 (5.73)	8.75 (2.64)	−6.5
PP	167 (60.37)	331.5 (50.98)	30 (25.40)	35.24 (10.63)	+296.26
MP	81 (30.00)	190.00 (29.13)	8 (5.73)	30.2 (9.11)	+159.8
RP	20 (7.41)	97.5 (14.97)	5 (4.09)	23.00 (6.93)	+74.5
NCH	2 (0.74)	32 (4.91)	72 (59.01)	230.59 (69.56)	−198.59
Total	270 (100)	651.1 (100)	122 (100)	331.5 (100)	+319.6

Notes: Net lease in area is defined as (area leased in–area leased out); figures in brackets are proportion to totals.
Source: Vijay (2012)

and leased out can be attributed to nonresident households owning land in the villages, who were not covered by the survey. The villages with the highest share of land under tenancy are in the irrigated paddy-growing districts of the delta region. The two villages in the Godavari delta region had approximately 45 percent of both the land leased-in transactions and the total leased-in area. The distribution of land leased in and leased out is presented in Table 8.7. While NCHs are the only net suppliers of land in the lease market, the poor peasants and middle peasants are the net demanders of land in the lease market. While all classes participate in the leasing in and out of land, poor peasants dominate leasing in, with 60 percent of the land leased in. On the leasing-out dimension, the NCH predominate, with 70 percent of the land leased out.

A tenant can be a landless labor household who may be a poor peasant entering the land lease market for subsistence, while a rich peasant leasing in land may have a market orientation. In the surveyed villages, 605 families do not own any land, and they normally would have been recorded as agricultural labor households. However, 115 of the 605 households (19 percent), lease in land and operate as pure tenant cultivators; 81.63 percent of pure tenant households are poor peasant households. Another 16.52 percent of the pure tenants are middle peasants, while 1.73 percent are rich peasants. The amount of land operated under pure tenancy by poor peasant households is almost 74 percent of the total land operated. In the two delta villages, 52.68 percent of the households do not own any land. Among the non-landowning households, 38.85 percent of the households operate as pure tenant cultivators and most of them are poor peasants. The non-landowning potential agricultural labor households join the production structure as pure tenants; a path households choose to hedge against uncertainties in income.

8.5.3 *Options Open to Poor Peasantry and Agricultural Labor Households*

For an NCH to derive "secure" yearly income, the demand for land for tenancy should be inelastic. In other words, if the agricultural laborer and poor peasantry do not have many options to earn income outside tenancy, the tenancy rates would be relatively stable and NCHs could derive a secure yearly income. This section provides information on options available to labor-supplying households to earn income.

In aggregate, nearly 70 percent of the rural households are agricultural labor households and poor peasantry. But if one sees the options available to these labor-supplying households to meet their livelihood, there are some interesting trends. Here I analyze three options: to lease in land, migrate to urban areas, or having long-term labor market contracts. The option of entering the nonfarm sector in rural areas was not presented in the table as few agricultural households reported this option. (Table 8.9). In the case of canal-irrigated villages, entry into the land lease market looks to be the most important option to these households. In Mentipudi and Kothapalli villages, around 35 percent of the agricultural labor households and poor peasantry households enter the land lease market. In both the villages nearly 28 percent of agricultural labor households and poor peasantry households leasing in land are pure tenants. There were thirty-four households reporting long-term contracts and they were pre-dominantly in two villages: Seethampet village (ten households) and Nagaram village (eight contracts). In these villages, potential labor-supplying

Table 8.9. Options open to agricultural labor households and poor peasantry in the surveyed villages

Villages	% of AGL	% of PP	% of AGL and PP	Options open to AGL and PP households			
				Share of PP and AGL in tenancy	Share of PP and AGL as pure tenants	Share of PP and AGL households experiencing migrations	No. of households with long-term labor market contracts
Mentipudi	17.78	36.67	54.44	34.69	28.57	6.12	1
Kothapalli	47.12	33.65	80.77	35.12	28.57	20.83	0
Seethampet	53.53	12.94	66.47	8.85	4.42	1.77	10
Arepalli	16.57	59.47	76.04	5.06	1.56	35.41	7
Tatiparthy	13.43	71.76	85.19	5.43	0.54	22.83	0
Chinnapur	22.69	27.78	50.46	12.84	1.83	14.68	3
Nagaram	22.22	45.03	67.25	6.09	0.87	20.00	8
Jonanki	21.19	50.33	71.52	29.63	14.81	37.96	2
B.Koduru	40.11	38.42	78.53	4.32	1.44	2.88	3
Total	27.63	43.87	71.50	13.53	7.49	20.69	34

Note: AGL refers to agricultural labor households and PP refers to poor peasantry.

Source: Field survey

households are entering into the land lease market. In the rest of the villages, tenancy is not an important feature of livelihood but migration is. In these villages the proportion of land owned by noncultivating households is relatively smaller when compared to the irrigated villages (Vijay 2012a), which restricts the supply of land in the land lease market.

Land in the land lease market is predominantly supplied by nonresident and/or noncultivating households staying in the village. If we assume that the supply of land is inelastic, the rental rates in the market would be defined by the demand-side factors. The households demanding land in the land lease market are rich peasantry, middle peasantry, and the poor peasantry. Individuals in these classes would compete to get access to land in the land lease market. If one assumes that the nonresident and/or noncultivating households are only interested in maximizing rental income, individuals who provide the highest rent would get access to land. A rich peasantry is by nature market oriented and also wants access to land in the land lease market to derive scale advantages. A poor peasant, on the other hand, has "surplus labor." The market is not able to provide sustained employment to these households. So poor peasants self-exploit by using the "surplus" labor from their own farm. These households without any options would keep increasing the demand for land lease for subsistence and by implication the rental value of land. In the surveyed villages it was found that the poor peasants/pure tenants are pricing out the rich peasant in the land lease market and are the main demanders of land in the land lease market (Vijay 2012a). The higher rent provides the economic basis for sustenance of the noncultivating households and does not appear to be a driver for higher accumulation in rural areas.

8.6 Conclusion

The agrarian structure is undergoing changes due to the important role played by NCHs owning land at one end of the spectrum, and the stability of the proportion of agricultural laborers at the other end. The tremendous increase in land prices is influencing the portfolio of erstwhile cultivators. In other words, higher returns with a higher uncertainty in nonfarm sector investment compared to lower levels of returns and lower uncertainty in agriculture influences the portfolio choices of the cultivator. This possibly is the source that generates the noncultivating peasant household that chooses a mix of higher returns in the nonfarm sector and a lower risk in the farm sector. The certainty in the income of NCH in farm activity is made possible by the demand for land in the agricultural sector, particularly from the agricultural labor households. The land hunger of these households and the need to hedge against the uncertainties of life make the land lease market a very active market. In all possibility, since the demand for land is reasonably high from this segment, the rent is also likely to be high, ensuring that the income of NCHs from agriculture is not unreasonable. This explains the decline in cultivators and the concomitant increase in NCHs owning land and the stable proportion of agricultural labor households.

The evolving rural economic structure is witnessing the emergence of a new class of landlords in the sense that they own land but do not cultivate the land themselves; rather they derive a ground rent. Unlike the earlier landlords who are from "landlord" families and own large tracks of land, these new landlords could be erstwhile cultivators who have diversified out of agriculture but continue their interest in land. At least in the case of Andhra Pradesh, and specifically in the case of West Godavari district, one is witnessing a revival of absentee landlords of the second category and revival of tenancy.

If this hypothesis on the increasing importance of noncultivating households owning land in rural areas is reasonably true, then the land question comes back to the center stage. On the one hand, NCHs own land but are not cultivating it themselves and deriving ground rent. On the other hand, there is a large segment of land-hungry agricultural laborers and the poor peasantry without any viable option for their livelihood. Instead they enter tenancy relations. Restructuring agriculture with an emphasis on land reforms appears to be a necessary policy intervention to solve the agrarian crisis.

Notes

1. I was assisting the late R. S. Rao and Dr. Mudunuri Bharathi as an associate in the project. The data is used with the permission of Dr. Bharathi.
2. One hectare is equal to 2.47 acres.
3. The relative strength of the absentee landlords/noncultivating nonresident households in the irrigated areas of Godavari district can be seen in terms of decisions taken by landowners to declare that they would not cultivate any kharif crop on their land in the year 2011 as they maintained that prices were not remunerative for paddy. Interestingly, this area is known to have nearly 60 percent of the land under tenancy. In addition to remunerative prices being one of the reasons for a crop holiday, another reason was a policy initiative to give identity cards to tenants and give tenant legal status. The absentee landlords felt that legalizing tenancy status would weaken their position and so they had used a crop holiday as an instrument to discipline the laborers/tenants (Vakulabharanam et al. 2011).

References

Chakravorty, S. 2013. "A New Price Regime: Land Market in Urban and Rural India." *Economic and Political Weekly*, 48(17): 45–54.

Directorate of Economics and Statistics. 2006. "Statistical Abstract of Andhra Pradesh—2006," Hyderabad.

NSSO (National Sample Survey Organization) 1986. "Households Assets and Liabilities in India," 37th Round, June 30, 1982.

NSSO (National Sample Survey Organization) 1998. "Households Assets and Liabilities in India," 48th Round, Report No. 419, June 30, 1991.

NSSO (National Sample Survey Organization) 2005. "Households Assets and Liabilities in India," 59th Round, Report No. 500, June 30, 2002.

Ramachandran, V. K, Rawal, V., and Swaminathan, M. 2010. *Socio-Economic Surveys of Three Villages in Andhra Pradesh*. New Delhi: Tulika Books.

Rao, R. S. and Bharathi, M. 2010. "Comprehensive Study on Land and Poverty in Andhra Pradesh: A Preliminary Report." In *In Search of Method*, edited by M. Bharathi, Hyderabad: B-1, Collective and Centre for Documentation, Research and Communication, pp. 323–67.

Rao, Y. V. K. and Subrahmanyan, S. 2002. "Development of Andhra Pradesh 1956–2001: A Study of Regional Disparities." N.R.R. Research Centre, Hyderabad.

Vakulabharanam, V., Purendra Prasad, N., Laxminarayana, K., and Kilaru, S. 2011. "Understanding Andhra Crop Holiday Movement." *Economic and Political Weekly*, 42(50): 13–16.

Vijay. R. 2012a. "Structural Retrogression and the Rise of 'New Landlords' in Indian Agriculture: An Empirical Exercise." *Economic and Political Weekly*, 47(5): 37–45.

Vijay, R. 2012b. "Rise of 'New Landlords: A Response." *Economic and Political Weekly*, 47(40): 82–3.

Vijay, R. and Sreenivasulu, Y. 2013. "Agrarian Structure and Land Lease Arrangements: An Investigation in Nine Villages in Andhra Pradesh." *Economic and Political Weekly*, 48(26–7): 42–9.

Vyas, V. S. 2003. *Indian Agrarian Structure, Economic Policies and Sustainable Development: Variation on a Theme*. New Delhi: Academic Foundation.

9

Land and/or Labor? Predicament of Petty Commodity Producers among South Indian Villages

R. V. Ramana Murthy

9.1 Introduction

Historically, agrarian transformation has been necessary for capitalist development. It entailed monetizing agriculture, development of wage labor, and production of market surplus. However, current Marxist discourse takes a clear view on the issue of capitalist transition in postcolonial nations, arguing that it is significantly different from the European experience (Sanyal 2007: 20). Capital may change relations of production in agriculture and eschew the resolution of the agrarian question by not discharging petty commodity producers from agriculture and absorbing them into the modern capitalist sector (Bernstein 2010). There are convincing reasons why transformations at work in post-twentieth-century capitalism could be significantly different in postcolonial countries such as India. A large share of the workforce locked in the agricultural sector, the dominance of a self-employed workforce, and a growing informal sector inside and outside agriculture are some of the peculiar manifestations of the transformation. The land question, besides the population factor, is obviously complicated by a large chunk of the workforce being unable to quit agriculture. The peasantry is transformed from the subsistence stage to petty commodity production under the compelling influence of the market. Land concentration in agriculture is declining in India and the share of smallholdings, in number as well as in share of land, is growing steadily. It is natural that nonfarm needs compel some diversion of agricultural land in the course of urbanization and industrialization. Land acquisition in such a context is resisted aggressively in some places, while it is accepted reluctantly in others. The centrality of land to the lives of farmers is also a matter of how much they depend on it. The gradual diversification of rural households to nonfarm activities can alter the centrality of land for different classes of farmers. Thus far, the land question depends on the nature of the agrarian transition.

In parts of South India, landless, marginal, and small farmers are acquiring land, either through sale or lease, in smaller parcels, forming a majority in the agrarian structure. It is an intriguing phenomenon to observe farming in the country being brought under petty commodity production. How well petty commodity production sits with capitalist development is an important question, both as a development question and a land question. This chapter seeks to answer the following questions: Why is the share of petty commodity producers increasing in the agrarian structure? How extensive is petty commodity production as the source of livelihood? Does the state make the capitalist transition less harsh through social welfare transfers that outweigh the effects of pauperization imposed by capitalist development? I shall provide an explanation, albeit partial, of these issues from evidence drawn from household data from a study conducted in rural Andhra Pradesh.

The rest of the chapter is divided into three sections. Section 9.2 discusses perspectives on the agrarian question. It begins with a discussion of petty commodity production (PCP) and its relationship to capitalist development in India. Section 9.3 presents the findings of field observations obtained from three village surveys in Andhra Pradesh on the profile and nature of PCP. It begins with the agrarian distress in the state covering patterns in land distribution, access to agricultural credit, and rural incomes, and ends with a brief discussion of state-disbursed benefits to the rural poor. Section 9.3 revisits the process of proletarianization and marginalization of petty producers.

9.2 The Agrarian Question

9.2.1 *Petty Commodity Production and the Agrarian Transition*

Marxist literature has traditionally treated petty commodity producers as a class progressively diminished by the logic of capitalist development, turning them into a proletariat in the larger course of transformation, referred to as the resolution of the agrarian question (Byres 1977). However, alternative courses of transition are possible, as outlined by Bernstein (2010), where dissolution of petty commodity producers may not be a necessary condition for capitalist transformation under globalization of capital. PCP has also survived for several historical reasons such as political space gained in anti-colonial movements, anti-feudal land reforms, and postcolonial conditions of capitalist development. PCP in agriculture can not only persist in developing economies but may grow in the share in production as nonfarm capitalist sectors clearly benefit in terms of accumulation from copious flows of market surplus at lower prices. The petty commodity producers' bid to survive through over-self-exploitation conveys that they are labor in disguise. Pauperization and poverty continue to be endemic features of their existence. However, unlike the way they were wiped out in the past, their transformation could be much harsher in the age of liberal democracy and hegemony of capitalist development.

Forms of precapitalist production relations persist in most parts of the world, including India. The uneven development of production forces and penetration of capital across time and space have created complex modes of production across the world. The renewed scholarship on the agrarian question has produced an extremely rich literature that explains not only the complexity of modes of production but also the persistence of PCP under neoliberal globalization (Akram-Lodhi and Kay 2010; Bernstein 2010; Lerche 2013). PCP denotes the existence of a substantial class of small farmers involved in agricultural production primarily using family labor for subsistence. Penetration of markets, generalized commodity production, rising pressure of consumption needs, and inadequate nonfarm employment opportunities often compel PCP households to cling to their small parcels of land while participating in the market. Marx (1954) in *Capital I*, part III, while observing the agrarian transition in England that displaced the free peasantry through the twin processes of primitive accumulation and differentiation, sufficiently cautioned against mechanical anticipation and repetition of such experiences. Kautsky ([1889] 1988), while endorsing the view on the final exit of this class in his classic piece *The Agrarian Question*, also vividly described the ability of PCP households to survive under extreme competition through over-exploitation of family labor and self-starvation. It is also now widely acknowledged that the resolution to the agrarian question in advanced Western capitalist countries occurred through colonialism, which facilitated the agrarian transition to capitalism through migration (Bagchi 2008: 85). Imperialism not only provided cheap raw materials and commodity markets but also an ability to relieve their surplus population, an option unavailable to most countries that are late entrants to capitalism (Bernstein 2010). Hence, PCP persists as a default response mechanism until a viable transition to the nonfarm sector becomes possible. The petty commodity producers are integrated with capitalist markets through production as well as consumption. Expansion of PCP is also unlikely to create a contradiction with the capitalist transformation as long as it cheapens agricultural production. What could be the implication of the preponderance of such petty producers in agriculture to the land question? How central is the land to the reproduction of petty commodity producers? These questions are addressed in this chapter.

Petty commodity producers stand hopelessly at the receiving end as markets play havoc with prices that fail to cover the costs of production, leaving a bare minimum of profit to meet the family's consumption needs. In other words, from simple reproduction it fails to graduate to expanded reproduction. The logic of self-exploitation works through the hopes of earning profits by undercutting costs, but gets beaten by competitive and manipulative markets. To survive, petty commodity producers tend to over-exploit themselves to make up the deficit. The PCP household manages to reproduce itself because: 1) it can survive with lower returns; 2) exploitation is internalized (it can accept lower implicit wages in bad times); and 3) it maintains low consumption levels. However, the independence of these households may be considerably undermined by the terms and conditions of interlocked contracts and insufficient access to credit markets to raise required working capital. An increased consumption culture and market orientation can also weaken its kinship network and cooperation, which further tends to weaken its coping mechanisms (Harriss-White 2012).

From having gained access to small parcels of land, petty commodity producers also benefit from the speculative prices of land. Land becomes the key asset to play in keeping the class position of the household. Thereby, it plays a vital role in reproduction of social relations of family, marriages, and social value. Land remains an important store of wealth and a source of security in old age. Thus, the PCP household strengthens the bourgeois property regime through its participation, which makes its class position totally ambivalent.

9.2.2 *Petty Commodity Production and Capitalist Development in India*

Having described petty commodity producers in capitalist development, we shall locate them in India's capitalist transition. India's path of capitalist development has perhaps seven distinct aspects. First, India adopted a liberal democratic political system well before it experienced substantial capitalist development of the Western variety. Second, the Indian state played a major role in shaping the nature of capitalist development and negotiating competing class interests for its growth. The property rights over land have evolved from a long process ranging from *Ryotwari* settlements (where ownership rights were given to peasants) to abolition of intermediaries and state distribution of village commons. Market forces, aided by developmental policies such as free electricity, fertilizer subsidies, procurement prices, market yards, subsidized credit, and microcredit, encouraged commercial farming to extract market surplus necessary for stable accumulation of capitalist sectors. Within the space created by markets and the state, PCP in India survives, the shortfall to subsistence and survival being provided through welfare transfers. Third, India's capitalist development is spearheaded by the state more through forcible mobilization of small savings than through agrarian surplus (Sanyal 2007:159). Fourth, as a late entrant to industrial development in the twentieth century, the technological conditions in India are unlikely to allow the absorption of a substantial share of the surplus labor in agriculture into the modern sector. Fifth, the option of large-scale international migration does not exist for a populous country such as India. Therefore, the majority of the workforce is locked in the space of PCP. Sixth, large-scale primitive accumulation, which could have eliminated PCP, is pre-empted by liberal democratic spaces and human rights discourses in the dominant ideologies (Chatterjee 2008). Finally, the poverty of the masses, including petty producers, caused by capitalist extraction is softened by the welfare populism of the state to the point of political management of dissent (Chatterjee 2008). The liberal bourgeois politics in its vast diversity provides incremental space for the common masses. The comprehensive analysis of political economy of capitalist development in India compels a synthesis of all these aspects.

A major lacuna in the debate on the mode of production in India that took place in the 1970s, in all its richness in capturing the microfoundations of capitalist transition,[1] is the scant attention paid to the role played by the state in resolving the agrarian question. Under postcolonial conditions of global capitalism and electoral democracies, class hegemony is largely negotiated by the state through its policies. A far more penetrating analysis on this aspect is carried out by Sanyal (2007), who made a departure from the standard base–superstructure model to a neo-Gramscian state–civil society model, which brings in the politics of a developmental state under bourgeois democracy much more explicitly while negotiating capitalist development. Sanyal argues that the "state–civil society" dichotomy is much more useful to understand the politics of the hegemonic state. Gramsci, in the Italian context, argues that a weak bourgeoisie is compelled to compromise with the feudal classes, which allows the alliance to form a hegemonic leadership over the masses. Hegemony results from consent given by the masses in exchange for incremental benefits provided by the state. This, by and large, becomes the model in most countries where bourgeois revolutions did not succeed. Therefore, the bourgeoisie could only use a gradualist strategy of progressively weakening the power of feudal classes through what Gramsci called "passive revolution" (Gramsci 1996: 202). Liberal democracies enabled such a transition and class compromise characterizes most of the countries where bourgeois revolutions have failed to take place.

Partha Chatterjee (2004, 2008), while taking this further, provides a caveat for understanding the case of postcolonial states. In developing countries, the civil society constitutes a minute share of the population, while a huge section of the population owning no property lies outside civil society, in what he terms "political society." Political society is one which does not have all the rights of citizenship, but has the right to franchise. Squatters, street vendors, urban slum dwellers, landless poor, and *dalits* (self-labeled untouchables of the caste system) who may till assigned lands constitute this political society (Chatterjee 2004: 27–47). Their means of survival does not often have legal entitlements, but the state patronizes them by protecting their otherwise illegal existence. Managing political society, for Chatterjee, is key to the politics of developmental states such as India. The competitive electoral politics gives some scope for the political society to negotiate incremental benefits from the state. He further argues that the state, as in the West, uses various governmental technologies such as statistics to indicate "development" through various programs and engage the population. This is further helped by international capital through funding various non-governmental organizations (NGOs), which help the capacity building of the state and thus extend *governmentality*.[2] By extending Chatterjee's logic, the small peasantry, pauperized by the capitalist system, qualify as a political society. Alleviation of the farmers' plight often becomes part of the election agenda. In past elections, much against the ruling establishment, debt waivers, compensation for families of farmers who commit suicide, occasional state procurement under farm lobby pressures, and so on have been part of governmentality. In addition, poor peasants become recipients of welfare in the form of employment programs, public distribution, pensions, housing loans, cash-transfers, and so on.

There is an impression that agriculture has lost its importance in the contemporary economic growth discourse. This view is taken further by suggesting that agriculture has become decoupled from the larger economy. The issue needs to be put into perspective. In India, the issue of agrarian constraint to capitalist growth was recognized during the planning era, *circa* 1960. It was held that the modern sector's growth suffered an accumulation crisis during the mid-1960s, when no palpable agrarian surplus was forthcoming. Worse, the supply shortfalls in food grain production in the economy posed an inflationary barrier to growth.[3] Following the failure to achieve institutional reforms, the Indian state resolved the production problem through the technological means of the green revolution. Institutional interventions ensured the necessary market surplus for the capitalist sector, which in turn ensured the viability of the farmer and productivity growth for the sector. In the neoliberal era, there are tendencies to withdraw the different forms of intervention that involve subsidies and other public expenditures. The absence of intervention tends to produce crisis for the petty producers, while medium and big farmers are better placed to diversify and capitalize on new opportunities. As production trends are unaffected so far, and agrarian distress is internalized by the small peasantry, it makes neoliberal reformers feel they have nothing much to worry about. Petty commodity producers of smaller holdings are a proletariat in disguise, whose income portfolio gets distributed between small profits from farming and allied activities and wage in farm and nonfarm sectors. Proletarianization underlies the marginalization and pauperization of PCP households. Social and kinship networks are often employed to escape the severity and survive. While neoliberal policies push them to the brink through withdrawal of subsidies and price supports, PCP households have become below-poverty population subjects of the state. A slew of welfare transfers began figuring in state policy.

9.3 Results of Three Village Surveys in Andhra Pradesh

9.3.1 *Small Peasantry and Agrarian Distress in Andhra Pradesh*

An important change that has occurred in the structure of operational landholdings in the last sixty years (1950–2010) is that the share of small and marginal farmers has grown from 56 percent to 84 percent and the share of the area under them has increased from 18 percent to 49 percent. The share of medium farmers has gone down from 35 percent to 16 percent and the share of their landholding has declined from 54 percent to 45 percent. The share of big farmers has shrunk to less than 1 percent, with landholding of around 6 percent (Table 9.1). Big farmers have divested 31 percent of their land, which is acquired by the marginal and small farmers. Thus, the structure is now overwhelmingly dominated by a class of "small/marginal-producer-farmers," a section that is the most vulnerable. Further, there is a growing tenancy in canal-irrigated areas, as a section of medium farmers who played a crucial role during the green revolution are leaving agriculture as their principal occupation, and have begun leasing out their lands since the late 1980s (Parthasarathy 2002: 45). There has been a steady rise in noncultivating households observed in the state of Andhra Pradesh and elsewhere (Vijay 2012: 37–45, and Chapter 8, this volume). They lease their land mostly to landless laborers and marginal farmers. Tenancy, which usually takes place through oral agreements, is unrecognizable under the law; hence, tenants do not have access to institutional support. While there is no reliable official data on the extent of tenancy, several primary studies have indicated that 70–80 percent of cultivators in coastal Andhra are tenant farmers cultivating more than 50 percent of the land in each village (Vijay 2006; Ramachandran, Rawal, and Swaminathan 2010). The large holdings of more than 50 acres have been subdivided over the past sixty years due to property inheritance customs, reducing most of the large landowners to medium farmers. While there is large-scale diversification of erstwhile large farmers into allied activities such as fisheries, poultry, dairy, orchards, and various nonfarm activities such as finance, commission agents, rice milling, pesticide-fertilizer-seed dealerships, and transport business, their progeny have largely become urbanized. Some of

Table 9.1. Class-wise relative shares of operational holdings in Andhra Pradesh (in %)

	1956–7		1980–1		2005–6	
Farm holding	Holding	Area	Holding	Area	Holding	Area
Marginal	38	8	51	13	62	23
Small	18	10	22	17	22	26
Medium	33	44	25	50	16	45
Large	9	38	2	20	<1	6

Notes: Figures are in acres. The classification is taken from the NSS classification, which defines marginal farm holding as owning or operating less than 2.5 acres (one hectare), a small farm operating 2.5–5 acres (1–2 hectares), medium as 5–20 acres (2–8 hectares), and large farms as operating more than 20 acres (8 hectares).

Source: Various National Sample Surveys (NSS)

those who remained in the village became rich/progressive farmers undertaking investment-intensive farm activities.

Interestingly, increasing marginalization of the agrarian structure in the state, however, did not reflect any decline in growth or productivity. When we consider the last fifty years, there are four broad phases of food grain growth in Andhra Pradesh, which has three major regions, namely, coastal Andhra, Rayalaseema, and Telangana. The compound growth rates of area, production, and yield in the three regions for different sub-periods is given in Table 9.2. We can observe that first, during the period 1973–81, high yielding variety (HYV) rice seeds were introduced, largely in coastal Andhra and to some extent in Telangana. Rayalaseema experienced a shift away from food grains to oil seeds. Then, in the second phase (1981–91), there is a decline of area in the latter two regions of Rayalaseema and Telangana, while growth in production was sustained through yield. Andhra Pradesh was a rice-deficit state during this period (Ramanamurthy and Mishra 2012). In the third phase, 1991–2001, the area stabilized in Telangana. In the fourth phase, 2001–11, the area and yield under food grains, mostly rice, increased. The expansion of *rabi* (winter) crops under borewells contributed to growth in production. The production of food grains in the state grew over 3 percent in the post-liberalization period, which is not only higher than the national growth rate, but higher compared to the pre-liberalization period in the state, including the green revolution period. The yield growth stayed well over 3 percent during 1983–2000, even though there was a slight decline in the last decade. The expansion of borewell irrigation not only led to a rise in production of food grains, but also a diversification of cropping patterns toward non-food grain crops such as cotton in Telangana and groundnuts in Rayalaseema. However, this agricultural growth since 1997 has been accompanied by the phenomenon of unabated suicides among small and marginal farmers (Galab, Revathi, and Reddy 2010), mostly among cotton farmers. Studies have noted a

Table 9.2. Area, production, and yield of food grains in Andhra Pradesh (in %)

Area	Coastal Andhra	Rayala Seema	Telangana	Andhra Pradesh
1973–4 to 1982–3	0.14	−2.68	−0.76	−0.77
1983–4 to 1990–1	−0.44	−7.18	−2.86	−2.43
1991–2 to 2000–1	0.39	0.23	0.35	0.36
2001–2 to 2010–11	0.38	1.53	1.12	1.13
1973–4 to 2010–11	0.07	−1.95	−0.88	−0.67
Production				
1973–4 to 1982–3	3.78	−1.12	3.22	2.86
1983–4 to 1990–1	0.33	−4.15	2.29	0.53
1991–2 to 2000–1	3.22	1.18	4.58	3.55
2001–2 to 2010–11	3.99	3.50	5.87	3.08
1973–4 to 2010–11	2.81	0.52	2.68	2.11
Yield				
1973–4 to 1982–3	3.63	1.60	4.01	3.66
1983–4 to 1990–1	0.77	3.26	5.30	3.04
1991–2 to 2000–1	2.83	0.95	4.21	3.19
2001–2 to 2010–11	3.59	1.94	4.84	1.93
1973–4 to 2010–11	2.74	2.52	3.59	2.73

Source: Estimated from Department of Economics and Statistics data

general crisis of viability of farmers in the state growing food grains as well as non-food grains (Ramanamurthy and Mishra 2012).[4] The state shows a peculiar situation of immiserization that calls for a careful examination.[5] In spite of a preponderance of studies on agrarian distress, which have all singularly focused on neoliberal policies of deflation, decline of institutional credit, increased input costs, decline in public investment, and import competition as responsible for the agrarian distress (Deshpande 2010; Patnaik 2007; Reddy and Mishra 2010), there is not much engagement from an agrarian studies perspective. It is not possible to explain the crisis of the bottom of the agrarian structure without a nuanced analysis of agrarian classes.

9.3.2 *The Three Villages in the Sample*

In light of the above discussion of PCP, we shall now examine the empirical aspects of its reproduction from the study of three villages in rural Andhra Pradesh.[6] A primary survey was conducted in three villages in the state of Andhra Pradesh, namely, Achampet, Pulimaddi, and Kalavapamula, which are located in the three regions of the state, Telangana, Rayalaseema, and coastal Andhra. While this regional classification is based more on the cultural-historical background, each is also different in terms of irrigation by source, ownership structure, and economic development. While Telangana and Rayalaseema are dominated by borewell irrigation, the coastal region has predominantly canal irrigation (in use for more than a century). The basic objective of the survey was to focus on the conditions of petty commodity producers in agriculture. We give a brief description of conditions in the three villages in the following.

Achampet is a smaller Telangana village in Medak district with about 250 households and population of 1,100 people. It has 560 acres of arable land, most of which is irrigated through borewells. The traditional landlords, belonging to the Reddy community, have sold off all their land, which has been acquired by marginal and small farmers belonging to backward castes and *dalits*. The government has given small pieces of land to *dalits* as a land redistribution measure. Thus the village is an example of the rise of a new class of farmers from lower castes as owner-cultivators from "below," most whom were previously farm laborers.

Pulimaddi is a dry village in Kurnool district of the Rayalaseema region of the state. The village has an arable land of 1,500 acres, one-third of which is owned by the Reddy community. The rest is owned by backward caste and *dalit* farmers. The village has a history of factional feuds, which ended two decades ago.[7] The main landlord family that had once dominated the village has suffered a huge economic decline; since then others in the Reddy community began focusing on children's education and diversification of cropping patterns using tractors, which also promoted them to rich farmers from a status of big farmers. The foot soldiers of these warring factions, mostly *dalits*, won some social space in the process and managed to acquire land through wage labor. While a few big landed households still exist, the majority of households have turned into small, semi-medium, and medium farmers. The village grows dry crops such as chickpea under extreme water scarcity. The small peasantry in the village can be said to have emerged from "above" and "below."

Kaluvapamula is an irrigated village in the developed Krishna district. The village has a population of 1,780 with 448 households and 1,900 acres of cultivated land. The village has more than a hundred-year history of irrigation, with capitalist farming after the green revolution. Sugarcane and paddy are the principal crops. The traditionally dominant Kamma farmers own more than 80 percent of the agricultural land. After being the beneficiaries of the green revolution in 1970–85 they began to diversify into nonagricultural professions without giving up their land ownership. They lease their lands to tenants from backward castes and *dalits*, who are either mostly landless or marginal farmers.[8] Thus, the landless in the village have emerged as the petty commodity producing tenant farmers in the village.

9.3.2.1 THE SAMPLE HOUSEHOLDS

Primary information on farm size, crops grown, costs and returns, employment, allied activities, nonagricultural employment and income, credit, and welfare transfers were collected from 454 households, distributed as 145, 154, and 155 households in Achampet, Pulimaddi, and Kalavapamula respectively. The data was collected February through April 2013. The sample, identified on a stratified random basis, covered 35 percent of the households, where 35 percent of caste households as well as all farm size classes were covered. Indian village society consists of a social hierarchy of castes and a class hierarchy. Economic class power is concentrated in upper castes and declines steadily down the caste hierarchy. This approach helps to examine the cross section of class and caste in the study villages.

9.3.2.2 LAND OWNERSHIP

In terms of land ownership, there are some crucial differences among the villages. There is a visible decline of landlordism in the three villages.[9] In Telangana and Rayalaseem villages, namely Achampet and Pulimaddi, which are both largely unirrigated, the traditional dominant landlord castes have sold off a considerable share of their land in the last three decades, which was purchased by *dalits* and backward caste farmers. Now landed households constitute 91 and 69 percent respectively (Table 9.3). In Achampet, the government has distributed completely uncultivable wasteland of 2 acres per head; in Pulimaddi some poor households belonging to *dalits* and backward castes managed to buy small patches of land in the 1980s and 1990s. In the third village, Kaluvapamula, which is in the coastal heartland Krishna district, landed households constitute only 31 percent and landless 69 percent, as traditional landlords have retained almost all their land and they lease it (because agricultural land is very expensive in this village compared to the other two villages). In irrigation-rich villages like Kaluvapamula, one does not hear about any land distributed, as observed for backward dry land villages where common lands or uncultivable wastelands were available for distribution.

Our study sample, when seen in terms of class distribution, has about 74 percent marginal, small, and semi-medium farmers, who constitute the petty commodity producers in agriculture (Table 9.3), characterized not only by the smallness of their landholding, but also their dependence on family labor for cultivation and simultaneously hiring out as wage labor within the agricultural and nonagricultural sectors. The petty commodity producer class constitutes 70–80 percent in each of the villages in the form of owner-cultivators, owner-cultivator-tenants, and landless tenants. In Achampet and Pulimaddi, the majority of farmers are owner-cultivators and owner-tenants, whereas in Kalavapamula, a majority of farmers are landless tenants and owner-tenants.

While depeasantization commonly happens through pauperization, migration, and proletarianization, a process of *peasantization* of landless also occurred in different ways in these three villages. In Achampet it occurred through a combination of land purchases by landless labor and the distribution of

Table 9.3. Agrarian structure in sample villages in Andhra Pradesh

	Achampet	Pulimaddi	Kalavapamula	Total
Land ownership (in %)				
Landed households	91	69	31	67
Landless households	9	31	69	33
Size classification (in %)				
Marginal	58.9	17.9	45.3	42.6
Small	28.8	23.6	12.8	23.1
Semi-medium	10.3	27.4	20.9	18.3
Medium	2.1	25.5	18.6	13.6
Large	0.0	5.7	2.3	2.4
	100	100	100	100
Type of farm household (%)				
Owner cultivators	65.0	33.8	13.2	32
Landless tenants	4.5	7.1	45.5	25
Owner tenants	13.0	27.9	12.4	18
Noncultivator households	13.0	7.8	5.0	11
Landless	4.5	23.4	24.0	14
Total	100	100	100	100

Source: Field study

wasteland by the government. In Pulimaddi, some of the landless households managed to purchase land through wage incomes in the 1980s and 1990s, when land prices were within their reach.[10] As Table 9.3 demonstrates, 30 percent of small and marginal households were landless agricultural labor two generations ago. Such purchases have become less frequent due to soaring land prices in the recent period. In Kalavapamula, which is a canal-irrigated village, *peasantization* occurred through tenancy when the upper caste farmers from medium and large households left farming for better paying nonfarm activities. Except for a few sales to those within their own caste within and outside the village, they continue to own and control the land. *Dalits* and backward castes now cultivate this land as tenants.

The majority of current land owners in all three villages reported that they obtained the land through inheritance (Table 9.4). Some marginal farmers received land during the previous generation. Roughly one-tenth of land was inherited through dowry, although many refused to divulge actual figures. Land ownership through distribution of wasteland by the government is observed in Achampet, which formed a small share, about 3–4 percent in ownership. No such land distribution was seen in the other two villages.

The caste-wise distribution of land in the three villages conveys that while *dalits* and backward castes own land to some extent in the two villages of Achampet and Pulimaddi due to the dilution of the share held by the traditional upper caste (Table 9.5), the Kamma dominance in land ownership continues in the coastal village of Kalavapamula. The land price in the village

Table 9.4. Sources for land acquisition (in %)

	Achampet	Pulimaddi	Kalavapamula
Inherited	76.8	69.4	80.0
Dowry	10.8	10.3	12.2
Purchased	10.3	16.8	7.8
Government	3.1	4.5	0.0
Total	100	100	100

Source: Field study

Table 9.5. Caste–class distribution of households (in %)

	Owner cultivators	Tenants	Non-cultivators	Landless
Achampet				
Upper castes	67	0	27	6
Backward castes	65	19	13	3
Scheduled castes	65	19	10	6
Pulimaddi				
Upper castes	33	33	15	19
Backward castes	32	36	10	22
Scheduled castes	40	41	0	19
Muslims	25	12	13	50
Kalavapamula				
Upper castes	18	53	24	6
Backward castes	20	49	4	27
Scheduled castes	6	79	0	15
Scheduled tribes	10	10	0	80
Muslims	0	50	0	50

Source: Field study

is between Rs. 150,000 and Rs. 250,000 per acre, and has become a major speculative asset. At these prices there is no possibility of *dalits* and backward castes buying land to become even marginal or small farmers in this village. The average holding of land is highly skewed in all the three villages, where upper castes hold two to eight times more land than scheduled caste and backward caste households. The peasant differentiation in the village has a history of more than a century, as canal irrigation has continued since the colonial period. The emergence of the rich peasantry, whose surplus financed professional education of children and even sent a couple of dozen abroad. Compared to the other two villages, the coastal Andhra village of Kalavapamula has the highest inequality in land distribution. Land concentration in the upper caste is stark compared to backward castes and scheduled castes (Table 9.6). Thus, upper castes are likely to urbanize faster and *dalits* and backward castes may stick to the land for a longer time.

Table 9.6. Average size of landholding across castes (in acres)

	Achampet	Pulimaddi	Kalavapamula
Upper castes	5.9	12.19	9.68
Backward castes	2.8	5.0	1.78
Scheduled castes (Dalits)	3.2	4.5	1.38
Scheduled tribes	–	–	1.0
Muslims	–	3.0	0

Source: Field study

9.3.2.3 AGRICULTURAL CREDIT

A decade ago, there was great concern about declining institutional credit to the farm sector (Reddy and Mishra 2010). However, a recent Planning Commission Report suggested a manifold increase in agricultural credit at the aggregate level.[11] Our sample study also confirmed this increase in credit flow. However, the credit flow continues to be extremely unequal. The lion's share of institutional credit goes to medium and large farmers.

We found that there is a substantial coverage of formal credit in all three villages. Further, substantial self-help group (SHG) credit also flows though disbursal of loans to women's groups, which marks a huge change in terms of politics of organizing the poor by the state. The institutional credit reaches farmers in the form of crop loans from banks (commercial and cooperative banks), loans from mortgaging family gold, and SHG loans to women. From the data (Table 9.7), considerable improvement can be seen over the last six years. The total institutional credit in the three villages now stands at 68.6 percent while non-institutional credit is around 37.4 percent. The shares of institutional credit in Achampet, Pulimaddi, and Kalavapamula are 53.5, 73.4, and 68.6 percent respectively. Such inter-village differences are quite possible as far as bank credit is concerned. Why? In all three villages, credit distribution appears to be highly unequal among different classes. In Achampet, marginal and small farmers, who constitute 86 percent, received only 37 percent share in bank credit, 8.5 percent in Pulimaddi, and 24 percent in Kalavapamula. In Kalavapamula, 80 percent of farmers belong to one caste, namely *kammas* and most of them are noncultivating landlords who take the bank loans and re-lend to tenant farmers. Tenant farmers, denied bank credit, are forced to depend on commission agents and private moneylenders for credit.

For marginal and small farmers, SHG loans now constitute 40–50 percent of their total loans. Even though it is not known whether they use them only for agricultural purposes, there is a considerable amount of money being pumped into poor households through these loans. For tenant farmers, these loans matter a lot. Further, for tenant farmers in Kalavapamula, the cooperative

Table 9.7. Institutional and non-institutional credit (in %)

	Institutional	Non-institutional
Pulimaddi		
Landless	88	12
Marginal	34	66
Small	44	56
Semi-medium	58	42
Medium	70	30
Large	76	24
Noncultivators	78	22
Achampet		
Landless	47	53
Marginal	56	44
Small	38	62
Semi-medium	56	44
Medium	57	43
Noncultivators	50	50
Kalavapamula		
Landless	51	49
Marginal	58	42
Small	65	35
Semi-medium	40	60
Medium	48	52
Large	100	0
Noncultivators	43	57

Source: Field study

bank in the village has distributed Joint Liability Group (JLG) loans up to Rs. 5,000 per farmer, which is insufficient to meet cultivation costs. Having no access to bank credit, tenant farmers continue to borrow from moneylenders at high interest rates. The other aspect that dries up the small farmers is the private irrigation cost in dry land areas. In Achampet, we found that of the 250 borewells dug so far, over 200 went dry, leaving debts for the small farmers.

9.3.2.4 FARM COSTS AND RETURNS

To understand the distress of petty producers, an examination of the data on farm costs and returns is necessary. We examined nine principal crops grown in these three villages: paddy, cotton, maize, groundnut, horse gram, sorghum, sunflower, tobacco, black gram, and sugarcane. The returns on paddy cultivation have slightly improved after a five-year run of poor returns during 2007–12.[12] We observed that in the surveyed year, paddy market prices and procurement prices barely covered Cost A1 (which is only paid-out costs) and not Cost A2 (which includes rental cost). Except in Pulimaddi, where superfine rice is grown, in the other two villages, returns over Cost A1 are too meager and returns over Cost A2 are negative.[13] Similarly, for cotton, maize, sunflower, tobacco, *rabi* paddy,[14] and sorghum, the returns over Cost A1 are

Table 9.8. Farm costs and returns in Andhra Pradesh, 2012–2013 (in INR)

Village/Crop	Price (Rs. /Quintal)	Yield (Quintal/acre)	Cost of production		Return per Quintal	
Achampet (Medak District)			Cost A1	Cost A2	Over cost A1	Over Cost A2
Paddy Kharif	1,018	19.4	788	1,149	230	−130
Paddy *Rabi*	1,025	17.8	1,014	1,407	11	−382
Maize	1,189	14.3	803	1,293	385	−103
Pulimaddi (Karnool District)						
Bengal Gram	3,537	6	2,338	3,536	1,199	1
Cotton	3,777	4.7	3,316	4,835	461	1,057
Groundnut	3,426	7.1	2,076	3,081	1,350	345
Sorghum	1,324	15.8	677	1,129	647	195
Paddy Kharif	1,600	20.6	816	1,162	783	437
Sunflower	3,264	5.6	2,888	4,163	375	−899
Tobacco	5,425	7	3,766	4,786	1,658	638
Kalavapamula (Krishna District)						
Paddy Kharif	1,081	24.5	840	1,093	241	−12
Paddy *Rabi*	926	26	857	1,096	68	−169
Sugarcane	2,157	42.7	1,428	1,912	729	245
Black Gram	3,562	4	1,774	1,774	1,787	1,787

Source: Field study

too meager and returns over Cost A2 are negative in most cases (Table 9.8). Thus it is clear from this table that farming in the state has become largely unviable. Except for large farmers in the irrigation-endowed regions who still cultivate on their own, the income from farming of most crops is either barely enough to sustain consumption or even negative depending on the market and production conditions. Higher investment on private irrigation, when coinciding with successive dry spells of monsoons, often prove to be catastrophic, resulting in suicides of petty producers. Such lack of viability of farming in the state has been noted by several other scholars since 1999 (see Ramachandran, Rawal, and Swaminathan 2010; and Ramanamurthy and Mishra 2012).

9.3.2.5 RURAL HOUSEHOLD INCOMES

An examination of the income portfolios of households across the farm-size classes throws light on the nature of the transition. We suggest two important trends, namely petty commodity producers are diversifying their incomes from farm to nonfarm sources; second, they diversify more toward wage employment. In contrast, medium/big farmers also diversify, but proportionately less and more toward business/regular salaried sources, suggesting a growing class differentiation. Farm income, here, constitutes direct income

Table 9.9. Average net agricultural, nonagricultural, and total incomes of rural households (in INR and %)

	Agriculture and Allied Activities		Nonagriculture		Total	
	Rs.	%	Rs.	%	Rs.	%
Landless	18,946	23	62,863	77	81,809	100
Marginal	29,833	33	60,710	67	905,402	100
Small	26,616	13	184,606	87	211,221	100
Semi-medium	49,580	35	92,214	65	141,794	100
Medium	122,113	22	423,497	78	545,609	100
Large	266,558	34	497,845	66	764,403	100
Noncultivators	34,416	19	148,064	81	182,479	100

Source: Field study

from selling agricultural and allied produce, wages, and rental incomes. Farm business income refers to the income that comes directly from selling agricultural products. According to our survey data, there is little differentiation between the class of marginal, small, and semi-medium households (all owning less than 10 acres/4 hectares), while business income is distinctly high for medium and large farmers (above 10 acres and 20 acres respectively). These two broad groups suggest the increasing differentiation between capitalist farmers and petty commodity producers. This is visible in terms of the sources of income of the households. The average annual farm income for marginal, small, and semi-medium farmers is Rs. 7,341, Rs. 12,171, and Rs. 25,135 respectively. For medium and large farmers, it is Rs. 99,353 and Rs. 226,813 respectively, which are four to six times higher than the income earned by the PCP households. Marginal, small, semi-medium, and medium households draw only 33, 13, 35, and 22 percent of their family income from agriculture and allied activities, while large farmers draw 79 percent (Table 9.9). However, the PCP households, landless, marginal, and small, while deriving a major portion of their incomes from nonfarm activities, draw this more from wage labor, while medium farm households are able to diversify into regular employment, trading, and self-employment. Thus, petty producers in agriculture hang on to agriculture but it is not their principal source of income. This means that there is a considerable growth of nonfarm wage employment within and outside the village where petty commodity producers and the landless proletarian class participate, indicating a visible process of proletarianization.

Farm business income. Within the farm income of marginal, small, and semi-medium farmers, we found that 30–40 percent of income is earned through animal husbandry while agricultural wage income constitutes 74–82 percent (Table 9.10). The farm business income for PCP households ranges from

Table 9.10. Average household net income from agricultural activities (in INR and %)

	Farm business income	Animal husbandry	Rental	Wage	Total
In Rs.					
Landless	0	4,667	0	14,278	18,946
Marginal	7,341	10,880	0	11,611	29,833
Small	12,171	6,377	0	8,067	26,616
Semi-medium	25,135	8,998	870	14,578	49,580
Medium	99,353	13,480	0	9,281	122,113
Large	265,813	409	0	337	266,558
Noncultivators	0	2,835	21,861	9,720	34,416
In %					
Landless	0	25	0	75	100
Marginal	25	36	0	39	100
Small	46	24	0	30	100
Semi-medium	51	18	2	29	100
Medium	81	11	0	8	100
Large	99.7	0.2	0.0	0.1	100
Noncultivators	0	8	64	28	100

Source: Field study

Rs. 7,000 to Rs. 25,000, whereas for medium and large farmers it is between Rs. 99,000 and 265,800. However, the productivity differences are not much because the cost factor for big farmers is nullified by higher costs. Their higher incomes are simply due to larger scale of operations in agriculture and diversification into nonagricultural activities. For noncultivating landowning classes, rental income constitutes a major source of income in the village (see Chapter 8 in this volume). Thus, for the petty producer class of tenants, their farm income is negative, while their principal livelihood is derived from wage labor, supplemented by dairy and animal husbandry. The tiny pieces of land, despite generating occasional losses and low incomes, still remain important for the PCP households since land as an asset supports the allied activities such as dairy, in addition to contributing to social prestige for the rural poor.

Nonfarm incomes. There is a marked decline of traditional nonfarm activities (basically traditional services such as pottery, handlooms, laundry, arts, leatherwork, etc.) in the villages,[15] replaced by a range of new nonfarm wage and self-employment activities that require circulation between the village and other villages/town/city. Business opportunities are realized by those who can invest the capital. They are mostly upper castes who are medium and large farmers and some farmers from backward castes of similar class standing. *Dalits* are still unable to access this business sphere. This is the reason that they want to become farmers, for the social prestige it had historically that they never enjoyed, even if it does not give adequate income. They supplement income from animal husbandry and wage labor. Trading, regular

Table 9.11. Net income from nonagricultural activities (in INR and %)

	Trade	Casual	Regular	Self-employment	Other*	Total
In Rs.						
Landless	5,295	7,898	16,465	13,097	20,108	62,863
Marginal	18,967	1,953	5,869	19,190	14,731	60,710
Small	47,865	2,194	8,892	14,460	51,196	124,607
Semi-medium	23,970	3,919	43,598	11,534	9,194	92,215
Medium	53,591	118	15,616	6,625	347,546	423,496
Large	0	0	245,839	252,006	0	497,845
Noncultivators	15,006	6,469	74,132	1,778	50,679	148,064
In %						
Landless	8	13	26	21	32	100
Marginal	31	3	10	32	24	100
Small	37	1	5	8	49	100
Semi-medium	26	4	47	13	10	100
Medium	13	0	4	2	81	100
Large	0	0	25	75	0	100
Noncultivators	10	4	50	1	35	100

Note: * Includes income through traditional activities, wage income from migration, and National Rural Employment Guarantee Act (NREGA).

Source: Field study

employment, profitable self-employment activities, and long-term migration are generally monopolized by the landed and upper caste households, which enable them to diversify their activities.

The average annual net non-farm income for marginal, small, and semi-medium households in the sample villages is Rs. 60,710, Rs. 124,607, and Rs. 92,215 respectively (Table 9.11). For medium and large households, it is Rs. 423,496 and Rs. 497,845 respectively, four times that of PCP households. Therefore, nonfarm incomes are clearly differentiating the rural residents. Overall, there is a huge increase in nonfarm income sources and a decline of farm incomes for petty producers. Even in farm income it is the allied activities that fetch them a little more money than farming, where much of the surplus is squeezed out by moneylenders, traders, landlords, and input suppliers.

9.3.2.6 SOCIAL WELFARE TRANSFER

Rural distress became an important political agenda of the Congress-led United Progressive Alliance (UPA) in parliamentary elections in 2004, much to the discredit of the Bharatiya Janata Party (BJP)-led National Democratic Alliance. Backed by several NGO activists, the UPA government mulled several programs for different population groups, such as farmers, agricultural labor, rural women, students, the old, and the handicapped. The civil society groups campaigned for action addressing falling employment levels and rising malnutrition after the directives of the Supreme Court on rotting stocks in

Table 9.12. Average welfare transfers to households per annum in three villages (in INR)

	Achampet	Pulimaddi	Kalavapamula	Average
Public distribution system (PDS)	4,701	4,881	4,408	4,663
Pensions	2,612	4,306	3,088	3,335
Integrated child development services (ICDS)	587	924	748	753
Midday meals	2,599	2,693	2,131	2,474
Scholarships	1,600	1,875	1,600	1,692
Mahatma Gandhi national rural employment guarantee scheme (MGNREGS)	5,739	9,975	3,315	6,343
Total	17,838	24,654	15,290	19,261

Source: Field study

godowns (warehouses) of the Food Corporation of India.[16] The political society's conditions are thus addressed by a slew of measures, hailed as pro-poor by commentators. Further, an innovation is that some of these measures became constitutional rights and a permanent feature of governance. There are more than a dozen actively implemented schemes that involve in-kind and cash transfers in Andhra Pradesh. These include the Public Distribution System (PDS), Mahatma Gandhi Rural Employment Guarantee Act (MNREGA or NREGA), old-age pensions, financial help for housing, student scholarships, state medical insurance (*Arogyashree*), midday meal, Integrated Child Development Services (ICDS), National Insurance and Family Benefit Scheme, 3 percent interest farm loans, and agricultural loan waivers.[17] These schemes now form a major part of the electoral discourse through which the parties seek to take credit for introducing social security to the poor. We have collected data on six major welfare schemes covering in-kind and cash transfers received by the households and computed the total average annual welfare transfer to households. We found that almost all the poor (even well-off) households in the villages are covered under major schemes. The PDS and NREGA programs are implemented relatively better in South India in general.

The average annual welfare transfer for the three villages is given in Table 9.12. MNREGA tops the list of welfare transfers, conferring an average annual wage income of Rs. 6,343, followed by Rs. 4,663 for PDS, and Rs. 3,335 for pensions. Midday meals and scholarships rank next. The total average annual welfare transfer for households in the three villages is about Rs. 19,261, which constitutes 24 percent and 15 percent of the annual income of the landless and marginal farmers respectively. Thus welfare transfer constitutes an important share, if not the principal share, of petty commodity producers' subsistence reproduction. The extent of legitimacy that the state can derive is telling by the fact that UPA assumed power for a second time.

Further, the legitimacy that it gained is enormous, as the causes of distress are no longer blamed on the state, but are relegated to the personal sphere of stakeholders who are mostly farmers. The distress of petty commodity producers is addressed through the politics of welfare. However, any reduction of welfare measures could have expected outcomes.

9.4 Concluding Remarks

The agrarian question in a postcolonial context presents a complex trajectory in capitalist transformation. The small peasantry, in spite of adverse conditions of production, may persist in farming. Suicides, pauperization, unemployment, and the precariousness of the class are all the outcomes of conditions of PCP. PCP can coexist with a capitalist system, as long as PCP households manage to produce the market surplus needed for the system. Until alternative opportunities to migrate arise, these households are compelled to hang on. Diversification into nonfarm activities is inevitable in the course of capitalist development; it is already happening with gradually increasing participation in nonfarm activities, widening the income portfolio. Our evidence suggests that for the majority of farm households, who are essentially petty commodity producers, the share of agricultural income is less than what is earned from nonfarm sources. Their diversification is also proceeding toward more wage labor. Yet, the ambiguous family labor-based petty commodity production can persist until the final resolution. Until then, one still finds a preference for land ownership, with the added reasons of speculative holding and reproduction of social status. The precariousness that stems from lower returns, high risk, lack of access to institutional support in credit, technology, and marketing is matched somewhere by rising wages and alternative opportunities. Meanwhile, the precariousness is addressed by the state in terms of welfare transfers. Thus, diverse processes enable the existence of petty commodity producer households. As these changes take place, land is becoming less central to their lives.

Notes

1. The issue of agrarian transition and mode of production in India was debated vigorously in various issues of *Economic and Political Weekly* and *Social Scientist* in the 1970s. Characterization of the mode of production in Indian agriculture got pegged between two positions, that is, dominantly precapitalist and semi-feudal, with emerging capitalist relations at one end (Bhadhuri 1973; Patnaik 1973) and dominantly capitalist relations with persisting subtle semi-feudal relations (Desai

1984; Rudra 1978). There were other equally compelling views that capitalist relations had already entered under colonial rule (Banaji 1975; Frank 1996), while some others held the opposite view, that colonial rule introduced feudal relations and blunted growth of production forces by unequal exchange and drain of surplus against formation of potential capitalist relations (Prasad 1987). The debate, as stated by Alice Thorner (1982), remained inconclusive from the diverse positions taken by the Marxist political economists and the practitioners.

2. "Governmentality" is a Foucauldian term that refers to a process of dividing populations into groups that are amenable to statistical measurement, using census and other methods, and fixing governmental programs to the targeted groups, thereby gaining biopower over the groups. The participation of members in the program gives the state a positive power over the population besides the sovereign power it already has. This power over groups is a discursive power that keeps them fractured (Foucault 2007).
3. Ashok Mitra (1977) argued that the kulak class prevails over the public policy by influencing agricultural price policy, which caused the terms of trade to shift in favor of agriculture, affecting industrial accumulation.
4. Andhra Pradesh stands second in India beside other states such as Maharashtra, Karnataka, UP, Punjab, Haryana, and Kerala. Over two and half lakh (250,000) farmers have committed suicide between 1995 and 2011. Most of the victims are small and marginal farmers and many belong to backward classes and scheduled castes (Sainath 2012). Several scholars who have analyzed the farmer suicides contend that these suicides are the legacy of the economic reforms (Deshpande 2010; Mishra 2009; Nair and Menon 2009; Parthasarathy 2002; Revathi et al. 2009; Singh 2010).
5. The contemporary agrarian crisis that began in the mid-1990s is much more than a mere outcome of neoliberal policies adopted since 1991 as contemplated by some scholars (Reddy and Mishra 2010). But the neoliberal reforms have certainly accentuated the crisis. Ecological strain resulting from intensive monocropping and extensive use of groundwater has imposed externalities (Reddy et al. 2001).
6. The study was done in 2012–13, during which Telangana state was not yet separated from Andhra Pradesh.
7. The history of factional fights in Rayalaseema villages dates back to pre-British times. The village chieftains who were the warlords in the Vijayanagar Empire passed the unfettered legacy into modern times. Thomas Munroe, the Collector of Bellary, had suppressed these local warlords, called "poligars," ruthlessly. Still, some of them survived as factionists who dominate village politics. While much of the factionism has died down, it survives in a small belt in Karnool-Cuddapah districts. A last active phase saw the induction of these into the two bourgeois parties in the 1980s. Some of them graduated into big time politicians, one of them became even became a chief minister, to name Rajasekhar Reddy (see Balagopal 2005).
8. Kamma farmers managed low wage rates by shifting to piece rates in the 1980s, but were unable to control them due to the 1989 Scheduled Caste and Scheduled Tribes Atrocities Prohibition Act. Then leasing became a better option, where risk could be transferred to the tenant and the moral hazard problem avoided.

9. Landlordism here refers to dominance of big landlords over agricultural labor toward submission for economic and other services, not necessarily feudal in nature.
10. The actual stories of land purchase are tortuous ones, where *dalit* families remained joint families, invested the surplus from the farming to acquire further land, while wage incomes are used for maintaining consumption. It is difficult to differentiate the source of savings, which can range from underconsumption to rising real wages to state policy of public distribution.
11. Report of Working Group on Institutional Finance, Cooperatives and Risk Management for twelfth Five-Year Plan (2012–17), Planning Commission (p. 13).
12. In 2013 the situation had slightly improved as both procurement prices as well as market prices have gone up.
13. Commission on Agricultural Costs and Prices provides six categories of prices, according to different cost concepts: (i) Cost A1 (all fixed and variable paid out costs excluding rent), (ii) Cost A2 (Cost A1+rent), (iii) Cost A2+ (FL), (iv) Cost B1 (Cost A1 + Interest on value of owned fixed assets excluding land), (v) Cost B2 (Cost A2+imputed rent on own land), (vi) Cost C1 (Cost B1+ imputed family labor); (vii) Cost C2 (Cost B2+imputed family labor); (viii) Cost C2* (Cost C2 using Minimum Wages if they are higher than market wages); (ix) Cost C3 (cost C2*+10 percent managerial input over C2*). Cost A2+FL that includes rent as well as family labor is relevant cost for tenant farmer, Cost B1 and B2 is relevant for small and marginal farmers, and cost C1 and C3 are relevant costs for supervisory landlord farmers. The respective cost of cultivation is converted into cost of production per quintal by applying the yield.
14. *Rabi* refers to the winter crop, which spans from January to April, while the monsoon crop during June–November is called *Kharif*.
15. Activities such as pot-making, ironwork, carpentry, washing, and weaving are mostly non-existent. There is a rise of new nonfarm sector activities such as finance, commission agents, pesticide/seed dealers, trading, hotel, petty shops, grocers, vegetable vending, cloth merchants, food vending, transport, photographers, motor mechanics, television mechanics, phone recharging, electricians, winding works, plumbers, masons, construction workers, television cable operators, drinking water suppliers, photocopying centers, stationary shops, private school teachers, bus/truck/van/auto/harvester/poclainer/tractor drivers, MNREGA mate workers, technical assistants, SHG organizers, nurses, ANM (auxiliary nurse midwife) workers, Anganwadi workers, mid-day meal workers, tailors, liquor shop, medical shops, and so on, which are the new job activities emerging in the village and outside.
16. See "Supreme Court Orders on the Right to Food: A Tool for Action," available at: <www.sccommissioners.org/CourtOrders/tool_for_action.pdf> (accessed September 12, 2016).
17. The various social welfare schemes and programs in the state are as follows: Under Public Distribution System, 20 kg rice is given to below poverty line (BPL) families at Rs.1 per kg; Congress government in the state also opened a fee reimbursement scheme for the poor in all streams of higher education; *Arogyashree* is a medical

reimbursement scheme under which every member of designated below-poverty households is entitled to private medical care (the program later became a public–private health insurance model); National Family Health Benefit Scheme is another life insurance scheme for the poor whereby Rs. 50,000 is paid outright in every accidental death of the rural poor; the Integrated Child Development Services provides a nutrition supplement to infants up to five years and pregnant mothers, providing an egg every day to each child; the mid-day meal scheme provides lunch to school-going children; 1.3 percent loan scheme is an interest-subsidized loan scheme against the usual bank interest of 7 percent; the SHG program is one of the biggest in the country where 13 million women are covered under microcredit (each receiving a loan ranging from Rs. 5,000 to Rs. 25,000 depending on the group life); MNREGA (Mahatma Gandhi Rural Employment Guarantee Act), launched in 2005, gives an average employment of sixty-seven days to one member of rural families in Andhra Pradesh.

References

Akram-Lodhi, H. and Kay, C. 2010. "Surveying the Agrarian Question" (Part I and II). *Journal of Peasant Studies*, 37(1–2): 177–202, 255–84.

Bagchi, A. 2008. *Perilous Passage: Mankind and the Global Ascendancy of Capital*. Plymouth: Rowman and Littlefield.

Banaji, J. 1975. "India and the Colonial Mode of Production." *Economic and Political Weekly*, 10(49): 1887–92.

Bernstein, H. 2010. *Class Dynamics of Agrarian Change*. Sterling, VA: Kumarian Press.

Bhaduri, A. 1973. "A Study in Agricultural Backwardness under Semi-Feudalism." *Economic Journal*, 83(329): 120–37.

Byres, T. J. 1977. "Agrarian Transition and Agrarian Question." *Journal of Peasant Studies*, 4(3): 258–74.

Chatterjee, P. 2004. *Politics of the Governed: Reflection on Popular Politics in Most of the World*. New Delhi: Permanent Black.

Chatterjee, P. 2008. "Democracy and Economic Transformation in India." *Economic and Political Weekly*, 43(16): 53–62.

Desai, A. R. 1984. *India's Path of Development: A Marxist Approach*. Mumbai: Popular Prakasan.

Deshpande, R. S. 2010. "Agrarian Transition and Farmers' Distress in Karnataka." In *Agrarian Crisis in India*, edited by D. N. Reddy and S. Mishra, New Delhi: Oxford, pp. 199–229.

Foucault, M. 2007. *Security, Territory, Population: Lectures at the Collège de France 1977–1978*, edited by M. Senellart, translated by G. Burchell. London: Palgrave Macmillan.

Frank, A. G. 1973. "On Feudal 'Modes,' Models and Methods of Escaping Capitalist Reality." *Economic and Political Weekly*, 8(1–26): 36–7.

Frank, A. G. 1996. "India in the World Economy, 1400–1750." *Economic and Political Weekly*, 31(30): PE50–PE64.

Galab, S., Revathi, E., and Reddy, R.P. 2010. "Farmer's Suicides and an Unfolding Agrarian Crisis in Andhra Pradesh." In *Agrarian Crisis in India*, edited by N. Reddy and S. Mishra, New Delhi: Oxford University Press, pp. 164–98.

Gramsci, A. 1996. *Selections from the Prison Notebooks*. New Delhi: Orient Blackswan.

Harriss-White, B. 2006. "Poverty and Capitalism." *Economic and Political Weekly*, 41(13): 1241–6.

Harriss-White, B. 2012. "Capitalism and the Common Man: Peasants and Petty Commodity Production in Africa and South Asia." *Agrarian South: Journal of Political Economy*, 1(2): 109–60.

Kautsky, K. [1889] 1988. *The Agrarian Question*. London: Zwan Publications.

Lerche, J. 2013. "The Agrarian Question in Neoliberal India: Agrarian Transition Bypassed?" *Journal of Agrarian Change*, 13(3): 382–404.

Marx, K. 1954. *Das Capital*, vol.1, part III. Moscow: Progress Publishers.

Mitra, A. 1977. *Terms of Trade and Class Relations: An Essay in Political Economy*, London: Routledge.

Mishra, S. 2009. "Agrarian Distress and Farmers' Suicides in Maharashtra." In *Agrarian Crisis in India*, edited by N. Reddy, D. Reddy, and S. Mishra. New Delhi: Oxford University Press, pp. 126–63.

Nair, K. N. and Menon, V. 2009. "Distress, Debt, and Suicides among Farmer Households: Findings from Village Sudies in Kerala." In *Agrarian Crisis in India*, edited by N. Reddy, D. Reddy, and S. Mishra. New Delhi: Oxford University Press, pp. 230–60.

Parthasarathy, B. 2002. "Changing Agrarian Structure and Nature of Transition in Post-Green Revolution Period." In *Development of Andhra Pradesh: 1956–2001. A Study of Regional Disparities*, edited by Y. Krishna Rao and S. Subrahmanyam. Hyderabad: N.R.R Research Centre, pp. 25–59.

Patnaik, U. 1973, "Central Committee Resolutions on Certain Agrarian Issues." *Social Scientist*, 13: 65.

Patnaik, U. 2007. *Republic of Hunger and Other Essays*. New Delhi: Tulika Publications.

Prasad, P. H. 1987. "Towards a Theory of Transformation of Semi-Feudal Agriculture." *Economic and Political Weekly*, 22(31): 1287–90.

Ramachandran, V. K., Rawal, V., and Swaminathan, M. 2010. *Socio-Economic Surveys of Three Villages in Andhra Pradesh: a Study of Agrarian Relations*. New Delhi: Tulika Books.

Ramanamurthy, R. V. and Mishra, R. 2012. "Unremunerative Pricing of Paddy: A Case of Andhra Pradesh." Study No. 38, DRG Study Series, Reserve Bank of India. Available at: <https://rbi.org.in/scripts/PublicationsView.aspx?id=14604> (accessed September 12, 2016).

Reddy, N. and Mishra, S. (eds) 2010. *Agrarian Crisis in India*. New Delhi: Oxford University Press.

Rudra, A. 1978. "Class Relations in Indian Agriculture–III." *Economic and Political Weekly*, 13(24): 994–1004.

Sainath, P. 2012. "Farm Suicides Rise in Maharashtra, State Still Leads the List." *The Hindu*, July 3.

Sanyal, K. 2007. *Rethinking Capitalist Development: Primitive Accumulation, Governmentality, and Post-Colonial Capitalism*. London: Routledge.

Singh, K. 2010. "Agrarian Crisis in Punjab: High Indebtedness, Low Returns, and Farmers' Suicides." In *Agrarian Crisis in India*, edited by N. Reddy, D. Reddy, and S. Mishra. New Delhi: Oxford University Press, pp. 261–80.

Thorner, A. 1982. "Semi-Feudalism or Capitalism: Contemporary Debate on Classes and Modes of Production in India." *Economic and Political Weekly*, 17(51): 1993–2009.

Vijay, R. 2006. "Agrarian Structure and Agrarian Relations: Illustrations from Village Studies." *Studies in Humanities and Social Sciences*, 2(1): 22–56.

Vijay, R. 2012. "Structural Retrogression and Rise of New Landlords in Indian Agriculture: An Empirical Exercise." *Economic and Political Weekly*, 46(5): 37–45.

10

Land Reform in Kerala and West Bengal

Two Stories of Left Reformism and Development

Anirban Dasgupta

10.1 Introduction

This chapter examines the extent to which land reforms created the conditions of capital accumulation in West Bengal and Kerala, the two states in postcolonial India that attempted to bring about major land-related institutional change. The idea that land reforms can stimulate a process of relatively egalitarian capitalist industrialization by increasing the productivity of agriculture and equalizing the distribution of primary productive assets, has been discussed frequently in the development literature (Griffin, Khan, and Ickowitz 2002; Kay 2002). In India, West Bengal and Kerala are the only states where such a process could have materialized in principle, as the initial condition of significant land reforms was satisfied in these cases. At a time when land reforms are no longer high on the development agenda, our study is a reminder of the only attempts at serious redistribution of assets in independent India and how that has affected the course of accumulation in their local contexts.

Land reforms in both Kerala and West Bengal were brought about through progressive legislative acts that were the results of long-standing peasant movements led by the left political forces. Although both the nature and impact of these interventions have been different for the two states, they share common ground in attempting an alternative model of development in the context of postcolonial India. The claim here is not that the left governments had a clear framework of capital accumulation for their states before them when they brought about the land reform legislation in 1957 (reformulated in 1969 for Kerala) and 1977. Political literature of the left parties indicates that the primary driving force behind land reforms was to provide the peasantry with some relief from the exploitative system of agricultural production they were part of. Nevertheless, being unique for undertaking serious land-related interventions, it would not have been unreasonable to expect the two states to follow a distinct development trajectory in comparison to the other parts of India where land reforms were mostly limited to the abolition of the colonial intermediaries in the land revenue system.[1]

Given the above background, the specific aim of this chapter is to compare the development paths of these two states from the vantage point of their agricultural sector. In doing so, we want to understand if the process of capitalist accumulation was strengthened or made more equitable in these two states with the stimulus of land reform. We start by asking, how did changes in agrarian relations brought about by the land reform legislation affect agricultural production as well as the different classes of the agricultural population? Further, in what ways did the outcome of the reforms in agriculture impact the broader dynamics of capitalist accumulation in these states? Although serious land reform is no longer considered a policy option or a political imperative in the Indian context, this comparative study illustrates how important local institutional factors can be in mediating the relationship between land tenure systems, agricultural growth, and the extent of capitalist development in the overall economy. Apart from the historical lessons from these specific experiences for the Indian left, this chapter seeks to clarify the understanding of the role of agriculture, and thus the place for land, in development in India.

We will engage with two distinct theoretical literatures in analyzing the cases in question: the political economy literature of the Byres–Bernstein school that has consistently tried to adapt the classical Marxian concepts to contemporary realities of agrarian change (Bernstein 2004; Byres 1991) and the more mainstream, policy-oriented literature on agriculture-led development credited to John Mellor and others (Dunham 1991; Mellor 1995). A political economy analysis is essential to understand the nature of accumulation that had evolved in agriculture, starting from the land reforms and their implications for the different agrarian classes. The framework of farm–nonfarm linkages popularized by Mellor is useful to evaluate how the land reform programs affected changes in agricultural production and how this in turn impacted the nature and pattern of growth across the rest of the economy.[2]

While there is a rich literature on land reform for both Kerala and West Bengal individually, the present study differs significantly in its coverage as well as conceptualization. First, it offers a comparative analysis of the two cases and thus hopes to tease out important differences between the two states in terms of the design, implementation, and impact of the reforms as well as on the wider development pattern of the two states.[3] Second, it uses the insights from the comparative analysis to understand whether the demand for land reform has any relevance in contemporary India in making capitalist accumulation more robust as well as equitable. This can help clarify the often muddled discourse on inclusive growth in India from an explicit political economy position. Lastly, this chapter hopes to contribute indirectly to the literature on the politics of the Indian left as a comparative analysis of two major examples of left reformism attempted in postcolonial India. A critical rethinking of left interventions in the past becomes even more relevant today when the conventional left political formations have been weakened significantly in their erstwhile strongholds of Kerala and West Bengal.

As for any other subnational analysis for India that involves policies implemented at the state level, one has to remember that constituent states in the Indian republic do not function as independent economic entities. While agriculture and land are state subjects in the Indian constitution and state governments do play a very important role in shaping the agricultural sector, agricultural production takes place in a wider macroeconomic context including the trade regime, which clearly constitutes a set of national-level variables. Having said that, it remains a testament to India's vibrant democracy that different so-called "models" of development followed by different state governments (often from different political persuasions) are hotly debated in Indian public life.[4] There is also clear evidence to show that state-level policies and emphases significantly affect the real development experiences of ordinary citizens in various ways (Kohli 1987).

The rest of the chapter is structured as follows: Section 10.2 reviews the theoretical literature in order to present a conceptual framework for understanding how the agrarian reality in an economy shapes overall development and what role land reforms can play in facilitating this function. Section 10.3 introduces the Kerala and West Bengal land reforms and provides the basic data regarding implementation of reform measures. Section 10.4 analyzes the post-land reform development trajectories in the two states. Section 10.5 attempts to evaluate the extent to which land reforms were able to bring about a distinctive pattern of development by resolving the agrarian question, especially in its accumulation dimension. The concluding section discusses the limits of left reformism in the Indian context by highlighting how land reforms were not sufficient to stimulate capitalist accumulation in these two states. It argues that in both cases there was a disarticulation of the accumulation dimension of the agrarian question, a conclusion that resonates with developments in contemporary Indian capitalism.

10.2 Land and Capitalist Development: Some Theoretical Issues

The role of agriculture in the creation of value in the economy has been a persistent theme in political economy since the writings of the Physiocrats in the middle of the eighteenth century. Marx and Marxian writers following him such as Engels, Lenin, and Kautsky, wrote extensively on the so-called "agrarian question," which relates broadly to the advent of capitalist production in agriculture (and the changes in class structure that it brings about in the countryside) as a precursor to the establishment of industrial capitalism.

For our purposes in this chapter, we will begin by discussing the contemporary reading of the agrarian question and its relevance in today's globalized world. Akram-Lodhi and Kay (2010) present a useful characterization of the different "problematics of the agrarian question" based on writings of the distinguished agrarian scholars Terry Byres and Henry Bernstein. According to them, the three major problematics are those related to accumulation, production, and politics. The central focus of the accumulation problematic is the role agriculture plays in facilitating industrial accumulation by producing surplus in the form of food, nonfood commodities, as well as labor. The production problematic is most clearly derived from the classic version of the agrarian question (following Kautsky and Lenin) and concerns itself with the development of capitalism as the dominant production system in agriculture (from feudalism or other precapitalist forms) and analyzes the factors that determine the speed at which this capitalist transition takes place in a given context.[5] The politics problematic seeks to understand the changes in the rural class configuration that evolves through the class conflict accompanying agrarian transition.

It is useful to pause and reflect on how forms of state-mandated institutional changes in land relations, or what are called "land reforms" more generally, may affect the resolution of the agrarian question in terms of accumulation, production, or politics. Most progressive land reform legislation, whether a result of sustained political struggle or that of administrative exigency, would have a strong anti-landlord bias geared toward weakening the hold of traditional landlords on agricultural production including their domination in the credit, labor, and commodity markets. To the extent such reforms are implemented with seriousness, they will clearly change the class configuration in the agrarian landscape along with a shift in the dominant production systems in agriculture. Depending on the specifics of the land reforms undertaken and the historical context, the exact nature of agrarian transition will differ. For example, in Punjab, the postindependence land reform in the early 1950s was limited to the abolition of the revenue intermediaries that existed in the colonial system. This led to a system of production dominated by large-scale capitalist farmers who were invested in agriculture as a profitable enterprise to the detriment of the smaller tenants or self-cultivators (Gill 1989). On the contrary, in the case of both South Korea and Taiwan in the second half of the last century, where land reforms came in the form of conferring long-term cultivation rights to the tenants and eventually transferring ownership rights, a system of small peasants using family labor emerged as the dominant form of production in the process completely eliminating the erstwhile landlord class (a significant section of these being Japanese) (Griffin, Khan, and Ickowitz 2002). Needless to say, the vast experience of actually implemented land reforms across the world over the twentieth century can provide many other examples of agrarian transition.

The key question for us in this chapter is to understand how the accumulation dimension of the agrarian question is impacted by the advent of land reforms. A slightly different way of raising the same question is to ask: How does land reform change the trajectory of overall development in an economy through its impact on the agrarian production process? At the outset it is important to recognize that the accumulation process that we are interested in analyzing is not detached from the dimensions of production and politics; in fact it is only through a change in these latter dimensions that land reforms will have an effect on the larger economy through accumulation.

A schematic representation of a so-called "virtuous path" that follows from the logic of redistributive land reform may be the following (see Griffin, Khan, and Ickowitz 2002; Storm 2015: 683): confiscatory land reform leads to the weakening (or possible elimination) of the landlord class and gives rise to a dynamic system of peasant-based agriculture, given a system of providing complementary inputs such as credit, technology access, and so on is in place. This change in the agrarian structure results in growth and surplus generation in agriculture. A part of this surplus is transferred out of agriculture for industrial accumulation (both in rural and urban areas) through either state-directed taxation measures or some mode of incentivizing investment outside agriculture. As a result, the nonagricultural sector (industry and services) grows at a higher rate than agriculture (additionally supported by the demand generated from a more prosperous agriculture) and labor moves out of agriculture along a Lewisian path, completing the process of structural transformation. Of course, the entire process can, in principle, happen without land reforms as well but if we have land reform there is the added advantage of starting with a more equitable distribution of resources that has the potential of generating a more broad-based or "inclusive" growth process.

John Mellor and his associates had the above schema in mind when they advocated an agriculture-led growth strategy (Mellor 1995). There were two caveats, though. First, as mentioned before, land reform per se was not essential. As long as an economy could generate significant growth in agriculture (even through technological means *à la* green revolution instead of institutional changes), the process can work out. Growth thus generated may not be particularly shared or inclusive but since its magnitude can potentially be substantial, trickle-down effects can make a significant dent in poverty. Second, there were crucial linkages between agriculture and the nonagricultural sector that needed to be strong for this strategy to be successful. This included demand-side consumption linkages as well as supply-side production linkages both upstream and downstream.

It is worth repeating that land reforms that were instituted in Kerala and West Bengal were not policy interventions with explicit development objectives in a conventional sense but were natural steps taken in culmination of long-standing demands and struggles of the small peasants and landless laborers who constituted an important support base of the left political movements led by the mainstream communist parties. Rather than a longer-term revolutionary move to further the cause of socialism, land reform was seen as a means to provide short-term "relief" to the exploited peasantry in what was characterized as a semi-feudal agrarian system (Bhaduri 1973). Left leaders have repeatedly said that given the constraints of running a state government within the larger national governance apparatus dominated by the interests of big capital, no fundamental change in the economic or political system was possible. The objective instead is to "provide relief to the people and strengthen the democratic movements" (Karat 2007).

However, if one looks carefully, it is not difficult to detect an alternative vision of development among enthusiasts of the left governments at the initial stages. Take the case of Ashok Mitra, distinguished left-wing economist and intellectual as well as the finance minister of the first left front government from 1977. Introducing a set of articles in the journal *Social Scientist* in early 1978 (Mitra 1978: 7) on the opportunities ahead of the new government, he says:

> The reconstruction albeit partial, of the income structure in the villages [brought about by land reforms] should lead to an enlargement in the demand for goods and services, which could immediately give rise to the possibility of a planned expansion of cottage and small industries. This is an area where the Left Front government would have to apply itself with drive, skill and imagination. The prospect of a linkage between the demand newly emerging in the countryside and some urban industries, supplying either inputs for agriculture or finished commodities for rural consumption, would be equally relevant. True, this series of additional demands by themselves could have but a marginal impact on the overall industrial situation. But even were there to be a slight shift in demand for the products of the organised industrial sector, to that extent a few thousands [sic] could be additionally absorbed in jobs. This additional employment might in turn lead to a further strengthening of demand across-the-board, thereby contributing, howsoever modestly, to general industrial growth.

Ashok Mitra clearly indicates a framework of an industrial accumulation process kick-started through the restructuring in agrarian relations brought about by land reforms. In spite of important differences in the details, in its basic tenets this "model" is not very dissimilar to the experience of Taiwanese development through agriculture–industry linkages discussed in Mao and Schive (1995).[6] Chandra (1978a, 1978b) charts the possibilities of a new form of industrialization in West Bengal not dependent on investment from big capital, whether foreign or domestic (equally applicable for Kerala with minor changes) with the advent of the new government in 1977. The strategy presented is that of small-scale, labor-intensive industrialization including rural industrialization that works in tandem with the interventions in land tenure so as to exploit the possibilities of the linkages between agricultural growth and industrial development to the fullest.[7]

This discussion highlighted the key role of the agricultural sector in enabling industrial accumulation and the specific function of land reform in this context in generating a more equitable growth outcome. With this theoretical understanding we move on to the details of the land reform programs and their implementation in Kerala and West Bengal.

10.3 Overview of the Land Reforms

In Kerala, the process of land reform started with the election of the first Communist Party-led state government in 1957. One of the first policy moves of this government was passage of the Kerala Agrarian Relations Bill (KARB) in 1959. The KARB had relatively strict provisions of handing over ownership rights to the actual tillers.[8] There was also a land ceiling fixed at 15 acres of double-cropped paddy (or equivalent), going up to 25 acres for larger families. The enactment of KARB led to political chaos with steep opposition from all the non-left political parties including the Congress Party, which represented the interest of the landed class. Although other issues against the Communist government were raised in the sustained movement by the opposition, it was essentially focused on the anti-land ownership provisions of the KARB. Finally this movement led to the dismissal of the state government by the central government in 1959, which invoked the controversial Article 356 of the constitution (Raj and Tharakan 1983). The newly elected coalition government by the Congress Party introduced the Kerala Agrarian Relations Act (KARA) in 1960 with some minor changes to the earlier bill. However, KARA could not be put into operation due to other legal obstacles and was replaced with the Kerala Land Reforms Act in 1963. Subsequent versions of the bill diluted the progressive provisions of the original bill.

The newly elected left government passed the Kerala Land Reforms Amended Act (KLRAA) in 1969, which was ultimately implemented in 1970. A ceiling limit of 20 acres for a family of five was declared in the act. However, rubber, tea, and coffee plantations, private forest, nonagricultural land, and land owned by religious, charitable, and public educational institutions were excluded from the act. The act very importantly also made provisions for abolishing tenancy. According to KLRAA, from the beginning of 1970, all land under tenancy automatically accrued to the state as the de jure owner while the de facto rights of such land were given to the sitting tenants (Raj and Tharakan 1983). By the late 1970s, most tenants were issued purchase certificates by the state government on verifying their status as genuine lessees according to KLRAA. Lastly, there was also an important provision for ownership rights of homestead plots given to the rural poor at subsidized rates with the government paying half the purchase amount on behalf of the beneficiaries.

In terms of land redistribution, by 2007, about 31,043 hectares of land had been distributed among 165,149 individual beneficiaries accounting for between 1–2 percent of the net cropped area, with merely 0.19 hectares per beneficiary (Agravala 2010). The land to the tiller program was much more successful, with almost 2.8 million tenants being conferred ownership rights on roughly 600,000 hectares of land (Appu 1996, cited in Ramakumar 2006). Up to 1996, 528,000 households also received ownership of homestead plots (Ramakumar 2006).

The land reform program in West Bengal proceeded in two phases. The first phase took place between 1967 and 1970 under two United Front governments, which was a coalition of different left parties including the CPI (M) and a breakaway group of the Congress Party called the Bangla Congress. The CPI (M) controlled the key ministry of land resources and played an active role in land redistribution during this period. The ministry also passed an amendment to the previous land reform clause with stricter provisions. Instead of 25 acres per person, the ceiling was revised to 12.3 acres of irrigated land and 17.3 acres of unirrigated land for households of up to 5 persons (Sengupta and Gazdar 1997).[9] The main focus of land reform in this phase was appropriation of ceiling surplus land and its distribution among poor and landless peasants. The left parties, especially the CPI (M) and the CPI (ML) (the Naxalite or the ultra-left faction that was not a part of the government),[10] were active in identifying ceiling surplus land including benami land (land held under a bogus name or name of relatives to avoid ceiling provisions) held by the richer peasants and landlords as well as putting pressure on the local government functionaries to use the provisions of land ceiling legislation and appropriate the ceiling surplus land.

After the fall of the United Front government there was a lull in the implementation of land reform. The second phase started in 1977, when the CPI (M)-led left front government (LFG) was voted to power. The main facets of the LFG land reforms included a) land redistribution, and b) enforcing tenancy rights and decreasing the rental share for sharecroppers through the registration of tenants in the Operation Barga program.[11] The registration was carried out initially though state-wide camps organized by the state land reform department where share-tenants or *bargadars* were encouraged to register themselves. This bureaucratic exercise was actively supported by the CPI (M) through their extensive networks in rural West Bengal.

Although roughly 205,444 hectares of land was supposed to have been redistributed among 2,102,529 individuals between 1977 and 2011 (Government of West Bengal 2012), it constitutes only 4 percent of the net cropped area with per capita land redistributed being a meager 0.1 hectare. The tenancy reform program was relatively more successful than land redistribution, with about 1.5 million sharecroppers constituting approximately 75 percent of the total number recorded by the state. There is mixed evidence on the extent to which tenants managed to retain their crop share mandated by Operation Barga. The security of tenure guaranteed to registered tenants has also not been absolute, with a statewide survey carried out in 2000 reporting that 14.4 percent of *bargaders* or share-tenants lost possession of their land,[12] with this proportion going up to 30 percent in some districts (Dasgupta 2010).

10.4 Post-Land Reform Economies in Kerala and West Bengal

We next present the broad trajectory of growth and development in Kerala and West Bengal over the last three decades to decipher how (and to what extent) the land reforms, which were largely complete by the early 1980s, shaped this trajectory. Instead of presenting a detailed account of the development experience that is outside the scope of this chapter, we will start with an analysis of agricultural performance following the land reforms. This allows us to understand the direct impact of the reforms, if any, on economic production. Subsequently we will discuss some other salient features of the respective state economies that have evolved post-land reforms and are important from the perspective of this chapter.

Any balanced account of Kerala's land reform would admit that the agrarian relations in the state were significantly impacted by the state's land tenure policies. The traditionally powerful landlord caste/classes such as the Namboodiris and Nayars were weakened, while a large section of the tenants and agricultural laborers gained ownership rights, albeit over small landholdings. The agricultural laborers also gained in terms of increasing wages and better social security provisions (Ramachandran 1997). However, this empowerment of the peasantry and workers did not lead to agricultural growth in the subsequent period. Kannan and Pushpangadan (1988) present a comprehensive analysis of data on yield, area, and production for both food and nonfood crops to show that the period contemporaneous with the land reform process (as well as the next few years) 1975–6 to 1985–6, witnessed stagnation in agriculture with negative rates of growth in output. The main factor explaining this stagnation is the rapid rise in real wages of agricultural labor through unionization, which in the absence of any major technological breakthrough outweighed any increase in labor productivity. This study also contains an important finding that the growth in tertiary sector service activities, which was the main growth sector in the economy in the ten years under question, was being driven by external remittance income mainly from the Gulf region and was thus not related to the working of the primary and secondary sectors. This disjuncture continued to be a pattern of the Kerala economy for years to come.

Kerala, as an exemplary case of advances in social development among Indian states, is well known in the development literature. Its success in achieving human development indicators comparable to advanced Western economies on some fronts despite modest economic performance is justifiably heralded for the effectiveness of public action in bringing about progressive development outcomes (Dreze and Sen 1995, 1997). In some accounts, the policies of the subsequent left governments are identified as the major determinants of this remarkable record, along with historical factors. While it is true that the presence of left political movements as well as left-induced policies have had a far-reaching impact on the social landscape by empowering the poor and marginalized, when it comes to land reform in particular, its economic impact on the growth process in the Kerala economy is ambiguous.

Kannan (1998) makes an important contribution analyzing the Kerala development pattern from the perspective of labor. The presence of powerful labor organizations in the state since the early 1980s, due to the predominance of left politics, shaped the pattern of Kerala's development. Wage rates for various categories of labor increased well beyond the cost of living adjustments as well as increases in labor productivity and were accompanied by nonwage benefits related to pensions and health care. Kannan (1998) shows that the bargaining power of labor was also bolstered by the phenomenon of labor migration to the booming Gulf economies from the late 1970s. This relative advantage that labor as a class started to enjoy, led them to oppose technological change in industrial production due to the apprehension of job loss. However, such tactics resulted in a profit squeeze for employers and ultimately in loss of employment due to capital flight out of the state. The case of agriculture was no different, with major resistance by organized labor associations to modern technology, forcing landowners (often small farmers themselves) to shift to less labor-intensive crops. Kannan (1998: 64) explains the problem succinctly:

> Between the mid-1970s and early 1990s, the area under paddy registered a decline of nearly 30 percent and by the mid-1990s of 40 percent! This decline in area

> aggravated an already deteriorating situation of employment of agricultural labourers. On a conservative estimate the decline in employment (excluding harvesting and post-harvesting operations), due to a decline in area, was around 35 million man days between the mid-1970s and early 1980s and around 50 million man days by the mid-1990s. Moreover, the crop cultivating agricultural sector stagnated since the mid-1970s to the mid-1980s thereby imposing an additional constraint in enhancing employment in the economy as a whole.

While from the demand side, employment was being curtailed due to the relatively high price of labor, the high quality of public infrastructure in health and education as well as the social empowerment of the traditionally marginal sections also meant that the supply of labor was being qualitatively changed. An educated group of young job-seekers emerged whose expectations regarding employment were very different from the previous generation. There was thus a mismatch in the labor market resulting in open unemployment rates that were consistently much higher than in other parts of India. Notwithstanding warnings that the remittances from abroad cannot be relied upon in the long term, the Kerala economy has continued to depend critically on remittances and has shown itself to be remarkably resilient to the impact of the recent financial crisis. Detailed studies indicate that the loss of foreign employment for Kerala workers had been minimal during 2008–9 (Zachariah and Rajan 2010). The buoyant service sector, which currently contributes about 78–79 percent of state domestic product (Reserve Bank of India, 2015: table 7) is largely financed and sustained by this remittance income, making the economy vulnerable to external conditions.

A recent analysis characterizes Kerala as an "emerging nonagricultural economy" (Kannan 2011: 64) with barely 14 percent of state domestic product being contributed by agriculture. In recent years the secondary sector, led by construction, has recorded robust growth rates along with the service sector. Agriculture is largely stagnant, growing at merely 0.5 percent annually in the decade 1997–8 to 2007–8. Interestingly, agriculture still provides employment to about 33 percent of the labor force in 2009–10, indicating that the process of structural transformation is far from complete.

Unlike Kerala, there was a spurt in agricultural production in West Bengal from the early 1980s, after a long period of stagnation that goes back to the preindependence period. Several authors have credited the land reforms with this turnaround although there are dissenting voices (Rawal and Swaminathan 1998; Sen and Sengupta 1995) that identify the role of technological changes including better irrigation facilities for this success. Notwithstanding the reason, this period of high growth in agriculture has been tapering off from the 1990s and the agrarian economy in West Bengal has been in a constant crisis ever since. This crisis is clearly evidenced in the cost of cultivation data reported in Khasnabis (2008), which shows a clear deterioration of net income in paddy cultivation (still the major crop in the state) from the early 1990s to the middle of the 2000s.

There have been significant changes in the agrarian structure of West Bengal over the three decades since the land reform program started. While the inequality of land distribution has remained largely unchanged (with a slight increase in some periods), there has been a significant decrease in tenancy from 1971–2 to 2002–3 with a significant increase in the last decade ending in 2013. In keeping with the overall trends in the country, the percentage of employment in agriculture has fallen in West Bengal as well but the latest data from 2011–12, still shows 56 percent of workers, according to their usual status, engaged in agriculture and related activities (National Sample Survey Organization (NSSO) 2013). The corresponding share of agriculture in the state domestic product is between 17 and 18 percent. Another significant change concerns the relative decline of self-employed workers in agriculture (i.e., owner cultivators) in comparison to agricultural laborers. In the 2001 census, the agricultural workers as a group outnumbered cultivators for the first time in West Bengal. In the latest 2011 census, this trend is expected to continue and be reinforced further (distribution of workers by occupation from the 2011 Census is not available yet). This implies that contrary to the original objective of creating a small peasant-based economy with most agrarian workers having access to cultivable land, we have a situation of significant proletarianization emerging in the economy without dynamic accumulation.

Industrial development in West Bengal has been stalled since the 1960s due to a series of structural factors including regional development policies of the central government such as inadequate issuance of industrial licenses and bank credit (see Mitra 2007). The LFG has not been successful in reversing this stagnation during its rule with the situation worsening in the last three decades (Bag 2011; D'Costa 2005). Organized manufacturing has been woefully inadequate in providing productive employment to the labor that is rendered surplus in the agrarian sector. The slack has been picked up largely by the unorganized sector (manufacturing and services) with its well-known vulnerabilities. According to a recent estimate, more than 90 percent of total employment generated in the nonfarm sector in 2007 was in the unorganized sector (Government of India 2010: 91).

It is in this context that the LFG made a desperate attempt toward industrialization beginning in 2006. This industrialization drive depended primarily on attracting corporate capital by offering incentives, including (but not limited to) land. This move led to violent protests (most prominently in Singur and Nandigram) where the government made efforts to acquire farmland for industrial production. It is perhaps not an overstatement to say that in spite of many other sources of discontent with the LFG during its long tenure in West Bengal, it was the state's land acquisition drive for industrialization that crystalized the discontent against the LFG and eventually led to its defeat in the 2011 elections. The political leaders of the Left Front and its supporters have come up with their own justification for the current move toward industrialization. One of the main points emphasized in this respect is that industrialization is the obvious next step to the government's intervention in agriculture, building on the benefits of land reforms. Thus, land acquisition was justified merely as a continuation of the earlier policy of land redistribution of the state rather than a major change in course (Bhattacharya 2007).

10.5 Land Reform and the Agrarian Question in the Long Run

For both Kerala and West Bengal, more than three decades have passed since the implementation of the land reforms.[13] It is reasonable to expect the major impact of these reforms to have materialized over this period. This is not to suggest a priori, that land reforms have played an important role in this context. That remains to be analyzed. Moreover, there must have been other factors, not directly related to land reform, which played a role as well.

To initiate the discussion, let us recall the three aspects of the agrarian question following Bernstein: those related to accumulation, production, and politics. Given that land reforms were primarily motivated to resolve the political aspect of the agrarian question, that is, to reconfigure the rural class structure in support of the oppressed classes of poor peasants and laborers, Kerala and West Bengal have both shown substantial progress. The stronghold of large landlords has been weakened in local factor and output markets, the weaker sections of the agrarian population have been empowered both economically and politically, and feudal exploitative practices have been largely eliminated. However, there are still reasons for concern. First, there is evidence, as well as statements from left political leaders, that the traditional landholding class has transformed itself into a class of powerful rural traders and businessmen (Mishra 2007; Mishra and Rawal 2002) while retaining their control of land, albeit in smaller amounts. In some cases, middle or large peasants have joined their ranks showing some mobility in the rural class structure. It is interesting, however, that this process has not created a class of rural capitalists who invest substantial surplus in production-related activities in agriculture or outside it.

Second, while tenancy has been regulated to a large extent, it is clearly not eliminated. In Kerala, legal tenancy has been prohibited through the KLRAA legislation of 1969 but recent studies (Nair and Menon 2006) point toward the existence of underground lease arrangements, even though it is not captured in the NSSO macro surveys. In West Bengal there are reports of decline in share tenancy contracts recorded under Operation Barga and an increase in short-term leases for *boro* or summer rice (Sharma 2006). There is a common argument made now by researchers and policy makers that legalizing tenancy in Kerala with safeguards against large-scale contract farming (Nair and Menon 2006) and relaxing the regulations around tenancy in West Bengal are necessary (Hanstad and Nielsen 2004a). They argue that this would make for smoother operation of lease and land sale markets and ensure more efficient allocation of cultivated land among able and willing farmers. We feel that such relaxation of tenancy regulations (including its legalization in Kerala) should only happen after analyzing in detail the characteristics of lessees and tenants in the new lease arrangements and ensuring that the original motivation behind the tenancy reforms of providing small farmers with long-term access to land, is not compromised by such moves.

Third, questions of systematic exclusion of specific sections of the population from the land reform process on the basis of class, caste, and gender need to be brought to the fore. In the case of both West Bengal and Kerala, critics have pointed out the sacrifice of the interests of landless agricultural laborers on the altar of "peasant unity" as practiced by the left parties in designing and executing the land reform programs (Rammohan 2008; Rudra 1981). There is also serious criticism about the Kerala land reforms being inadequate or exclusionary from the perspective of lower caste Dalits, as well as women (Kodoth 2004; Rammohan 2008).

The production problematic, as we have mentioned before, focuses on the development of capitalism in agriculture with "the dispossession of pre-capitalist predator landed property and the peasantry" (Bernstein 2006: 451). The exact characterization of the production process in terms of the extent of penetration of capitalism presents severe conceptual problems. How does one identify capitalistic production in agriculture? Is the extensive presence of wage labor or its increasing proportion in the agricultural population sufficient to make this claim? What about the simultaneous presence of wage labor and self-cultivation in the same production process? This chapter does not enter this debate.[14] It would suffice to say that a classical process of dispossession of the peasantry from their land and the ubiquitous use of wage labor in production has not happened in either Kerala[15] or West Bengal. This is only to be expected in states where the land reform process was geared toward empowering the smaller peasants and landless laborers with secure access to land and preventing large-scale proletarianization.

Lastly and most importantly, we now focus on the accumulation problematic of the agrarian question. The accumulation dimension of the agrarian question is concerned with both the generation of surplus in agriculture within a specific agrarian structure as well as with the transfer and use of this surplus to further industrialization or overall development of the economy in question. In other words, it presents an understanding of the role agricultural accumulation plays in industrial accumulation, typically in a capitalist context. However, the same can also be applied in the context of so-called socialist accumulation as in the case of the Soviet Union. Land reform is not a prerequisite to study this interlinkage but clearly major changes in land tenure are likely to have implications for the resolution of the accumulation dimension of the agrarian question.

If we review our understanding of the development processes at work in the economies of Kerala and West Bengal over the post-reform period, we do not find any evidence of a satisfactory resolution of this problematic. In the case of Kerala, land reform was not successful in bringing about any significant change in surplus generation in agriculture. The stagnation of agricultural production in the period immediately following the implementation of reforms suggests that. We have discussed at least one reason behind this stagnation related to the overpricing of labor referred to earlier in the chapter (Kannan 1998). It is possible that there are other factors as well. But irrespective of the cause, it can be unambiguously said that land reforms were not able to transform agriculture into a dynamic surplus-generating sector. Given this failure, it is not surprising to see that industrialization has not picked up in the state and the service sector growth in the state is largely driven by external stimuli such as remittances. Of course, one can argue that alternative sources of surplus required for industrialization could have been made available. Corporate-controlled big capital could have brought in investment without any local level linkage with agriculture. Remittance funds potentially could also have been a source of capital investment. The fact that these options did not work out as well for Kerala is a complex matter, which is beyond the scope of this chapter.

One may ask at this point, for an economy like Kerala's, which has barely a third of its labor force engaged in agriculture producing only 12–13 percent of state income, is there any use focusing on an agrarian question for the state, whatever may be its characteristic? The answer to this justifiable question lies in understanding who is still dependent on agriculture and why they continue to be in the sector in spite of the miserable returns. We would like to assert that agriculture in its current stagnant, crisis-ridden state is the "parking lot" of the most vulnerable section of the rural population who do not have any other alternative source of sustaining their livelihoods, including migration.[16] The agrarian question will remain valid as long as they are trapped in agriculture (see Chapters 8 and 9 in this volume).

The West Bengal case provides a study in contrast to some extent. The discontinuous jump in agricultural production from the early 1980s points to the possibility that the reform measures, especially the security of tenure provided to sharecroppers, might have had a positive effect on agricultural production. Although there is evidence that cultivation of *boro* or summer rice using high yielding variety (HYV) seeds, technology, and small-scale irrigation has led this growth, one can make an argument that the reforms and the presence of a pro-peasant government facilitated this transformation in agriculture by providing incentives to the small cultivators to use labor and non-labor inputs more intensively (Banerjee, Gertler, and Ghatak 2002). However, this growth spurt did not last long and by the early 1990s there was a crisis in agriculture that has continued since then in varying degrees. There are also external reasons related to macro policies pursued at the national level that have precipitated this crisis, such as liberalization of input markets and a sharp reduction in public investment in agriculture. But one question remains: Was there substantial surplus generated in agriculture during the first eight to ten years of high growth in the 1980s, and if so, was this surplus invested in industrial accumulation? There is no clear answer to these questions but based on data on industrial growth, one can surmise that even if surplus was generated it was not used for any successful industrialization effort in the state itself.

There is also a related question of whether the possibility of generating high-volume surplus in West Bengal agriculture exists, given the extreme fragmentation of land. The limited land redistribution program also has not been successful in creating a class of independent peasant producers. Because of the very scarce supply of land above the land ceiling, the average amount of land redistributed during the LFG regime was barely 0.11 hectares per beneficiary or about a quarter of an acre. The land reform beneficiaries have not been able to live off their land in the post-reform period. This is demonstrated in the study by Chakraborti (2003), which finds that more than 90 percent of new land recipients did not find year-long employment on their own land. The corresponding figure for *bargadars* is 83.5 percent. Most of these people join the ranks of agricultural laborers who work for larger landowners for wages.[17] It is not surprising, thus, to find that there has been an unprecedented increase in the number of agricultural laborers in the rural population of West Bengal. Indeed, laborers constituted a majority of the agricultural population for the first time in 2001. The crisis in the agrarian sector is summarized by recent data that show that while the share of agriculture has dropped from 34.45 percent to 26.42 percent between 1999–2000 and 2004–5, its share in the workforce has remained almost stagnant (47.76–46.35 percent within the same time interval).

The rupture of the accumulation aspect of the agrarian question is also visible in the rather ambiguous official stance of the CPI (M) regarding its move toward industrialization in the mid-2000s. On the one hand, there was an attempt to justify the strategy to attract industrial capital on the basis of the success of land reform. In this account, the emphasis is on the spectacular growth in agriculture since the inception of the reforms and how that growth has been translated into increased purchasing power of the rural peasantry and created the basis for broad-based industrial growth (Bhattacharya 2007). On the other hand, there is recognition of the severe crisis looming in the agrarian sector, which makes it necessary to look beyond agriculture for economic growth and employment. The blame for the crisis is put squarely on the agricultural policy pursued by the national government for how they have made agriculture unviable for small producers by eliminating subsidies as well as downsizing public investment in agriculture.

While it is true that the broad agenda of neoliberal development followed by the Indian state has had definite implications for the crisis in agriculture in terms of the sharp reduction in public investment (Reddy and Mishra 2009), it is not clear how the state of the small farmer is different in West Bengal compared to other states despite three decades of agrarian reforms.[18] Moreover, there is also the question regarding the best response in the face of the crisis. The LFG's move toward corporate-led industrialization ostensibly to generate employment has justifiably come under considerable attack in this respect (Bhaduri 2007).

10.6 Conclusion

We will finish this discussion with a final comment about the nature of the agrarian question itself, especially in the context of the accumulation dimension. Bernstein (2006) has noted that in the current phase of capitalist globalization, the agrarian question with respect to capital may have become less relevant. Particularly, in the contemporary world of internationally integrated networks of agricultural production, distribution, and consumption, the national-level intersectoral linkages between agriculture and industry bolstered by a home market have ceased to be a prerequisite for capital accumulation to continue (Lerche 2013). Bernstein instead emphasizes the agrarian question of labor, which broadly corresponds to the crisis of social reproduction faced by people dependent on agriculture and yet not able to make a living out of it exclusively.

If we look at the two cases of Kerala and West Bengal, the land reforms notwithstanding, there is hardly any linkage between the agricultural sector and the process of large-scale capital accumulation, either within the state or elsewhere in the country. Even the visions of development that are being sought either explicitly or implicitly through some form of capitalist accumulation, do not subscribe to any agriculture-linked process in a serious way. However, there are two aspects pertaining to agriculture that continue to be important. First, the question of food. Although, imports pertaining to select commodities are significant (for example, edible oils), India produces most of its domestic food consumption. Policy makers and analysts clearly recognize a food constraint for the Indian economy, although this specific term is seldom used. Of course, individual states are not (and need not be) self-sufficient in food production and thus Kerala is consistently a food-deficient state while West Bengal has a food surplus at least with regard to rice, the main food grain. The second aspect, in line with Bernstein's position on the agrarian question of labor, is related to the question of survival and livelihoods for the population that is stuck in agriculture due to a stalled structural transformation. Both Kerala and West Bengal have a challenge in this regard, albeit to varying extents: more than half the labor force in West Bengal and one-third in Kerala.

Irrespective of one's position in the debate about the contemporary nature of the agrarian question, the Kerala and West Bengal cases demonstrate a definite disarticulation of the accumulation dimension of the original agrarian question. The caveat is that it has been largely the case for India as a whole, but this should not divert attention from the fact that these two states are exceptional in postcolonial India in what they attempted in terms of land reforms. That none has been able to chart a successful path of equitable development using land reforms demonstrates the limits of left reformism as practiced in India at the level of constituent states.

In terms of bolstering a process of capitalist development on the one hand, land reforms have clearly not worked in India, as the experience of the two states that seriously attempted such reform, shows. On the other hand, states such as Punjab and Haryana, hotbeds of capitalist development, did not have any land reform. This indicates that land reform is not a prerequisite for high rates of capitalist accumulation. However, if one interprets the original left ideology of providing relief to the poor as one of ensuring a more humane trajectory of capitalist development, land reform seems to have provided some relief to the rural poor but at the cost of robust capitalist expansion or economic growth. Some versions of Marxism will probably decry the idea of humane capitalism as utopian and the agenda of land reform as mere populism which slows down the inevitable transition to capitalism and the eventual struggle for socialism. Conversely, from a liberal reformist perspective, equitable access to productive assets like land is undoubtedly an effective way of making the growth process more inclusive. The fact that some particular attempts with land reform have failed is not reason enough to abandon the agenda itself. Therefore, the ultimate lesson from the Kerala and West Bengal experience discussed here is contingent on who is taking the lesson.

Notes

1. Jammu and Kashmir and Tripura might be considered exceptions to some extent.
2. With some qualifications, the Mellor framework can be used effectively as an operationalization of the more theoretical understanding of the agrarian question by Bernstein and Byres, especially in its accumulation dimension
3. There is little comparative work on Kerala and West Bengal. Nag (1983, 1984, 1989) compares mortality, fertility, and access to health services in rural Kerala and West Bengal, Desai (2001) compares the left party formation and its role on social reforms in the two states and Harriss and Törnquist (2015) is an overview comparison of the development experiences of the two states. None of them look at land reforms and agriculture specifically.
4. For example, the huge media attention given to the Bihar and the Gujarat models of development under the chief ministership of Nitish Kumar and Narendra Modi respectively.
5. One can also incorporate the cases of implanting noncapitalist forms of production systems like those attempted in the USSR, China, or Vietnam within this framework.
6. Ashok Mitra denied any discussion of the East Asian experience as a role model for development in either the party or the government at the time of the first Left Front Government (personal interview, 2013).
7. Although there are numerous academic papers on the so-called Kerala model of development, they basically talk about the Kerala development experience of achieving high levels of human development without corresponding high economic growth. I have not been able to locate writings by intellectuals or party activists in Kerala elaborating a blueprint of alternative development following the land reforms but I suspect that they are there.
8. This so-called "land to the tiller" provision had a major loophole. Even tenants who merely supervised cultivation were considered tillers and therefore qualified for ownership (Raj and Tharakan 1983).
9. This ceiling went up to 17.3 acres and 24.2 acres of irrigated and unirrigated land respectively for larger families.
10. The Indian Communist movement split along ideological lines in 1964 into the pro-Soviet Communist Party of India (CPI) and the pro-China Communist Party of India (Marxist). Subsequently, in 1967, a more radical faction of the CPI (M) broke away to form the Communist Party of India (Marxist-Leninist).
11. According to the West Bengal Land Reform Act (amended) 1972, the legal basis of the Operation Barga, *bargadars* were entitled to 75 percent of the produced crop except in cases where the landlord provided all non-labor inputs. In such cases the tenant's share was reduced to 50 percent of the produce.
12. The survey does not mention specific causes for losing possession of land. Micro level studies report informal transactions in Barga land through which tenants surrendered their rights to their landlords for a price (Rawal 2001; Hanstad and Nielsen 2004b).

13. One can argue that beyond 1983, West Bengal still had redistributed some land but the scope of this is negligible compared to what had taken place before.
14. The debate on mode of production in *Economic and Political Weekly* with seminal contributions by Ashok Rudra and Utsa Patnaik can be consulted in this regard by interested readers. See Chakrabarti and Cullenberg (2003) for a different take on the mode of production debate.
15. With the exception of the plantation sector, which was exempted from the land reform legislation.
16. An apt term quoted by Rajesh Bhattacharya (2010: 201) from Dandekar (1992: 54–5).
17. Some of these cultivators may also earn a livelihood in the rural nonfarm sector. Recent figures do show a sharp increase in the level of rural nonfarm employment (NSSO 2013: 81).
18. I am consciously limiting myself to the economic conditions facing the agricultural population; the question of political empowerment in a Left-ruled state may of course be raised.

References

Agravala, P. K. 2010. *Land Reforms in States and Union Territories in India*. New Delhi: Concept Publishing Company.

Akram-Lodhi, A. H. and Kay, C. 2010. "Surveying the Agrarian Question (Part 2): Current Debates and Beyond." *Journal of Peasant Studies*, 37(2): 255–84.

Appu, P. S. 1996. *Land Reforms in India: A Survey of Policy, Legislation and Implementation*. Delhi: Vikas Publishing House.

Bag, K. 2011. "Red Bengal's Rise and Fall." *New Left Review*, 70(July–August): 69–98.

Banerjee, A. V., Gertler, P. J., and Ghatak, M. 2002. "Empowerment and Efficiency: Tenancy reform in West Bengal." *Journal of Political Economy*, 110(2): 239–80.

Bernstein, H. 2004. "Changing before Our Very Eyes: Agrarian Questions and the Politics of Land in Capitalism Today." *Journal of Agrarian Change*, 4(1–2): 190–225.

Bernstein, H. 2006. "Is There an Agrarian Question in the 21st Century?" *Canadian Journal of Development Studies*, 27(4): 449–60.

Bhaduri, A. 1973. "A Study in Agricultural Backwardness under Semi-Feudalism." *The Economic Journal*, 83(329): 120–37.

Bhaduri, A. 2007. "Alternatives in Industrialisation." *Economic and Political Weekly*, 42(8): 1597–1601.

Bhattacharya, B. 2007. "On Industrialisation in West Bengal." *The Marxist*, 23(1): 1–8.

Bhattacharya, R. 2010. "Capitalism in Post-Colonial India: Primitive Accumulation Under Dirigiste and Laissez Faire Regimes." PhD Dissertation, Department of Economics, University of Massachusetts, Amherst.

Byres, T. J. 1991. "The Agrarian Question and Differing Forms of Capitalist Transition: An Essay with Reference to Asia." In *Rural Transformation in Asia*, edited by J. Breman and S. Mundle, New Delhi: Oxford University Press, pp. 3–76.

Chakrabarti, A. and Cullenberg, S. 2003. *Transition and Development in India*. London and New York: Routledge.

Chakraborti, A. 2003. *Beneficiaries of Land Reforms: The West Bengal Scenario*. Kalyani: State Institute of Panchayats and Rural Development.

Chandra, N. K. 1978a. "Industrialization and the Left Movement: On Several Questions of Strategy in West Bengal: Part One." *Social Scientist*, 7(1): 3–32.

Chandra, N. K. 1978b. "Industrialisation and the Left Movement: On Several Questions of Strategy in West Bengal: Part Two." *Social Scientist*, 7(3): 57–75.

D'Costa, A. P. 2005. *The Long March to Capitalsim: Embourgeoisment, Internationalisation and Industrial Transformation in India*. Basingstoke: Palgrave Macmillan.

Dandekar, V. M. 1992. "Forty Years after Independence." In *The Indian Economy: Problems and Prospects*, edited by Bimal Jalan, New Delhi: Penguin Books India, pp. 33–84.

Dasgupta, A. 2010. "Land Acquisition to Consolidate Land Reforms? The Case of West Bengal." In *Development, Equity and Poverty: Essays in Honour of Azizur Rahman Khan*, edited by A. Dasgupta, L. Banerjee, and R. Islam. New Delhi: Macmillan, pp. 453–77.

Desai, M. 2001. "Party Formation, Political Power, and the Capacity for Reform: Comparing Left Parties in Kerala and West Bengal, India." *Social Forces*, 80(1): 37–60.

Dreze, J. and Sen, A. 1995. *India: Economic Development and Social Opportunity*. New Delhi: Oxford University Press.

Dreze, J. and Sen, A. 1997. *Indian Development: Selected Regional Perspectives*. New Dehli: Oxford University Press.

Dunham, D. M. 1991. "Agricultural Growth and Rural Industry: Some Reflections on the Rural Growth Linkages Debate." *ISS Working Paper Series/General Series*, 114: 1–22.

Gill, S. S. 1989. "Changing Land Relations in Punjab and Implications for Land Reforms." *Economic and Political Weekly*, 24(25): A79–A85.

Government of India. 2010. "West Bengal Development Report." Academic Foundation for The Planning Commission, New Delhi.

Government of West Bengal. 2012. "Economic Review." Bureau of Applied Economics and Statistics, Calcutta.

Griffin, K., Khan, A. R., and Ickowitz, A. 2002. "Poverty and the Distribution of Land." *Journal of Agrarian Change*, 2(3): 279–330.

Hanstad, T. and Nielsen, R. 2004a. "West Bengal's Bargadars and Landownership." *Economic and Political Weekly*, 39(8): 853–5.

Hanstad, T. and Nielsen, R. 2004b. "From Sharecroppers to Landowners: Paving the Way for West Bengal's Bargadars." RDI Report on Foreign Aid and Development No. 121, Rural Development Institute, Seattle.

Harriss, J. and Törnquist, O. 2015. "Comparative Notes on Indian Experiences of Social Democracy: Kerala and West Bengal." CSES Working Paper No. 29, Centre for Socio-Economic and Environmental Studies, Kochi.

Kannan, K. P. 1998. "Political Economy of Labour and Development in Kerala." *Economic and Political Weekly*, 33(52): L61–L70.

Kannan, K. P. 2011. "Agricultural Development in an Emerging Non-Agrarian Regional Economy: Kerala's Challenges." *Economic and Political Weekly*, 46(9): 64–70.

Kannan, K. P. and Pushpangadan, K. 1988. "Agricultural Stagnation in Kerala: An Exploratory Analysis." *Economic and Political Weekly*, 23(39): A120–A128.

Karat, P. 2007. "Bamponthi Sarkarer Bhumika abong amader bojhapora" [The Role of the Left Front Government and Our Understanding], *Marxbadi Path* (*The Marxist Path*), May.

Kay, C. 2002. "Why East Asia Overtook Latin America: Agrarian Reform, Industrialisation and Development." *Third World Quarterly*, 23(6): 1073–1102.

Khasnabis, R. 2008. "The Economy of West Bengal." *Economic and Political Weekly*, 43(52): 103–15.

Kodoth, P. 2004. "Gender, Property Rights and Responsibility for Farming in Kerala." *Economic and Political Weekly*, 39(19): 1911–20.

Kohli, A. 1987. *The State and Poverty in India: The Politics of Reform*. New York: Cambridge University Press.

Lerche, J. 2013. "The Agrarian Question in Neoliberal India: Agrarian Transition Bypassed?" *Journal of Agrarian Change*, 13(3): 382–404.

Mao, Y. K. and Schive, C. 1995. "Agricultural and Industrial Development in Taiwan." In *Agriculture on the Road to Industrialization*, edited by J. W. Mellor, Baltimore, MD: Johns Hopkins University Press, pp. 23–66.

Mellor, J. W. 1995. *Agriculture on the Road to Industrialization*. Baltimore, MD: Johns Hopkins University Press.

Mishra, S. K. 2007. "On Agrarian Transition in West Bengal." *The Marxist*, 23(1): 1–22.

Mishra, S. K. and Rawal, V. 2002. "Agrarian Relations in Contemporary West Bengal and Tasks for the Left." In *Agrarian studies: essays on agrarian relations in less-developed countries*, edited by V. K. Ramachandran and M. Swaminathan, New Delhi: Tulika Books, pp. 329–55.

Mitra, A. 1978. "This Opportunity Has to Be Converted into a Challenge. . . . " *Social Scientist*, 6(6–7): 3–8.

Mitra, A. 2007. "The Oblivion of Idealism." Available at: <http://old.himalmag.com/component/content/article/954-the-oblivion-of-idealism.html> (accessed June 26, 2016).

Nag, M. 1983. "Impact of Social and Economic Development on Mortality: Comparative Study of Kerala and West Bengal." *Economic and Political Weekly*, 18(19–21): 877–900.

Nag, M. 1984. "Fertility Differential in Kerala and West Bengal: Equity-Fertility Hypothesis as Explanation." *Economic and Political Weekly*, 19(1): 33–41.

Nag, M. 1989. "Political Consciousness as a Factor in the Accessibility of Health Services: A Case Study of Rural Kerala and West Bengal." *Economic and Political Weekly*, 24(8): 417–26.

Nair, K. N. and Menon, V. 2006. "Lease Farming in Kerala: Findings from Micro Level Studies." *Economic and Political Weekly*, 41(26): 2732–8.

NSSO (National Sample Survey Organization). 2013. "Key Indicators of Employment and Unemployment in India." Report No. NSS KI 68/10, New Delhi.

Raj, K. N. and Tharakan, M. 1983. "Agrarian Reform in Kerala and Its Impact on the Rural Economy: A Preliminary Assessment." In *Agrarian Reform in Contemporary Developing Countries*, edited by A. K. Ghose, London: Routledge, pp. 31–90.

Ramachandran, V. K. 1997. "On Kerala's Development Achievements." In *Indian Development: Selected Regional Perspectives*, edited by J. Dreze and A. Sen, New Delhi: Oxford University Press, pp. 205–356.

Ramakumar, R. 2006. "Public Action, Agrarian Change and the Standard of Living of Agricultural Workers: A Study of a Village in Kerala." *Journal of Agrarian Change*, 6(3): 306–45.

Rammohan, K. T. 2008. "Caste and Landlessness in Kerala: Signals from Chengara." *Economic and Political Weekly*, 43(37): 14–16.

Rawal, V. 2001. "Agrarian Reform and Land Markets: A Study of Land Transactions in Two Villages of West Bengal, 1977–1995*." *Economic Development and Cultural Change*, 49(3): 611–29.

Rawal, V. and Swaminathan, M. 1998. "Changing Trajectories: Agricultural Growth in West Bengal, 1950 to 1996." *Economic and Political Weekly*, 33(40): 2595–602.

Reddy, D. N. and Mishra, S. 2009. *Agrarian Crisis in India*. New Delhi: Oxford University Press.

Reserve Bank of India. 2015 "Handbook of Statistics on Indian Economy." Available at: <https://www.rbi.org.in/scripts/annualPublications.aspx?head=Handbook+of+Statistics+on+Indian+Economy> (accessed on June 26, 2016).

Rudra, A. 1981. "One Step Forwards, Two Steps Backward." *Economic and Political Weekly*, 16(25—26): A61+A63+A65+A67-A68.

Sen, A. and Sengupta, R. 1995. "The Recent Growth in Agricultural Output in Eastern India, with Special Reference to the Case of West Bengal." Paper presented in the Workshop on "Agricultural Growth and Agrarian Structure in Contemporary West Bengal and Bangladesh," Centre for Studies in Social Sciences, Calcutta.

Sengupta, S. and Gazdar, H. 1997. "Agrarian Politics and Rural Development in West Bengal." In *Indian Development: Selected Regional Perspectives*, edited by J. Dreze and A. Sen. New Delhi: Oxford University Press, pp. 129–204.

Sharma, H. R. 2006. "Liberalising the Lease Market." *Economic and Political Weekly*, 41(8): 696–8.

Storm, S. 2015. Structural Change. *Development and Change*, 46(4): 666–99.

Zachariah, K. C. and Irudaya Rajan, S. 2010. "Impact of Global Recession on Migration and Remittances in Kerala." CDS Working Paper no. 432, Centre for Development Studies, Trivandrum.

11

How Much Land Does a Capitalist Need?

Historical Patterns of Land Acquisition and Indian Industrialization

Mircea Raianu

11.1 Introduction

Of the many recent struggles over land acquisition for private industrial use across India, those involving Tata companies have been among the most protracted and violent. Tata Steel's attempts to set up new plants in Odisha (Gopalpur, Kalinganagar) and Chhattisgarh (Lohandiguda), and especially Tata Motors' proposed Singur factory, have become emblematic of a particular mode of neoliberal accumulation, resisted by innovative forms of popular politics (Roy 2011). In each case, variations in industry-specific resource demands and relationships with state governments are important factors in shaping the contours of the land acquisition process. But despite the geographically disparate and relatively decentralized nature of the Tata Group's constituent companies, they do share a common operational logic, especially in formulating rehabilitation and resettlement policies (cf. Pati 2012).

Historically, the Tatas' early involvement in large dam projects and steel production sets them apart both from Jawaharlal Nehru's "temples of modern India," and from the latest wave of private acquisitions, led by new players such as POSCO, Vedanta, and Jindal. The activities of these companies, including Tata Steel, are seen by many activists as a form of extractive "internal colonialism," by analogy with the old East India Company. The losers in this process are mainly *adivasi* (tribal and/or aboriginal) populations in eastern India, who refuse to be converted into industrial laborers or to accept the loss of their land (Padel and Das 2010). However, other theoretical approaches to the "land question" today place far less emphasis on private capital, given that around 90 percent of acquisitions in India since independence have been carried out on behalf of state projects. States are not necessarily acting purely at the behest of capital, but pursuing their own developmental or public welfare objectives. According to this view, recent highly visible episodes of resistance, such as Singur or Kalinganagar, attract disproportionate attention because of the appealing narrative of the "giant private corporation against the little man," but cannot represent the problem in its totality (Chakravorty 2013: 59, 122–3). Such diverse appeals to the past suggest the need for critical reflections on the role of land acquisition in the longer trajectory of Indian industrialization.

The historian's task is to answer the difficult question of which pasts we should look to for a clearer understanding of the present crisis described by the editors of this volume as an "incomplete transition," in which large-scale capitalist development in India depends upon land acquisition at increasingly higher prices while the commodification of land proceeds without the conversion of agricultural populations into wage laborers. As the editors note, we must examine the manifold conflicts over the "circulation of land as a commodity," rather than seeking an originary moment of "primitive accumulation" (Chapter 1, this volume). Expropriation of land in India today is mainly done by domestic rather than foreign capital, with the state playing an enabling role. This tendency is not new, but its origins cannot simply be read back in time without an analysis of the legal and political circumstances in which it arose.

When and how did the involvement of private capital in land markets restructure agrarian social relations, and under what conditions was the state called upon to support this process through legislation or administrative fiat? I argue that the entry of Indian capital in the industrial sector during the late colonial period was a key moment in the evolution of the land acquisition process, which is easily overlooked when taking the postindependence era as a starting point. This moment made possible two seemingly contradictory but mutually constitutive trends: the legal designation of private capital as capable of fulfilling a "public purpose," and the increasingly direct involvement of the state in resource capture and management for the purpose of industrial development.

By "capital," I mean Indian-owned companies with incentives to undertake expansion into extractive and land-intensive industries in the early twentieth century, when British managing agencies largely set the terms of India's relationship with the global economy. The Tatas, more than any other firm, were able to leverage both the legitimacy of *swadeshi* (self-sufficient) enterprise and the necessary technical expertise to expand in this area. By "state," I mean specifically the colonial state at different levels (central and provincial), whose gradual abandonment of agrarian rent seeking and assumption of a "developmental" mantle would lay the foundations for centrally planned industrialization under Nehru. One marked institutional continuity between the colonial, post-independence, and contemporary iterations of the "land question" in India is the importance of competition across multiple nodes of state power to attract capital and secure control over resources, taking place today primarily between state-level governments.

The chapter seeks to substantiate this argument by drawing connections and comparisons between the establishment of two major Tata enterprises that obtained the status of proto-national industries before 1947: the Tata hydroelectric power companies in the Western Ghats, and the Tata Iron and Steel Company (TISCO) at Jamshedpur. The decision to focus on one interconnected set of companies in different sectors is driven by the availability of uniquely rich archival sources on the Tatas that allow historians to explain precisely when and how capital becomes "selectively territorialized" (Ferguson 2005: 378–9). The method I have adopted here is multi-sited archival research, relying on newly accessible materials at the Tata Central Archives in Pune and the Tata Steel Archives in Jamshedpur, alongside official records from colonial state archives in Mumbai, Delhi, and London. In particular, I focus on the negotiations between capitalists (the Tatas), the colonial state, provincial governments, and princely states, as well as on the interpretation and application of land acquisition law. An approach based on specific case studies is best suited to the reconstruction of long-term patterns of land acquisition in the absence of aggregate data before 1947. Such an approach also brings to light fruitful contrasts between the evolution of different regional economies within India.

The analysis of each case study proceeds in two stages. First, I consider how the use of legal instruments of dispossession varied according to the geographical distribution of state power in a context of multiple and overlapping sovereignties, which in turn determined how the provisions of the Land Acquisition Act (1894) were put into force. Second, I trace the extent of lands acquired, the purposes for which they were used, and the resulting spatially bounded regimes of resource extraction and production specific to each case. This kind of detailed mapping allows for the assessment of the dynamics of land acquisition, both past and present, in a more contextually sensitive manner. Dam projects affecting settled cultivators required minimum acquisition and maximum displacement, while steel plants in more sparsely populated areas characterized by comparatively insecure tenures required maximizing the surface area to be acquired and minimizing the extent of displacement due to labor needs. Accordingly, the hinterlands created by the steel industry were always more closely integrated with the site of production than those of the hydroelectric companies, which were left in comparative isolation.

However, in both cases the promise of proletarianization held out by capital as an incentive to part with the land—transforming the peasants of Mulshi or the *adivasi* residents of Chotanagpur into industrial laborers—did not happen on any significant scale. By contrast with the later Nehruvian aim of absorbing local communities into the new steel towns of Bhilai and Rourkela, Jamshedpur was "conceived as an entirely original space that should exclude them" (Sanchez and Strümpell 2014: 1279). The Jamshedpur model has arguably proven more resilient than the Nehruvian, neither pure "enclave" nor regionally integrated township with a stable and rooted working class. This points to some of the peculiarities of Indian capitalist development addressed in this volume, including the predominance of domestic over foreign capital and the persistence of surplus labor, whose origins I uncover in the pages that follow.

11.2 The Tata Hydroelectric Power Companies: Land Acquisition and Dispossession in the Western Ghats

The first recorded instance of a demand to put the Land Acquisition Act into force on behalf of a Tata hydroelectric project involved a concession at the Dudh-Sagar waterfalls in Goa. Despite the location of the falls within the boundary of Portuguese India, J. N. Tata assured the secretary of state in 1901 that the new undertaking would "utilize a portion of the power in British territory."[1] From the very beginning, this was in no way conceived as a public utilities company; instead, the aim was to start ancillary industries, "be the landlords of the place and have a yearly income from the property for the sale of water rights," and eventually get bought out by an American firm.[2] The government of Bombay was reluctant to "permit occupied lands & forests to be submerged for the benefit of Goa," since water for irrigation would flow into Portuguese territory, and suggested a renewable lease instead of permanent alienation. The government was also skeptical of the Tatas' monopolistic intentions with regard to water, and thought their works would require far more than the 85 acres applied for.[3] Ultimately, the matter was dropped as the Tatas' attention shifted away from Dudh-Sagar, but these negotiations set important precedents for safeguarding state prerogatives in future agreements.

In parallel, surveys were being undertaken for a more ambitious scheme at Lonavla in the Western Ghats, which would supply power to the tramways and mills of Bombay, with the tail water to be used for irrigation as far away as the Konkan coast.[4] A carefully crafted "Memorandum on Public Utility" spoke of immense benefits to the growth of Bombay industry, as well as the reduction of the "smoke nuisance," what one might today view as promoting "clean energy."[5] An application was submitted in late 1910 for land measuring 5,467.23 acres in total, with three proposed reservoirs at Lonavla, Shiravte, and Walwhan. For the government, the "public purpose" clause of the Land Acquisition Act in this case was invoked with the assurance that power would be sold without price discrimination. Tail water would not become the property of the company, which would instead be granted a share of the profits at a later date. Land would continue to be subject to revenue assessment at fixed rates, despite suggestions that the future profitability of the lands be considered in raising the rents. Although the company was held liable for the cost of alternative land for the displaced, in order to speed up acquisition the requirement to specify the exact plots required was waived (only names of villages were to be submitted). Finally, two clauses were borrowed from the recently completed agreement with the Tata Iron and Steel Company, referring to the necessity of returning the land if the works were not completed in a timely manner and to the protection of public rights of way, though even at this stage officials were unsure whether they could be enforced.[6] Each of these points was to prove a source of constant dispute and renegotiation during the following decade, as two new companies were registered for the construction of dams in the Andhra and Nila-Mula Valleys after World War I.

By 1919, the Tata power companies were embarking on the third round of land acquisitions, with the Andhra agreement recently signed and notifications under Section 40 of the Act about to be sent out to the villagers of Mulshi Peta in the Nila-Mula Valley. Reports of delayed or inadequate compensation, lands submerged without being measured, and works begun before the final settlements were reached, had begun to circulate among the Mulshi villagers, who duly sent their objections in a series of petitions to the government and began to organize in opposition to the project (Vora 2009: 31–6, 52–4). The petitioners rejected private capitalists' claims to fulfilling a public purpose, which was left completely undefined by the Land Acquisition Act. The substance of their legal argument was that the Act was "not intended for the wholesale acquisition of vast tracts covering hundreds of square miles, and causing the depopulation and submersion under water of numbers of villages." They pointed to a clause in the Andhra Valley agreement whereby the government received "an additional sum of money equivalent to the amount of compensation" plus the statutory 15 percent bonus for compulsory acquisition, as well as the profits on tail water and the rights to increase of assessment, as evidence of the unfairness of the proposed compensation.[7] The government responded that they made no profits from the additional market price of the land (collected without the bonus), which "goes to form a fund for the amelioration of local conditions," and that tail waters would be utilized "in the interests of industrial development."[8] The origins of this convoluted arrangement of "double compensation" lie in officials' recognition that similar land could simply not be offered in exchange as per the provisions of the Act, so "we should double the land values, pay half in compensation and keep the balance as a fund to use for the benefit of the expropriated hillmen." This was not done for purely altruistic or even paternalistic reasons, but as an attempt to extract a royalty commensurate to the "situation value" of lands that were about to appreciate due to their importance for the company, yet to which the government had renounced all claims. The provision was subsequently dropped in the Mulshi agreement, with all costs of resettlement and compensation to be borne by the company.[9] One of the many underlying conflicts exposed by the extraordinary resistance to the Mulshi project was over the character of the colonial state itself, as primarily a rent-seeking entity newly charged with promoting industrialization following the report of the Indian Industrial Commission in 1918. This required the state to partly abandon its role as the designated protector of a settled agrarian order and to actively engage with speculative pricing in land markets.

To return to the grievances of the Mulshi peasants and their leaders, one of the key points of contention was the extent of lands to be acquired. At the outset, the collector of Poona received conflicting information that suggested, as had been the case at Dudh-Sagar, that the Tatas were deliberately understating their requirements: "Tatas' say that the lake will only be 15½ square miles in extent: and almost in the same breath that they propose to acquire 40,000 acres. I am trying to get them to reconcile these figures."[10] To be sure, the majority of opinion in the valley was against acquisition entirely, symbolized by the cry of "Jan kinwa Jamin" (life or land) raised by Senapati Bapat's *satyagraha* movement (insistence on truth) (Vora 2009: 48, 84–5). But the particularities of land being submerged by dams were such that the Tatas sought to limit the amount of land to be acquired, while those willing to accept compensation or resettlement demanded the company acquire affected as well as submerged villages at the same rate. A petition from one K. B. Joglekar claimed the secretary to the governor had given the assurance in 1920 that lands outside the contours of the lake would be purchased, but he was curtly informed that no such requirement existed.[11] In fact, the commissioner had written to the Tatas pointing out that "while all the rice lands of the Cultivators' villages are taken up, their 'Workas' or 'Jirayat' lands which are useless to them are left untouched." It would only be fair that these lands be acquired as well. In response, the Tata Power Company Board directed that "care must be taken not to acquire any lands other than those which fall within the rule of 'injurious affection,'" and that compensation would be paid at market rates and not at the "excessive" acquisition rates (which had been agreed to only for the sake of expediency).[12] Eventually, lands in nine villages were purchased by private treaty at lower rates: "Rs. 25 to 30 for Jirayat lands as against Rs. 60/- being paid within the submerged area."[13] By virtue of its drastic effects on the geography of the surrounding region, the dam projects produced a distinct set of inequalities, as the provisions of the Land

Acquisition Act were not applied to degraded lands. The law, then as now, provided few mechanisms for redress that went beyond the alienation of land as a commodity. The character of land as a resource that could be easily devalued in multiple, often unpredictable, ways went unrecognized.

11.3 State Sovereignty and the Application of the Land Acquisition Act

This section steps back from the details of the Mulshi case, in order to examine the broader legal apparatus through which the state dealt with land claims in this period. The question of compensation was crucial because it could set a precedent in future acquisitions. Even as the company exploited the ambiguous and ad hoc nature of the process, the colonial state was repeatedly forced to clarify the principles which justified the operations of the Act in the context of clamors for legislative reform, and even confusion among its own officials. The standing instructions for acquiring officers in 1911, as the first Tata agreement was being negotiated, were to ascertain the "market value of the land" through laborious on-site investigations of sales. They were reminded that "'potential values,' such as may arise from the use to which the land is to be put, are expressly excluded from consideration." Service or *inam* lands could be acquired, but had to be compensated through a grant of equal lands in exchange or an annual payment.[14] The Mulshi petitioners were to be tragically disappointed in this respect. They believed the *sanads* (deeds granted to princely states in British India) held "from the time of the Peshwas," which included both the soil and revenue, were "absolute and for all times to come."[15] They also rejected compensation based on market value, proposing instead the "cost necessary to secure slightly superior lands in another locality." Since the lands sold on the market were often of the most inferior quality, they could not be a reliable guide in fixing the rates. The "real market value" according to the cultivators themselves was in fact three times greater than the proposed awards, ranging from Rs. 1,750 to Rs. 2,500 per acre (Vora 2009: 68–72).[16] The final amounts upheld by the District Court, by contrast, came to Rs. 550 per acre for *ambemohor* rice land, Rs. 350 per acre for *sal* (actually a forest seed containing oil) rice land, and Rs. 50 per acre for *jirayat* (dry crop) lands, plus the 15 percent bonus for compulsory acquisition.[17] They had been determined "unofficially" by the government, bypassing the Land Acquisition Act and thus "not strictly in accordance with the law."[18] Although higher and more speedily disbursed than in the Andhra Valley case, the Mulshi awards satisfied neither the proprietors nor the government.

A few years later, "political agitation" surrounding another dam project at Bhatgar forced another departure from established principles, erring "on the side of liberality" by paying similar rates for similar land. When an outraged surveyor pointed out the arbitrary nature of this decision, he was simply told: "The reasons for the enhancement of rates are not far to seek. They were chiefly political and due to the satyagraha movement which was in full swing then. Unless therefore the rates were attractive it was impossible to get the rayats (cultivators) to part with their lands without making trouble."[19] It must be emphasized that the Mulshi precedent was never codified, merely reproduced as a series of exceptions.

What these exceptions did reveal, however, was the peculiar interplay between the categorical assertion of state sovereignty that lies at the heart of the Land Acquisition Act, and the fragmentary and highly devolved nature of its implementation. The only legislative attempt to amend the Act around the time of the Mulshi struggle was a modest bill by assembly member J. Ramayya Pantulu, proposing to allow district courts to determine when the "public purpose" clause could apply to individual acquisitions (Krishnan 2012: 96–7). Of the many responses to the bill, which did not pass, that of the Development Directorate of Bombay Presidency stands out. In England, where approval of the legislature was necessary, compulsory acquisitions

> are reduced to a minimum, and may safely be, as titles are jealously guarded and the seller is normally a person with a title. In India with its present form of quasi-parliamentary government and a special branch of civil law which deals with benami transactions and a very haphazard method of dealing in land as though it were ordinarily a form of property transferable as freely as a roll of cotton cloth (thought it must be admitted that the law itself does not countenance such a procedure) it is absolutely necessary that a special collector must be appointed to extinguish all rights on the property.[20]

On the one hand, this statement may be interpreted as an unequivocal defense of the state bringing order to a system characterized by a multiplicity of tenures, with due allowance to the fact that the state's own land policies had sustained it for over a century. However, the lament over the liquidity of land reads as deeply ironic, considering how easily alienation and conversion of land (including the ostensibly secure *inam* titles) took place at Mulshi, and increasingly so thereafter. The government's willingness to allow the Tatas to undertake their own negotiations whenever possible and to secure exceptions in the calculation of rates, as long as its own fixed rents were secure, suggests that the massive wave of expropriations in the Western Ghats in the 1910s and 1920s were made possible at a particular historical juncture when ventures by private capital in the interior of the country may have exceeded the legal and political frameworks that could have contained it. With many key decisions taken ad hoc and through the exercise of sovereign discretion, a properly "hegemonic" legal regime governing acquisitions failed to emerge in this period.

11.4 The Tata Hydroelectric Power Companies: Regional Consequences and Conflicts

The Tata hydroelectric power projects eventually overcame the resistance at Mulshi, though not without heavy losses. As Vora shows in great detail, divisions grew between peasants and moneylenders, with the latter increasingly eager to settle with the company. By 1922, 80 percent of the Mulshi Peta had accepted compensation, even at the high point of Bapat's *satyagraha* (Vora 2009: 48–51, 152). As construction proceeded, the areas immediately surrounding the dams in the Western Ghats remained underserved by company-provided infrastructure. Pursuant to the clause on the protection of public rights of way inserted by the government into the agreements, the company initially sanctioned a modest expenditure on constructing and widening roads around the lake.[21] The Second Revision Settlement of the Mulshi Peta noted that "people living on the western side of the Dam are seriously handicapped or obstructed from visiting Poona, their chief market-town and even the places of weekly bazars in the Peta."[22] The collector wrote to the Tatas in 1930 pointing out that villages on the northern side of the lake were completely isolated, with ferries the only possible solution: "it is the greatest pity that all idea of constructing new roads in this area was not given up from the very start and all effort concentrated on supplying a satisfactory system of water transport." The villagers for whom new roads had been provided "will all be in a much worse position than they were before the dam was constructed."[23] Dire conditions also prevailed in the Andhra Valley, where rivers had become impassable at certain places. The company refused to erect bridges at these obstructed crossings, contending "that their liability was confined, in accordance with the terms of the agreements, to the provision of the amenities which had existed prior to the advent of their tail water," which, they added, was anyway the property of the government.[24] The promise of using the water for irrigation purposes remained unfulfilled (Vora 2009: 80). Up to the late 1930s, the government attempted to induce cultivators to take up irrigation through an exemption from water assessment, but found that due to a lack of capital and security of tenure, they were unable to do so.[25] The inhabitants of the valleys were caught between conflicting understandings of the rights and responsibilities of the company, for whom the provision of infrastructure was a matter best left to the government—yet another consequence of the ambiguous and ill-defined provisions of the initial agreements.

But what, then, of the prospect of proletarianization? This was one of the strongest incentives offered to the Mulshi peasants for acquiescing to the dam projects. Their counterparts in Andhra had agreed to serve as contract labor on the dam for seven years, and it was hoped this example would be emulated. A statement submitted by the company in advance of a Legislative Assembly debate in 1921 made the following claim:

> An agriculturist who earned a bare pittance of about Rs. 60/- to 80/- per year by putting in only 3 months of agricultural labour during the year could earn over 250 Rupees by working for a year on the works at the site of the dam. The obvious incentive to work would turn him into a daily labourer of regular habits, who would refuse to go back to the land at the end of the construction period and would join the ranks of industrial labour for his own and his country's good... [The] total income during the 6 or 7 years of construction which in many cases would greatly exceed the value of the land they were dispossessed of, apart from their rescue from the strangling grip of their money lenders.[26]

The Tatas were, in effect, promoting a sudden and irreversible change in the class structure of the valley, but without the creation of a long-term stable workforce on the site (by contrast with the strategy adopted at Jamshedpur). The worker, alienated from the land and appropriately disciplined, would be left to roam on his own in search of future employment, presumably in the factories of Bombay. The Settlement Report in 1926 optimistically declared that "the Tata works at the Mulshi camp have almost revolutionised the whole Peta," despite conditions being "at present unstable and in a state of transition." Wages of field and skilled laborers had quadrupled due to the Tata works. But the most positive result of the land acquisition process was that "a few agriculturists who were ever in slavish bondage of their Saokars [moneylenders] having paid off their debts from the amounts of awards are now free people entering on a new lease of life though under most trying conditions."[27] In response, the collector of Poona wondered whether "the demand for labour and food and the general stimulation of trade caused by the large camp and construction works is the only cause of the apparent prosperity of the inhabitants."[28] Indeed, most outward migration had taken place immediately after the acquisition, and the few remaining villagers resumed farming on inferior quality soils. Even the displaced who did make their way to the city could not benefit from ongoing links with the countryside, which otherwise played a crucial role in sustaining the livelihoods of the working classes of Bombay (Chandavarkar 1994). The Peta remained without irrigation and electricity until the early 1980s, when the Mulshi Dam Region Development Committee was formed. Its demands extended to the provision of roads, schools, and hospitals, as well as permanent appointments and reservations for affected persons in the company's works (Vora 2009, 158–62). This movement pursued a belated recognition that the company was tied to the land and people, despite all its attempts to disavow them. As with most subsequent large hydroelectric projects in India, labor demands were easily met by minimal casual recruitment from the areas around the dams, which remained isolated and impoverished (Hardikar 2013). Displacement for dams results in the destruction of land, which cannot be re-purposed or returned to the original owners. Provision of infrastructure or alternative land for cultivation thus becomes the claim of last resort for the displaced.

11.5 The Tata Iron and Steel Company: A Comparative Case

The case of steel plants presents, at first glance, a set of conditions radically different from the construction of dams. Land is not destroyed but converted to productive industrial use, and the conversion of agricultural populations into industrial workers appears more feasible. However, acquired land is almost never released, even when projects are later given up, and the demands of creating a secure township and housing a substantial labor force lead to the necessity of maximizing the surface area to be acquired. This section undertakes a brief comparative analysis of land acquisition for the steel plant at Jamshedpur, in order to point to some of the general features of the process not reducible to the characteristics of a specific project.

Around the same time as he was contemplating the Dudh-Sagar power project, J. N. Tata began the process of acquiring prospecting licenses for iron ore, coal, copper, and manganese in the Chanda District of the Central Provinces, with a view to setting up an iron and steel plant. He sent his son Dorabji and his nephew Shapurji Saklatvala as far afield as Hyderabad to pursue all possible leads on raw materials.[29] This venture, more than any other, would require the Tatas to negotiate with the numerous princely states then controlling vast tracts of land in central and eastern India. With investigations at Chanda yielding little result, especially with regard to coal of sufficient coking quality, a fortuitous series of discoveries were made of suitable iron ore deposits at Dalli-Rajara near Raipur, and coal beds at Jharia in Bengal. Saklatvala observed in 1904 that this would mark a significant shift in location eastward: "We shall not be able to manage easily two Iron-mining establishments at great distances in Dhullee and Chanda."[30] Having decided on the sources of raw materials, the next question to be resolved was the location of the plant. At this juncture, the most important intervention came from the princely state of Mayurbhanj near Bengal. The geologist P. N. Bose, then in the employ of the Maharaja, pointed to the extraordinarily rich iron ore deposits in the Gurumahisani District. The terms of the lease were favorable, so this became the preferred source.[31] This series of decisions reveals the ways in which a regime of extraction and production emerges across state spaces marked by multiple sovereignties (princely and British), and under the constraints of particular resource geographies.

For this reason, I pay close attention to the negotiations over Sini, a site located 170 miles inland on the Bengal–Nagpur Railway in the Seraikela state and ultimately not chosen, which presented difficulties similar to the falls at Dudh-Sagar. The colonial state grew concerned that the new company would evade their control, and the potential benefits it could reap from the industry would be reduced. The deputy commissioner of Singhbhum tried to persuade the Tatas to build exclusively in British territory:

> If the Sini site is adhered to and the Borgaon Tank constructed for its water supply, both the site and the tank will be in Native States, out of the direct control of British officers... Even after the construction the commencement of the Works, there are several details of administration where Native States are likely to be obstructive & troublesome in various ways.

The Tatas' response was to essentially play one power against the other:

> There are two alternatives open to us:- (1) To shift the site as they advise, or (2) to obtain an option from the Raja of Seraikela fixing definite price for the land to be occupied by the Works... it should be put before him that in case he wouldn't give us reasonable terms and a period of option, we would not establish the Works in his Territory but go over to British land.[32]

This was a risky strategy, because if the company bound itself to building in the Raja's domain, it would be giving up the possibility of using the Land Acquisition Act: "We might be thus placing ourselves at his mercy. The British Government could not help us much; there will be no land Acquisition Act to put into force; and the Raja may insist on his own price being paid."[33] As in the case of the hydroelectric companies, the primary aim of the Tatas' acquisition policy was to obtain security of tenure at the lowest possible rates, whatever the method. But it was not clear whether the flexibility of purchasing by private treaty could overcome the advantages provided by the weight of colonial law and bureaucracy.

The initial area acquired for the steel plant under Section 41 of the Land Acquisition Act was 3,564 acres at a rate of Rs. 13 per acre, with a total compensation amount of Rs. 12,000/- paid to the expropriated *raiyats*. Tata Iron and Steel Company (TISCO) informally agreed to provide alternative land for the displaced, but no explicit provisions were inserted into the agreement.[34] There were few protests, mainly because the company had deliberately avoided acquiring the more populated villages. A further 12,000 acres were added in 1917 for extensions of the plant, employee housing and leases to subsidiary companies. Borrowing certain terms from the agreement with the Tata Hydro-Electric Company, TISCO attempted to secure the right to permanently alienate land and to provide for future acquisitions.[35] The Maude Committee, appointed by the government of Bihar and Orissa in 1919 to investigate problems arising from the expansion of the steel town, stated in its report that more acquisitions were sure to follow, but difficulties would be encountered due to the rising price of land and insecurity of tenure in the remaining villages. TISCO held these villages on lease from the Midnapore Zamindari Company, which temporarily controlled the estate due to a succession dispute then pending in the courts. TISCO had "obtained all the rights of the superior landlord, but they were quite unable to control the *Pradhans* (village headmen) who assigned waste lands in the villages of Sakchi and Beldih for the erection of houses and thus exceeded their customary rights."[36] The preferred solution would be the further use of the Land Acquisition Act, the magic wand that could extinguish all existing property rights.

However, at this stage colonial officials had begun to doubt that the Act could legitimately be used to "clear up problems of doubtful ownership."[37] The Maude Committee argued that "the application of the Act has been unduly extended to cover cases in which land is required for private industrial concerns such as the Tata Iron and Steel Company," whose business was deemed "only indirectly useful to the public." Therefore, unless the Act was amended, the Committee's recommendation was the formation of a statutory Board of Works, a local government authority that would be far better equipped than the company to acquire land and assist the dispossessed tenants.[38] This was a clear challenge to the arrangement whereby the company exercised de facto sovereign rights over an expanding area in and around Jamshedpur, and it was prompted mainly by the legal question of land acquisition. The company itself had precipitated this impasse, by claiming that "we constitute a large Municipality, and as such we are a public body and entitled to the operation of the Land Acquisition Act."[39] This attempt to assuage the colonial state's doubts about the applicability of the "public purpose" clause soon backfired. Land became the main vehicle of a long struggle for political control over the surrounding region, which continues to this day.

Under conditions of frequent industrial unrest at Jamshedpur throughout the next two decades, no further action on the legal status of the township was taken. The question of TISCO's control over land resurfaced only after independence, when the Bihar Land Reforms Act of 1950 triggered a thirty-year court battle leading to a renegotiation of lease terms. The latest agreement was signed in 2005 with the newly formed Jharkhand government. Jamshedpur, now an urban agglomeration of more than 1 million people, has not been declared a municipality and services continue to be provided exclusively by Tata Steel through its newly formed subsidiary JUSCO (Jamshedpur Utilities and Services Company).

The process of land acquisition for the steel company produced a similar set of conflicts over the provision of infrastructure as in the case of the hydroelectric companies. The terms of the initial agreements (which, as I have shown, borrowed from each other) did not anticipate the effects of the drastic changes in the respective regions' geographies and the resulting divergence between the interests of state and capital. The companies were further drawn into the complexities of rural social relations within an agrarian system marked by multiplicity and uncertainty of tenures. They severed the bonds between moneylenders and peasants at Mulshi and sought to break the power of local village headmen at Sakchi, through a combination of legal proceedings, and the creation of new land and labor markets. The operations of both industries extended over large areas, yet were concentrated in a set of bounded spaces (dams, power houses, and transmission lines on the one hand, and factories, townships, and mines on the other). The key difference was the extent to which those boundaries were porous, and the steel company as producer, employer, and landlord found itself directly governing large populations that could not simply be displaced and freed as a roaming proletariat. Indeed, TISCO was forced to absorb and discipline labor on a semi-permanent basis, which tied it to the countryside and regional political economy in ways that were difficult to manage and predict.

11.6 Conclusion

The "land question" in India is now receiving due scholarly attention from a variety of disciplinary and theoretical perspectives. One persuasive revision of David Harvey's overarching model of "accumulation by dispossession" proposes an emergent neoliberal "regime of dispossession" in India fundamentally unlike its Nehruvian predecessor, with a shift from "land for production," as seen in state-controlled industrial estates and townships, to "land for the market" in special economic zones (SEZs) where agricultural land is indiscriminately converted into real estate (Levien 2013: 383–4). On the other hand, it has been argued that "accumulation by dispossession" does not apply at all in the Indian case, due to the preponderance of state projects throughout the history of land acquisition since the Nehruvian period, with the colonial state assumed to have been "not developmental" (Chakravorty 2013: 90–1, 129). Taking a longer view, I have sought to recover an earlier "regime of dispossession," marked by the limited assumption of control over spaces of resource extraction and production by private capitalists under the aegis of post-*swadeshi* industrial development. The colonial state *did* begrudgingly commit itself to this goal, but it was unable to reconcile its own fundamental character as an agrarian rent-seeking entity with the tasks it was to perform as a "land broker" for firms like the Tatas. The protracted negotiations over revenue assessment and the rights to water and land at Jamshedpur and in the Western Ghats reveal the frustrated desires of the colonial state to retain some measure of control and assert its own sovereignty, despite the substantive concessions it had given to the Tatas. It was a problem specific to this historical moment, with the Nehruvian state later taking upon itself the responsibility for industrialization, and the postliberalization state willing to make even greater concessions, such as allowing private capital to "capture the ground rent" from real estate (Levien 2013: 395).

If much has changed since 1991, we also find continuities between the present and the more distant colonial past of land acquisition. Negotiations of capitalists with multiple levels of sovereignty have always been an important factor in determining outcomes. The Tatas' playing the Raja of Seraikela against the British recalls the deal-making over the location of the Nano plant involving West Bengal and Gujarat, which has been described as the giving and taking of "geobribes" (Roy 2011: 267). Despite the simplification of formal tenures after the land reforms of the 1950s and 1960s, the Land Acquisition Act remains the best instrument available for resolving ambiguities in ownership and usufruct rights (Chakravorty 2013: 112). The broad definition of the "public purpose" clause of the Act has allowed the courts to routinely justify acquisitions (Wahi 2013: 51). As I have shown, this definition was first worked out as the colonial state faced new claims on behalf of private companies. The Tata case is instructive precisely because it acts at once as a precedent for, *and* a catalyst of, the state-driven mode of accumulation that took shape in the immediate postindependence period and that is now being radically reconfigured under conditions of rapid globalization and a fractured political landscape in India.

Notes

1. Tata Central Archives, Pune (hereafter TCA), FP-41. J. N. Tata to Richmond Ritchie, Secretary to Lord George Hamilton, May 10, 1901.
2. TCA, T53-DES-T20-IMPORTANT-EXTRACTS-DHARAVI-1- 0004, Edward Miller to Gostling and J. N. Tata, July 25, 1902.
3. Maharashtra State Archives, Mumbai (hereafter MSA), Revenue Department (Forests), 1911, No. 444, Vol. 116. Grant on lease to Messrs. Tata & Sons of certain forest lands in Dist. Belgaum for construction of reservoir.
4. TCA, T53-DES-T20-IMPORTANT-EXTRACTS-DHARAVI-1-0043-44. Macdonald Miller & Co. to Gostling and J. N. Tata, December 30, 1904.
5. TCA, T53-DES-T20-IMPORTANT-EXTRACTS-DHARAVI-2-0065-68. Memorandum on the Public Utility of the Bombay Hydro-Elec. Scheme, prepared by Dr. John Mannheim, October, 25, 1906.

6. MSA, Revenue Department (Lands), 1911, No. 625, Pt. I. Acquisition under Act I of 1894 of certain lands in the Poona and Kolaba Districts.
7. MSA, Revenue Department (B Branch), 1920, No. B-117, Pt. III. Nila Mula Power Supply Project, Deputation from inhabitants of thirty-seven villages of Mulshi Peta, pp. 73–83.
8. MSA, Revenue Department (B Branch), 1920, No. B-117, Pt. III. Nila Mula Power Supply Project, Deputation from inhabitants of thirty-seven villages of Mulshi Peta, p. 101.
9. MSA, Revenue Department (B Branch), 1920, No. B-117, Pt. VI. Nila Mula Power Supply Project, Correspondence relating to draft agreement under Section 41 of the L.A. Act, pp. 89–90, 295–303.
10. MSA, Revenue Department (B Branch), 1920, No. B-117, Pt. V. Nila Mula Power Supply Project, publication of press note and compensation terms.
11. MSA, Revenue Department (B Branch), 1920, No. B-117, Pt. III. Nila Mula Power Supply Project, Deputation from inhabitants of thirty-seven villages of Mulshi Peta, pp. 171–6.
12. TCA, T53-DES-T20-TP-BO-1-PG-025. Tata Power Board Meeting No.18, held on February 15, 1922.
13. TCA, Box 459, TEC/T20/TP/BO/1, Board Papers, Tata Power Co., 1921–4. Circular No. 40/22, September 8, 1922.
14. MSA, Revenue Department (Lands), 1911, No. 1636. Instructions for the guidance of officers acquiring lands under the Land Acquisition Act.
15. MSA, Revenue Department (B Branch), 1920, No. B-117, Pt. III. Nila Mula Power Supply Project, Deputation from inhabitants of thirty-seven villages of Mulshi Peta.
16. MSA, Revenue Department (B Branch), 1920, No. B-117, Pt. III. Nila Mula Power Supply Project, Deputation from inhabitants of thirty-seven villages of Mulshi Peta, pp. 91, 128–30.
17. Ambemohor is a particular kind of rice from the hills of Western Maharashtra, while sal is a forest seed that yields oil.
18. MSA, Revenue Department (B Branch), 1920, No. B-117, Pt. X. Nila Mula Power Supply Project, Rates of compensation to Mulshi Ryots.
19. MSA, Revenue Department (F Branch), 1924, No. 2591-24. Land Acquisition, Method of correctly estimating the values of land.
20. MSA, Revenue Department (Development),1922, No.79. Land Acquisition Act Amendment.
21. TCA, Box 459, TEC/T20/TP/BO/2, Board Papers, Tata Power Co., 1925–1933. Circular, December 23, 1925.
22. MSA, Revenue Department (B Branch), 1927, No. 9299-24. Condition of communications in Mulshi Peta, Report regarding the Second Revision Settlement of the Mulshi Peta, pp. 3–4.
23. MSA, Revenue Department (B Branch), 1927, No. 9299-24. Condition of communications in Mulshi Peta, Collector of Poona to Tata Power Company, January, 31, 1930.
24. MSA, Revenue Department (B Branch), 1924, No.180-24, Non-agricultural assessment, Kundley Valley Co. quarry purposes, pp. 275–8.

25. MSA, Revenue Department (B Branch), 1927, No. 8460-24, Tail water of Tata Hydro-electric works, exemption from water assessment.
26. MSA, Revenue Department (B Branch), 1920, No. B-117, Pt. V. Nila Mula Power Supply Project, publication of press note and compensation terms, Notes on Tata Power Co., p. 5.
27. MSA, Revenue Department (B Branch), 1927, No. 9299-24. Condition of communications in Mulshi Peta, Report regarding the Second Revision Settlement of the Mulshi Peta, pp. 10–12.
28. MSA, Revenue Department (B Branch), 1927, No. 9299-24. Condition of communications in Mulshi Peta, Report regarding the Second Revision Settlement of the Mulshi Peta, p. 49.
29. TCA, FP-36. D. J. Tata to J. N. Tata, September 25, 1902.
30. TCA, FP-41A. Shapurji Saklatvala to Messrs. Tata & Sons, Mining Department, Bombay, August 22, 1904.
31. TCA, B. J. Padshah to R. D. Tata, February 24, 1905.
32. TCA, B. J. Padshah to R. D. Tata, December 22–23, 1905.
33. TCA, FP-220. Tata Sons to R. D. Tata, December 21, 1905.
34. Tata Steel Archives, Jamshedpur (hereafter TSA), Land Acquisition File No. 5 (1908, Pt. I). From the Tata Iron and Steel Co. to the Government of Bengal, Revenue Department, January 3, 1908.
35. TSA, Land Acquisition File No. 12 (1918, Pt. I). Draft of the Preliminary Acquisition Agreement, General Manager's Letter no. 306, January 19, 1918.
36. India Office Records, British Library, London (hereafter IOR), Mss Eur E251/22. Papers of Sir Maurice Hallett, Report of the Jamshedpur Committee, pp. 20–1.
37. National Archives of India, Department of Commerce and Industries (Industries Branch), "A" Progs. Nos. 1–9, March 1919, Note by A. E. Gilliat in the Revenue and Agriculture Department, September 19, 1918.
38. IOR Mss Eur E251/22. Papers of Sir Maurice Hallett, Report of the Jamshedpur Committee, p. 22.
39. TSA, Land Acquisition File No. 12 (1918, Pt. I). Memorandum, June 11, 1918.

References

Chakravorty, S. 2013. *The Price of Land: Acquisition, Conflict, Consequence.* New Delhi: Oxford University Press.

Chandavarkar, R. 1994. *The Origins of Industrial Capitalism in India: Business Strategies and the Working Classes in Bombay, 1900–1940.* Cambridge: Cambridge University Press.

Ferguson, J. 2005. "Seeing Like an Oil Company: Space, Security and Global Capital in Neoliberal Africa." *American Anthropologist,* 107(3): 377–82.

Hardikar, J. 2013. *A Village Awaits Doomsday.* New Delhi: Penguin.

Krishnan, E. 2012. "Private Speculations and the Public Interest: N. C. Kelkar's Land Acquisition Bill." *Socio-Legal Review,* 8(2): 78–105.

Levien, M. 2013. "Regimes of Dispossession: From Steel Towns to Special Economic Zones." *Development and Change*, 44(2): 381–407.

Padel, F. and Das, S. 2010. *Out of This Earth: East India Adivasis and the Aluminium Cartel.* New Delhi: Orient Blackswan.

Pati, S. 2012. *Industrialization and Displacement: An Economic Impact.* New Delhi: Concept Publishing.

Roy, A. 2011. "The Blockade of the World-Class City: Dialectical Images of Indian Urbanism." In *Worlding Cities: Asian Experiments and the Art of Being Global,* edited by A. Roy and A. Ong, Hoboken, NJ: Wiley-Blackwell, pp. 259–78.

Sanchez, A. and Strümpell, C. 2014. "Sons of Soil, Sons of Steel: Autochtony, Descent and the Class Concept in Industrial India." *Modern Asian Studies*, 48(5): 1276–1301.

Vora, R. 2009. *The World's First Anti-Dam Movement: Mulshi Satyagraha, 1920–1924,* Ranikhet: Permanent Black.

Wahi, N. 2013. "Land Acquisition, Development and the Constitution." *Seminar*, 642: 49–54.

12

An Ethnographer's Journey through Land Grab for Capitalists by the Left Front Government in West Bengal

Abhijit Guha

12.1 Introduction

Historically, ethnography emerged as a method in the social sciences, particularly in anthropology, as a descriptive writing about a culture or society through the participation of the ethnographer in the daily life of a community in a specific site. Gradually ethnographers have moved from mere description of cultures to explanations and interpretations of human behavior in specific contexts (Nader 2011: 211–19). In course of time, explanations of societal phenomena were being sought by ethnographers through the positioning of their ethnographies in multiple sites and contexts wherein local events and global issues were often placed within a single ethnographic text. The value of this kind of multi-sited ethnography is more than one. *First*, detailed depiction of ground realities mediated by the author's personal encounters with the subjects, which are not found in macro-level theoretical expositions of the same problems. *Second*, positioning of the local actors and events in a macro-level context and theory and, *third*, generating a dialogue with the theoreticians who are not confined with the problems of a specific locale.

In line with the aforementioned trends of ethnography, I have attempted to place a case of land grab ("land acquisition" in the legal and governmental discourse) at the intersection of the local and the global that took place under a communist-led government in the West Bengal state of India. For this reason, I first go through the macro-level scenario of land grabs as depicted in the literature by academics and international funding agencies, which formed the global background of a local land grab in a particular ethnographic context.

Under the global capitalist market-driven regime, land grab in the literature (including media reports) is used as a blanket phrase to refer to large-scale commercial land transactions between countries involving transnational corporations for the production of food, biofuels, and timber (Borras, Jr. and Franco 2012: 34–59; Murphy 2013: 1–12). International funding agencies on the other hand prefer to use the term "compulsory acquisition" (FAO 2008: 5–17) and "land transfer" (World Bank 2011: xxv–xlv) in which people are displaced from their major means of production by the state. Without using the term "land grab," the United Nations Special Rapporteur on the Right to Food, in a 2009 report, however, emphasized putting in place proper institutional arrangements to ensure land rights, food security, pre-investment impact assessments, and transparency for the local populations prior to all kinds of large-scale land acquisitions or land leases by developing countries that are attracting international business investments under the impact of globalization (Schutter 2009: 12–15). An Oxfam discussion paper entitled "*Sleeping Lions*," on international investment treaties and access to food, land, and water also used the term "land grab" and pointed out succinctly that in the developing countries natural resources like land and water have "cultural and communitarian significance" that "exceeds their mere use and exchange value" (Perez, Gistelinek, and Karbala 2011: 5).

Academic attempts to look into the phenomenon of land grab from a global perspective have also started (Schutter 2011: 249–79). In a recent introductory article entitled "Towards a better understanding of global land grabbing," the authors viewed land grabs as a kind of land deal politics which was often lacking in the current debate. I quote from the authors:

> We will focus, ultimately, on the politics of land deals—something often lacking in the current debate—and therefore we embed the commercial act of exchanging land titles into a broader framework concerned with "land deal politics". Through this initiative, we hope to foster a dialogue with social movements, activists, policy makers, and concerned academics to produce data and debate potential implications. (Borras, Jr. et al. 2011: 209–16)

Under this general global background on the study of land grabs the chief aim of this chapter is to write the ethnography of land expropriation by the administrative wings of the state of West Bengal and peasant resistance against a pro-peasant communist party-led government in the West Bengal state in India.[1] The uneven path of the capitalist transition in West Bengal treaded by the then leftist government was marked by the contradictions between ideology and praxis, coupled with bureaucratic failures and peasant resistance against land grabs, which finally led to a stalemate toward capitalist transition. The ethnographic method of description, which turned out to be useful in revealing the troubled path of the stalemate involved the juxtaposition of typical anthropological field data with a variety of archival sources and my personal experiences of interaction with the politicians at the highest level of policymaking on the issue of reforming a colonial law, which was in vogue in India until the enactment of the new law, named the Right to Fair Compensation and Transparency in Land Acquisition, Rehabilitation and Resettlement Act, 2013.[2]

In Section 12.2 of this multi-sited ethnography, I briefly narrate my experience of interactions with the policymakers of the country (the members of the Parliamentary Standing Committee) in the setting of a committee room. My dialogue with the members revealed their attitude toward capitalist transition, which cut across political parties. Almost all of them were found to be interested in how to give land to the capitalists, who, according to the members, were already purchasing huge amounts of this natural resource from under the feet of the cultivators.

In Section 12.3, I first describe my field findings in some detail, which revealed some of the consequences of the dispossession of the peasants from their major means of production, that is, fertile agricultural land, by the government for two big private companies. This has placed my field area within the wider context of land acquisition in West Bengal. In this section I have also made an attempt to look into the administrative intricacies of land takeover, which were often hidden from the public view.

In Section 12.4 of my ethnographic narrative I described the story of protest by the peasants against a land grab by the state, which revealed the contradictions of capitalist transition. The peasant protest has a link with Section 12.2 of my ethnography wherein I describe my encounters with the members of the Parliament who, I believed were already alienated from the peasantry at the grassroots level. In Section 12.5 I briefly recount some anecdotal evidence from the field that illustrates the wide gulf in understanding of what the poor peasants of West Bengal aspire to and what the remote decision makers in New Delhi think about development.

12.2 Facing the Policymakers

I was among the Members of Parliament (MP) from different parts of the country at the big Parliament Library building, New Delhi, on June 17, 2008. I was feeling nervous, although I was invited as an expert to give suggestions on the reforms to be undertaken on the century-old Land Acquisition Act of India (1894) by which the government's power of *eminent domain* was used to acquire land for "public purpose" in lieu of monetary compensation given only to the land titleholders and calculated on the basis of the previous market price of land in the area.

The deputy chairman of the Standing Committee began the discussion by asking me to highlight the major points that I recommended to insert in the proposed bill. I talked at length trying to convince the MPs about those subjects that I thought were downplayed in the bill. I emphasized the recognition of local self-governments while getting consent of the affected people for land acquisition, protection of food security at the household level, corporate social responsibility, and avoidance of agricultural land from the scope of land acquisition for private, profit-making industries. The deputy, with a smiling face, reacted by saying that I had raised certain "basic issues and philosophy behind the Act and there was no dispute on the idea," and he continued quite emphatically, "the reality was, one could not avoid land expropriation since private companies were already purchasing huge chunks of land in the rural areas of the country." The chairman, who belonged to the Bharatiya Janata Party (BJP, now in power), also raised the same point. It seemed to me that the Standing Committee might not be interested in increasing the role of local governments, household-level food security, endangerment of land reforms, and all other issues on which I made my observation and published as a field anthropologist in cases of acquisition of huge chunks of fertile farmland in some of the villages of erstwhile Medinipur district of West Bengal where the peasants did not agree to sell their land to the big industrialists. My pro-peasant arguments did not seem to convince the policymakers who were already inclined in favor of land takeover by the industrialists through capitalist market mechanisms. The MPs, however, advised me politely to send the suggestions in script. Suffice it to say that I did my job on time. But my direct encounter with the policymakers ended at the Parliamentary Committee room. I will now go back to the story of my engagement with land grab by the government that I observed and describe my findings following the style of a multi-site ethnography (Buckley 2011; Gellner 2012: 1–16).

12.3 Dispossession of the Peasants in West Bengal by the Left Front Government

12.3.1 *Ideology and Praxis*

West Bengal is an agriculture-dependent state, which occupies only 2.7 percent of India's land area, although it supports over 7.55 percent of the Indian population and has a much higher population density than the national average (Census of India 2011). West Bengal had been ruled by the Communist Party of India (Marxist) (CPI (M))-led Left Front Government (LFG) for more than three decades (thirty-four years), which raised it to the position of the world's longest-running democratically elected communist government.

The LFG in West Bengal claimed its uniqueness among the Indian states not only by staying in power for thirty-four years through parliamentary democracy but also for the implementation of a pro-poor land reform program and decentralized local self-government with a fair amount of success (Mukarji and Bandopadhyay 1993; Lieten 1996).

In the late 1980s, and particularly in the wake of the liberalization in 1991, the focus of the development policy of the LFG radically shifted. The government, which was fully committed to land reforms, started to invite capital-intensive and technologically sophisticated private industrial entrepreneurs, including multinational corporations, to the state. Interestingly, the success in land reform in the state was first emphasized in the policy statement of the LFG as early as 1994 and was later cited in a report by the government as one of the justifications for huge industrial investment (Government of West Bengal 1994: 1–18; WBIDC 2000: 44).

However, contrary to what has been dubbed a "success" in order to justify industrialization, an earlier report of the government devoted to the evaluation of the *panchayats* (decentralized form of governance by elected members at the village level) in West Bengal observed that land reform was still an incomplete program (Mukarji and Bandopadhyay 1993: 41). In fact, during the early 1990s, the ruling LFG leaders argued that since land reform was a very successful endeavor in the state, raising agricultural production and the purchasing capacity of the peasantry, the state became the ideal ground for the establishment of capital-intensive heavy and medium industry. I call this the *industrialization-through-land reform argument*. The second line of argument came from more theoretically oriented Marxists of the ruling parties, who claimed that industries would be able to absorb the extra labor engaged in agriculture in disguised form and as well as that resulting from the introduction of mechanization in agriculture. The proponents of this argument also stated that agriculture, owing to land fragmentation caused by inheritance of property rights, and hikes in input costs, have already become nonviable for many small and marginal farmer families. This argument may be termed *employment-through-industrialization* (Mishra 2007: 1–22). The phrases "industrialization-through-land-reform" and "employment-through-industrialization" have been coined by me in an earlier publication in the journal *Social Change* in 2014 (Guha 2014: 205–28).

The West Bengal Human Development Report (WBHDR) published in 2004 also noted with concern the landlessness among rural households in the state. It noted a "disturbing feature," referring to the increase in landlessness despite land distribution and registration of *bargadars* (sharecroppers without ownership rights on the land) (Government of West Bengal 2004: 40). In spite of these findings, the government of West Bengal pushed its agenda of industrialization in the era of globalization ignoring its immediate and long-term effects on land reforms. This forms the wider context of the ethnography in which I will now place the local level events around a land grab in the villages and the land acquisition department of Paschim (West) Medinipur district in West Bengal.

12.3.2 *Dispossession in the Field*

I conducted anthropological field investigations with my students during 1995–7 and 2006–8 in some of the villages within the Kharagpur I block (the lowest administrative category of a district in West Bengal) of the erstwhile Midnapore district, which was subdivided into west and east Medinipur in 2002. The area was located in western Midnapore and was characterized by undulating lateritic soils. The rural people mainly subsisted on monocrop agriculture. The specific area of my fieldwork is situated on the bank of the river Kansai, the largest river of the district. Cultivation of rice (a staple of the district) in the villages under study depended primarily upon rainfall and no systematic irrigation facilities had been developed by the government during my fieldwork. The villagers residing on the southeastern bank of the river cultivated a variety of vegetables on the land adjoining their homesteads owing to a good supply of groundwater from traditional dug wells. But just on the west of the southeastern railway track the groundwater level was not very good for cultivation of vegetables. The main agricultural activity on the western side of the railway track centered on rain-fed paddy cultivation, which took about four to six months of the year. Land for four private industries had been acquired by the government on the western side during 1991–6. Among these four, three had already started production: 1) Tata Metaliks (manufacturing pig iron), 2) the coke oven unit of the Wellman Company (supplying coke coal to the Tata Metaliks), and 3) the Bansal Cement factory. The fourth was a proposed pig-iron plant named Century Textiles and Industries Ltd (hereafter CTIL) owned by the Birla group of industrialists for which fertile agricultural land was acquired in 1996 but no factory was built on this land until 2004 (Guha 2004: 4620–3). The fourth company finally withdrew its project because it foresaw stiff competition in the market under the liberalized economy. Interestingly, the West Bengal government had acquired agricultural land for all these industries despite the fact that a huge area of uncultivable undulating lateritic terrain ("wasteland" in the official jargon) lay just by the side of these agricultural lands on both sides of the railway track that extended almost up to the highlands on the bank of the Kansai. It may be that the companies did not choose these undulating lands since it would have been more expensive for them to level the ground for building industry. It should also be noted in this connection, that no members of the land-loser families were provided with a permanent job in those industries, although the local left political leaders tried to create the impression that at least one member of those families would be given employment (Guha 2014: 205–28).

The results of my fieldwork revealed the maladies of state-led capitalist development on many fronts. Foremost was the fact that land grabs by the government created not only food insecurity for the peasants but also endangered the land reform initiatives undertaken by the then pro-peasant LFG (Guha 2006: 155–73). In the absence of any viable industry-based employment, the dispossessed peasants were simply left to fend for themselves (Guha 2007a, 2013: 797–814; Majumder 2015). The monetary compensation provided to the land losers could not rebuild their household economy and this fact conformed to the findings of researchers in many countries of the world (Cernea 2008: 15–98).

The following events narrated by a peasant in the field area represented a typical case of a land-loser family affected by governmental acquisition of farmland in the study area. The family owned 0.54 acres of inherited land of which 0.22 acres was acquired by the government for the Tata Metaliks Company and their other piece of purchased land was also acquired for another company in the year 1991. The head of the family got compensation money and he saved some money in a nationalized bank. The remaining land which he possessed after the acquisitions could not provide the family food for the whole year and they had to buy food crops for two to three months of the year and this was the most adverse event in the life of the family (Majumder and Guha 2008: 121–33; Guha 2011b: 336–57).

12.3.3 *How a Land Grab Failed*

The land acquisition for CTIL is a typical example of a case of land grab, where a huge chunk of acquired agricultural land remained unutilized for more than five years (Guha 2007b: 3706–11). In Table 12.1 I have outlined the chronology of the CTIL land acquisition.

Compensation payments for the company started after June 1996 and continued up to March 1998 and then the payments came to a halt. The CTIL did not start any construction work for its proposed sponge and pig iron plant although buildings were constructed by the side of the land acquired for the company's security guards (Guha 2007a: 87–94). The company's managing director, B. K. Birla, in an interview to a reporter from *The Statesman* said that they would not proceed with the project as "the national market of pig iron has become very much competitive because of the entry of China and Australia in the field." He also said that this decision was conveyed to the state government. The state land and land reforms minister, Surya Kanto Mishra, on the other hand, told the reporter for *The Statesman*: "We are not finding any taker for the land" (*The Statesman*, November 18, 1999: 4).

Table 12.1. Succession of events showing land acquisition for CTIL

DATE	EVENT
July 2, 1992	Cabinet of the government of West Bengal approved the project for the acquisition of 1,137.78 acres of land by Act-II of 1948 in eighteen mouzas under Kharagpur Police Station.[3]
December 3, 1992	Assistant secretary writes to district collector to proceed.
October 28, 1995	The first public notification published in the newspaper. Acquisition by Act-I was made. Land amount reduced to 526.71 acres in 10 mouzas.
December 29,1995	The hearing of objections by the land losers completed.
January 10, 1996	Peasants prevented soil testing.
March 22, 1996	Mass deputation by the peasants at the district collectorate. Peasants demanded land for land or job and higher compensation.
April 2, 1996	Publication of agreement between the company and the government in the *Calcutta Gazette.*
April 10, 1996	Peasants submitted memorandum to boycott parliamentary election.
May 2, 1996	Poll boycott decision of the land losers published in *The Statesman.*
May 22, 1996	CTIL got possession certificate for 238.86 acres of land.
June 21, 1996	The report on the objection hearing at collector's office completed and all objections overruled.
May–July 1996	The peasants were not allowed to cultivate on the acquired land. By this time the payment of compensation started.
February 28, 1997	Land and Land Reforms Minister announces that state would reclaim unused land leased to industries.
April 11, 1997	CTIL got possession certificate over another 119.39 acres of land
March 12, 1998	*The Statesman* reported about the adverse effects of acquisition in Kalaikunda
June 21, 1998	Kalaikunda Gram Panchayat held meeting with the land losers about the land which remained unused by the Company
October 12, 1998–March 15, 1999	The order sheet of the land acquisition Dept. reveals the non-payment of compensation to land losers as the company has stopped placing funds since December 1996
April 19, 1999	Land Acquisition Office, Medinipur writes to the MD of WBIDC seeking advice over this hanging situation
November 18, 1999	The Company Managing Director tells the reporter of *The Statesman* that they are not interested to establish the plant the Minister also expresses his helplessness over the matter
December 7, 1999	Joint Secretary, Land and Land Reforms Dept. writes to Dist. Collector to send report on the Acquisition

In another detailed report, published seven months later by the same newspaper, it was stated that the state government still seemed to be undecided over the fate of the land acquired for CTIL (*The Statesman*, July 5, 2000: 4). The individual landowners, who possessed plots of land in this area, also stopped cultivation since they received acquisition notices in 1996. It may be of interest in this context to mention that in one of its publications, entitled *Destination West Bengal*, the West Bengal Industrial Development Corporation (hereafter WBIDC) listed the sponge and pig iron plant of Century Textiles as an "upcoming" industrial unit in the Kharagpur area (WBIDC 1999: 21). One important point about the acquisition of land for Tata Metaliks (a pig iron company which has started production) and Century Textiles Companies deserves mention. Huge chunks of fertile agricultural land were selected by the companies and the Cabinet Committee of the West Bengal government gave approval to this selection. My field observation in this area revealed the presence of a huge tract of undulating lateritic nonagricultural land on the western side of the southeastern railway track lying on the north bank of the Kansai river. During my fieldwork, the land losers of this area repeatedly pointed out that the government should have acquired the nonagricultural land for the industries instead of taking their agricultural land. When this point was raised before the officers and employees of the Land Acquisition Department of the district collectorate, they simply stated that it was the decision of the government, which the concerned department at the district level had to execute. Under the strong policy support toward industrialization no question of corporate social responsibility was raised in the then administration of the government of West Bengal (Guha 2011a: 79–98). The government neither requested that the companies ask for unfertile land for industries nor did the government declare tax relief for companies that would have saved the livelihood of peasants by avoiding fertile land for industries. In any case, the aforementioned episode of the failure of land acquisition for CTIL not only disillusioned the peasants about industrialization, but also created the ground for protest against the land grab for industries. In the next and the last section of the chapter, I will narrate the ethnography of the peasant resistance against the land grab.

12.4 Protests by the Peasants against the Land Grab

The protests launched by the landowning peasants of the Gokulpur-Amba (two of our study villages) against land acquisition took many forms, even though these did not last for a long period, as happened in Singur block during 2006–7 in the Hooghly district of West Bengal. Several peasants took up the statutorily available means and instruments to put up their objections against land acquisition under Section 5A of the colonial Land Acquisition Act, 1894 during December 1995. A government report dated June 21, 1996 vividly recorded the objections and described in detail how the latter were overruled by the district collector.

The objections submitted by 342 land losers contained the following points: first, the acquisition of agricultural land would affect the farmers seriously by throwing them out of employment; second, the land losers would not get compensation at the rate they expected; and third, the proposed acquisition was against public interest and beyond the purview of the land acquisition act. It was interesting to observe how the concerned officials of the Land Acquisition Department overruled all the objections raised by the peasants. Before rejecting the objections, the officials, however, recognized the severity and magnitude of the acquisition. To quote from the report: "It is a fact that since large quantum of land is being acquired and the people chiefly subsist on agriculture many people will be seriously affected in earning their livelihood and avocation" (Government of West Bengal 1996).

However, this was the only sentence in the whole report that upheld the interests of the peasants. The rest of the three-page report was devoted to justifying the acquisition. The arguments of the officials centered on the low agricultural yield of the lands, which were monocrop in nature. Furthermore, the report mentioned the merits of the location of the land that would provide important infrastructure facilities for industry, such as the nearby railway line and the national highway No. 6. The report noted that during the hearing of the objections, the petitioners could not "specify their individual difficulty in parting with the land," although in the same report it was stated "most of the objectors submitted that they have no objection if employment is assured to them, in the company in favor of whom acquisition is being done." It is not clear from the report why the authors of the same could not understand the nature of "individual difficulty" in parting with the land, which according to the report was their main source of livelihood.

Three points raised in the report were quite significant and revealed the insensitive way the government dealt with the acquisition, which was going to have a severe impact on the subsistence pattern of a group of rural cultivators in an area where the cultivation was predominantly dependent on rainwater and a single food crop was grown in an agricultural season. I now enumerate the points.

First, at one point the report mentioned: "It is worthwhile to point out that objections have been received only from 342 landowners for the acquisition of 526.71 acre which will affect at least 3000 landowners, if not more." It seemed the official position rested on the logic that since the overwhelming majority of the peasants would not face any difficulty (at least there was no record of objection under the Land Acquisition Act), there was no need to record any objection against this acquisition.

Second, after citing the location related advantages of the land, the officials overruled objections regarding the question of earning a livelihood by saying that the proposal had been approved both by the screening committee and by the state after considering all aspects. Incidentally, the screening committee for the approval of any project included the *sabhadhipati* (chairman) of the *panchayat samity* (the second tier of the statutory local self-government) and the Member of the Legislative Assembly (MLA) of the locality. It was obvious that these peoples' representatives, who were members of political parties of the LFG, would not object to a proposal that had already been approved by the cabinet and the concerned ministries of their own government.

Third, the report dealt with the point of "job for land" simply by saying that the Land Acquisition Act, 1894 did not provide any relief except compensation, but the government might take up the matter with the company, particularly for those peasants who would become landless and would be devoid of any source of earning a livelihood. After having overruled all the objections, the procedure for land acquisition marched ahead.

Besides recording objections within the legal framework of the Land Acquisition Act, the peasants of this area also took recourse to extra-legal means to fight against the acquisition of their agricultural land, which is narrated in the following paragraphs. The information on this part of the peasant protest has been collected from interviews with the leaders and participants of this movement as well as from press reports and the various written memoranda submitted by the villagers to the district and state administrations.

Two interesting incidents may be mentioned in this regard, which would throw some light on the reasons behind the popularity of this movement among the Kharagpur peasants. The first incident took place in the month of May 1995 when a local peasant leader with the help of other peasants sent a deputation to the Tata Metaliks Company (this company had started production during the early nineties) authorities demanding some compensation for the damage caused by the movement of trucks carrying goods for the company over unacquired agricultural fields (there was no crop in the fields at that time) of those cultivators. The trucks damaged the dykes of the fields (*ails*) and the soil. Under pressure from the peasants the company had to pay compensation in kind to seventy-five peasant families in the presence of the *pradhan* (elected head of the lowest tier, i.e., *gram panchayat* of the statutory local self-government) of Kalaikunda Gram Panchayat (the lowest tier of the constitutional local self-government). Some amount of fertilizer was given to those peasants whose lands were damaged by the movement of the trucks.

In the second incident, the peasants sent a deputation to the district administration about the damage caused to the unacquired agricultural fields of some peasants for putting up pillars to demarcate the lands acquired for Century Textiles Company in Kantapal, Mollachak, and other adjoining villages. Those cement pillars were anchored by digging up about 4 square feet of land to a depth of 3–4 feet. They became permanent structures right on the agricultural fields of the peasants whose lands were not acquired. About 24–5 such pillars were constructed in early 1996, which demarcated the boundary of the acquired land for CTIL. The peasants argued that cultivation of fields with plough and bullocks outside the area under acquisition was not feasible owing to physical obstruction created by the pillars (Guha 2007a).

The district administration had to agree with this demand of the peasants and arranged for a payment of Rs. 420 as compensation to those families affected by the construction of the pillars. This compensation payment continued for two years but with the decline of the movement, the administration discontinued this compensation.

Both these incidents revealed that under pressure from an intelligent and organized peasant movement, the company as well as the Land Acquisition Department had arranged compensation for peasant families having no provision under the existing legal and administrative framework (Guha 2007b: 3706–11, 2009: 41–53).

The movement reached its peak from the later part of 1995 up to April 1996 during which the farmers even resorted to violent means. In the first week of January 1996 hundreds of farmers in the Kalaikunda area stormed into the tent of the engineer who was conducting soil testing and a land survey on behalf of the CTIL A leading national daily reported on January 10, 1996:

> Land Survey and soil testing work in Mathurakismat Mouza in the Kalaikunda gram *panchayat* area of Kharagpur rural police station undertaken by Century Textiles—a Birla group of Industries—had to be abandoned following stiff resistance from villagers last week . . . The farmers also blocked Sahachak for nine hours yesterday . . . They also lodged a complaint with the police against the firm.
>
> (*The Statesman*, 1996, January 10: 6)

On March 22, 1996, the same national daily reported a mass deputation by a group of peasants of the Kharagpur region before the district administration (*The Statesman*, March 22, 1996: 6). In this deputation, the peasants demanded land for land or a job for the members of the land-loser families. They also demanded compensation of three lakh rupees (INR 300,000) per acre of agricultural land. After this deputation, about 100 farmers came to the district headquarters at Midnapore town on April 10, 1996 and submitted a memorandum to the district magistrate declaring that they would boycott the ensuing parliamentary election to protest against the acquisition of fertile agricultural land for the industrial projects. The peasants stated in their letter that this acquisition would disturb the local economy and destabilize the environmental balance of the region. This event was also reported in *The Statesman* (May 2, 1996: 6). It is important to note in this connection that neither the state nor district level Congress leadership (the Congress was the major opposition party) nor any MLA of this party showed any interest in supporting this movement of the peasants in Kharagpur region by raising their voice in the West Bengal Assembly. The local CPI (M) leadership and the elected *panchayat* members of this area not only remained silent about this spontaneous movement of the peasants but they also made every attempt to silence this agitation by labeling it a disturbance created by Congress to stall the progress of industrialization under the LFG. Without getting support from any opposition party, facing stiff resistance from the ruling left parties, and lacking a coherent organization, this localized peasant movement against land acquisition gradually lost its intensity. The land losers also made an attempt to organize themselves by refusing to accept compensation money for a very brief period under the leadership of a few local leaders but this effort too did not last long and the movement finally lost steam in the Kalaikunda region probably because of lack of any support from state-level political parties or human rights groups (Guha 2011b: 336–57).

12.5 An Anecdote from the Field

I started with a description of how the members of Parliament in India viewed capitalist transition by industrialization through land expropriation and narrated how, in the opinion of policymakers, land grab was seen as inevitable under the market forces. I then narrated how a land grab dispossessed the peasants and the way the administration looked at it and dealt with the acquisition. I also described why the government was unsuccessful in returning the land to the peasants when the company failed to build industry on it, which the government acquired for the capitalist, owing to unforeseen market forces unleashed by globalization. In effect, under liberalization, a leftist government could not provide rehabilitation to the peasantry which led the peasants to protest and bargain around land acquisition in Paschim Medinipur. I further observed that the peasants did not give away their lands under market forces as visualized by the parliamentarians in New Delhi. The peasants in my field area put up resistance, made intelligent bargains going outside the domain of the Land Acquisition Act, and finally lost their battle to the overarching state power.

It is also interesting to note that by 2006 the then Marxist government that was in power changed its age-old Leninist slogan "Land to the tillers" to "Agriculture is our foundation, industry our future" (Communist Party of India (Marxist) 2007). In fact, the then CPI(M) leadership argued in favor of huge capital investment in the state by saying that success in land reforms had created the ground for industrialization, although two important government reports during the Left Front regime had recorded very slow progress in the distribution of land to the landless and even a reversal of land reform benefits to an alarming level (Government of West Bengal 2004: 39–42; Mukarji and Bandopadhyay 1993: 41). Without reading these scholarly reports, the dispossessed peasants in my field area who depended on agriculture could not dream of a bright future in the upcoming industries, which only engulfed their land but could not provide any job to them. From my field observations and interviews I learned that almost all the peasants of the studied villages used to consume the paddy they grow in their land. I have not come across any peasant family who sold their paddy in the market. Purchasing rice for consumption was viewed by the members of a peasant family a humiliating task and was regarded as a dishonorable act for a *chasi* (peasant). Owning cultivable land was viewed as socially prestigious for the peasant families of this area. A "good peasant" in this area was one who could feed his family with the paddy grown in his field throughout the year. A popular maxim in this area, which I collected during my fieldwork was *Arthe maan/Khote dhan*. Freely translated from Bengali it meant: "Money gives prestige/Fertilizers yield paddy." In almost all our conversations, the members of the land-loser families always blamed acquisition of land by the government as the "root cause" (*mul karan* in Bengali parlance) of household level food shortages. They also expressed hopelessness whenever they talked about the number of months during which they purchased rice from the market for domestic consumption (Guha 2013: 797–814). Capitalist transition through land takeover by the state in my field area created a band of despondent peasants who instead of becoming an industrial proletariat still stuck to their peasant way of living in worse conditions than before. I will end my story with an anecdote from the field, which made a nice contrast with what the MPs at New Delhi implicitly communicated to me.

The event occurred near Kantapal village from where the huge chunk of land acquired for Century Textiles could be seen. I, with some of my students, was engaged in a discussion with the locals about the condition of the small dykes (*ail*) raised by the farmers to demarcate the plots of land possessed by different owners within the acquired area. Since no cultivation could be taken up for three successive seasons in the whole area it had turned into a grazing field and the dykes had started to break down. Two consequences of this situation followed. First, cultivators who still had unacquired land in the vicinity of the acquired area were facing difficulties in protecting their agricultural plots from the grazing cattle. Earlier there were other peasants who also shared the responsibility of driving out the cattle from the fields during agricultural season. Driving out the intruding cattle in paddy fields is always a collective affair in rural areas. After acquisition, the number of cultivators decreased in this area. Moreover, cows and buffaloes of the milkmen of the urban areas of Kharagpur town also ventured to exploit this huge chunk of land.

Second, after the breakdown of dykes the poorer people of the area, who used to collect a good quantity of small fishes of various types from those agricultural plots as a common property resource, were not getting any fish in those plots. In the discussion, three or four persons, including one middle-aged woman and an old man, were present. All of them were denouncing the government for the takeover of the fertile agricultural land for CTIL which had not yet been constructed. When the question arose, if people of this area had started to dislike the ruling party and the government, then why did they cast their votes at the *panchayat* and assembly elections to the same party every year? The reply came from the old man, which is reproduced here verbatim:

> Look babu, we poor people always have to ride on some animal almost blindfolded. After the ride for some time we start to realize whether it is a tiger or a bullock. But very often we have to twist its tail in order to keep it in proper direction. (translated freely from Bengali)

All of us, including the old man, burst into laughter but soon I realized that the joke symbolized the wide gap between the aspirations of the helpless local peasants in West Bengal and the distant policymakers in New Delhi.

12.6 Conclusion

Ethnography not only helps us to describe a culture or society, it also explains local events in a wider context. Accordingly, in this chapter an attempt has been made to situate the case of a local land grab by the West Bengal government within a global and regional context. Description of the local scenario in two sites (villages and administrative departments) helped me to understand the mechanisms of the failures of a leftist government which finally led to a stalemate in the path towards capitalist transition at least a decade before the Singur episode. The detailed description of how the governmental bureaucracy functioned to acquire land for the industrialists with a strong policy back-up on the one hand and the daily events around the visceral resistance put up by the land-losing peasants who did not want to become "willing sellers" on the other, revealed the contradictions toward the capitalist path of development and here lay the real value of this ethnographic journey. The outcome of this contradiction was as bad as a natural disaster since it not only pauperized the peasants without employment either in the industry or rehabilitation in agriculture but also distanced the policymakers from the peasants. The failure of the land grab could neither generate a labor force freed from agriculture, an industrial proletariat, nor created enthusiasm and hope for the capitalist investors. The Bengal leftist government's neo-Marxist theory of riding on the shoulders of land reform to achieve a successful capital-intensive industrialization finally proved to be a self-defeating exercise since the *praxis* sabotaged both past land reform and future industrialization. Generating capital and employment either through legal means or by the play of market forces seemed to be mere political rhetoric of the left political parties for contesting election battles in West Bengal. This is the macro-theoretical lesson I learnt from this multi-site ethnographic account on land grab in the state under liberalization.

Acknowledgments

I am indebted for this work to the villagers, my students, and the staff of the Land Acquisition Department of the erstwhile Midnapore district, West Bengal. I deeply acknowledge Achin Chakraborty, Director, Institute of Development Studies Kolkata, Anthony D'Costa of the Australia India Institute, University of Melbourne, and Mritiunjoy Mohanty of the Indian Institute of Management, Kolkata, for inviting me to present an earlier version of the chapter. I owe my special debts to Anthony for his critical comments on the successive drafts of the manuscript. However, I am solely responsible for any errors and omissions.

Notes

1. I have used the term "ethnography" not in the usual anthropological sense of the term, as the holistic description of a culture, but as a description of my own interaction as an ethnographer with land acquisition in a particular state of India in many field sites, including affected villages, government offices, archives, committee room of the Parliament house, and various public places outside the villages where I conducted the surveys over a long period of time. This may be called multi-site ethnography (Auyero 2004: 417–41; Burawoy 2013: 526–36; Gellner 2012: 1–16; Marcus 1995: 95–117).
2. This Act was passed on August 29, 2013 in the *Lok Sabha* (lower house of the Indian Parliament) and on September 4, 2013 in *Rajya Sabha* (upper house of the Indian Parliament) under the United Progressive Alliance (UPA) Government. The law, however, has not come into application since the new National Democratic Alliance (NDA) Government which has come to power in 2014 has restrained the application of the new law by amendments on the consent clause and the provision of social impact assessment through the promulgation of an Ordinance, a move which has been viewed by experts as the "weakening of the democratic and constitutional institutions"(Ramesh and Khan 2015: 124–30; Iyer 2015). The Act has provisions to provide fair compensation to those whose land will be taken away, intended to bring transparency to the process of acquisition of land to set up factories or buildings and infrastructural projects and assured rehabilitation of those affected for the first time in the history of the country (Cernea 2013).
3. Since Independence, besides the colonial Land Acquisition Act of 1894, there existed another State Act entitled *West Bengal Land (Requisition and Acquisition) Act, 1948*. The latter act is no longer applicable in West Bengal since March 31, 1993 by a

decision of the West Bengal Legislative Assembly. In fact, when this particular piece of legislation was first enacted in the State Assembly, it was stipulated that the Act had to be renewed in the Assembly by a majority decision every five years since this was a very powerful and coercive law. The government opinion was that the State of West Bengal, which had to receive millions of refugees from erstwhile East Pakistan just after independence, needed a huge amount of land for various developmental purposes. By *West Bengal Land (Requisition and Acquisition) Act* the Government could first requisition a particular piece of land for which the payment of compensation may not be made before the land takeover, while in the earlier Land Acquisition Act of 1894 the government could not take possession of any land without payment of compensation. In terms of political composition, both Congress and Left-ruled Governments, who were in power, continuously renewed the Requisition and Acquisition Act of 1948 in the State Assembly and applied it to acquire land for "public purposes" not only for the rehabilitation of refugees coming to West Bengal but also for other purposes, like the establishment of industries owned by private companies (Guha 2011a: 79–98).

References

Auyero, J. 2004. "When Everyday Life, Routine Politics, and Protest Meet." *Theory and Society*, 33: 417–41.

Borras, M. S., Jr. and Franco, J. C. 2012. "Global Land Grabbing and Trajectories of Agrarian Change: A Preliminary Analysis." *Journal of Agrarian Change*, 12(1): 34–59.

Borras, M. S., Jr., Hall, R., Scoones, I., White, B., and Wolford, W. 2011. "Towards a Better Understanding of Global Land Grabbing: An Editorial Introduction." *The Journal of Peasant Studies*, 38(2): 209–16.

Buckley, L. 2011. "Eating Bitter to Taste Sweet: An Ethnographic Sketch of a Chinese Agriculture Project in Senegal," Paper Presented at the International Conference on Global Land Grabbing, Organized by the Land Deals Politics Initiative (LDPI) in Collaboration with the *Journal of Peasant Studies* at the Institute of Development Studies, University of Sussex, April 6–8, 2011.

Burawoy, M. 2013. "Ethnographic Fallacies: Reflections on Labour Studies in the Era of Market Fundamentalism." *Work, Employment and Society*, 27(3) 526–36.

Census of India. 2011. "West Bengal Population Census data 2011." Available at: <http://www.census2011.co.in/census/state/west+bengal.html> (accessed March 28, 2016).

Cernea, M. 2008. "Compensation and Investment in Resettlement: Theory, Practice, Pitfalls, and Needed Policy Reform." In *Can Compensation Prevent Impoverishment?*, edited by Michael M. Cernea and Hari Mohan Mathur, New Delhi: Oxford University Press, pp. 15–98.

Cernea, M. 2013. "Progress in India: New Legislation to Protect Persons Internally Displaced by Development Projects." Available at: <http://www.brookings.edu/blogs/up-front/posts/2013/10/21-india-displacement-cernea/> (accessed March 2, 2016).

Communist Party of India (Marxist). 2007. "Thirty Years of West Bengal Left Front Govt." Available at: <http://cpim.org/content/thirty-years-west-bengal-left-front-govt/> (accessed March 1, 2016).

De Schutter, O. 2009. "Large Scale Land Acquisitions and Leases: A Set of Core Principles and Measures to Address the Human Rights Challenge." United Nations Special Rapporteur's Report on Right to Food, Geneva.

De Schutter, O. 2011. "How Not to Think of Land-Grabbing: Three Critiques of Large-Scale Investments in Farmland." *The Journal of Peasant Studies*, 38(2): 249–79.

Food and Agriculture Organization (FAO). 2008. *Compulsory Acquisition of Land and Compensation*, FAO Land Tenure Series 10. Rome: Food and Agriculture Organization of the United Nations.

Gellner, D. N. 2012. "Uncomfortable Antinomies: Going beyond Methodological Nationalism in Social and Cultural Anthropology." *AAS Working Papers in Social Anthropology*, 24: 1–16.

Government of West Bengal. 1994. "Policy Statement on Industrial Development," published on September 23, 1994, Calcutta.

Government of West Bengal. 1996. "Report on the Hearing of Objections for Land Acquisition for Century Textiles and Industries Ltd. (Unpublished)." Land Acquisition Department, Govt. of West Bengal, Midnapore.

Government of West Bengal. 2004. "West Bengal Human Development Report," Development and Planning Department, Govt. of West Bengal, Calcutta.

Guha, A. 2004. "Land Acquisition in a West Bengal District." *Economic and Political Weekly*, 39: 4620–3.

Guha. A. 2006. "Dispossession of Peasants by Industrial Projects." In *The Economics and Politics of Resettlement in India*, edited by Shobhita Jain and Madhu Bala, New Delhi: Pearson Education, pp. 155–73.

Guha, A. 2007a. *Land, Law and the Left: The Saga of Disempowerment of Peasantry in the Era of Globalisation*. New Delhi: Concept Publishing Company.

Guha, A. 2007b. "Peasant Resistance in West Bengal: A Decade before Singur and Nandigram." *Economic and Political Weekly*, 42: 3706–11.

Guha, A. 2009. "Peoples' Strategies of Survival around Land Assets: Decisions under Contrasting Situations." *Decision*, 36(1): 41–53.

Guha, A. 2011a. "Challenges to Corporate Social Responsibility: A Case Study from Kharagpur, West Bengal." *Sociological Bulletin*, 60(1): 79–98.

Guha, A. 2011b. "Special Economic Zones, Land Acquisition and Civil Society in West Bengal." In *Resettling Displaced People: Policy and Practice in India*, edited by Hari Mohan Mathur, New Delhi: Routledge and Council for Social Development, pp. 336–57.

Guha, A. 2013. "The Macro Costs of Forced Displacement of the Farmers in India: A Micro- Level Study." *The European Journal of Development Research*, 25(5): 797–814.

Guha, A. 2014. "Generating Public Consciousness around Land Acquisition for Private Industries." *Social Change*, 44(2): 205–28.

Iyer, R. 2015. "When Amendment Amounts to Nullification." *The Hindu*, January 15. Available at: <http://www.thehindu.com/opinion/lead/when-amendment-amounts-to-nullification/article6789569.ece/> (accessed March 3, 2016).

Lieten, G. K. 1996. *Development, Devolution and Democracy*. New Delhi: Sage Publications.
Majumder, A. 2015. "Socio-Economic Impact of Land Acquisition for Industries among the Peasants of Paschim Medinipur, West Bengal: A Test Case." Unpublished PhD thesis, Vidyasagar University, Midnapore.
Majumder, A. and Guha, A. 2008. "A Decade after Land Acquisition in Paschim Medinipur, West Bengal." *Journal of the Indian Anthropological Society*, 43: 121–33.
Marcus, G. E. 1995. "Ethnography in/of the World System: The Emergence of Multi-Sited Ethnography. *Annual Review of Anthropology*, 24: 95–117.
Mishra, S. K. 2007. "On Agrarian Transition in West Bengal." *The Marxist*, 23(2): 1–22.
Mukarji, N. and Bandopadhyay, D. 1993. "New Horizons for West Bengal's Panchayats," Department of Panchayats, Government of West Bengal.
Murphy, S. 2013. "Land Grabs and Fragile Food Systems: The Role of Globalization," Institute for Agriculture and Trade Policy, Minneapolis.
Nader, L. 2011. "Ethnography as Theory." *HAU: Journal of Ethnographic Theory*, 1(1): 211–19.
Perez, J., Gistelinek, M., and Karbala, D. 2011. "Sleeping Lions: International Investment Treaties, State-Investor Disputes and Access to Food, Land and Water," Discussion Paper, Oxfam. Available at: <https://www.oxfam.org/sites/www.oxfam.org/files/dp-sleeping-lions-260511-en.pdf> (accessed March 2, 2016).
Ramesh, J. and Khan, M. A. 2015. *Legislating for Justice: The Making of the 2013 Land Acquisition Law*. New Delhi: Oxford University Press.
WBIDC. 1999. "Destination West Bengal," Government of West Bengal, Calcutta.
WBIDC. 2000. "West Bengal: Industry News Update," Government of West Bengal, Calcutta.
The World Bank. 2011. *Rising Global Interest in Farmland: Can It Yield Sustainable and Equitable Benefits?* Washington, D.C.: The World Bank.

13

Land and Dispossession

The Criticalities in the Hills of Northeast India

Gorky Chakraborty and Asok Kumar Ray

13.1 Introduction

In the transition of an economy from agricultural- to nonagricultural-led growth, analyzing the role of the state in dealing with the issue of land can be an interesting area of inquiry. While this "transition problematique" has been an inseparable part of the development agenda of economies worldwide, there are differences in the process and its manifestations among nations and within them as well. While the differences between the technologically advanced and developing economies have been dealt with in the introduction to this volume, several contributors to this volume discuss various scenarios from different parts of India. We analyze the role of the state in its dealing with the issue of land in the hill areas of Northeast India (NEI). This chapter reflects on the contested nature of the land question in the hills of NEI due to the development agenda of the state. In other words, it deals with the complex ways by which institutional forces result in dispossession of the hill dwellers from their land, which is essentially collective in nature. In short, it locates in NEI, the central theme of this volume, that is, the social and political contradiction between development as an outcome and dispossession, which otherwise has been the hallmark of capitalism so far.

While appropriation and subsequent dispossession from land by the state as a consequence of its development agenda is not different between "mainland" India and NEI, we emphasize that there are dissimilarities regarding the process, which stem from a difference in understanding on the part of the state and that of the communities inhabiting the hills in the region. We argue that this gap between the perception of the community (under customary rights) vis-à-vis the colonial and postcolonial state (under the positive laws) regarding community land, influences the jurisprudential differences that lead to different understandings on property rights. This subsequently augments the process of dispossession through appropriation by the state. Nuanced understanding of the process of appropriation through various institutional measures highlights how the state has dealt with the land question in the hill areas of the region. This similar yet unique history of dispossession in the hills helps us to understand the process of social transformation in NEI. In other words, we discuss the scenarios where the state, in the name of pursuing its development agenda, defines property rights in a way that helps it to appropriate land that subsequently leads to dispossession of the hill communities from their community land.

NEI comprises eight states (see Figure 13.1). More than two-thirds of the area of the region is covered by hills, the rest encompasses three stretches of plains namely, Brahmaputra and Barak (in Assam) and Imphal (in Manipur). NEI covers 8 percent of India's geographical area and is home to 3.1 percent (40 million, census 2011) of the country's population. NEI shares more than 90 percent of its borders with other nations: China (Tibet) to the north, Myanmar to the east, Bangladesh on the southwest border, and Bhutan to

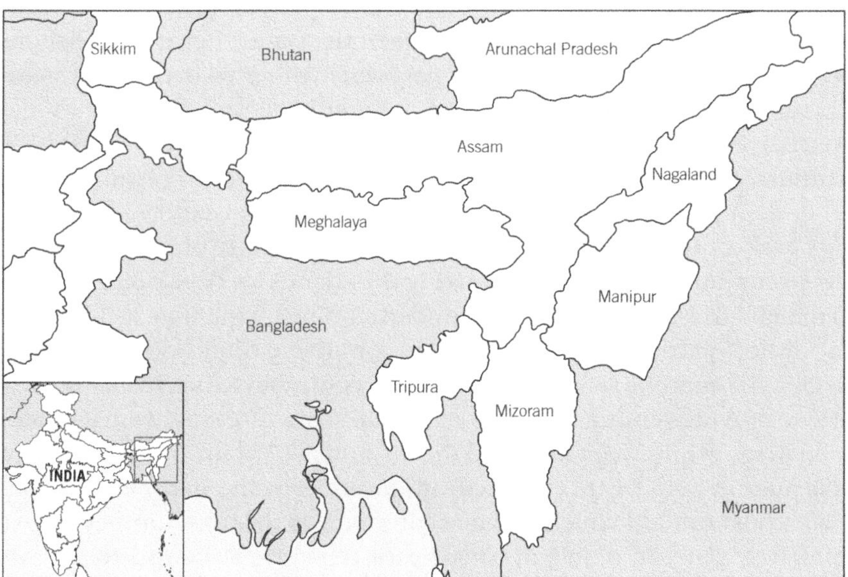

Figure 13.1. Northeast India
Source: OUP (accessed March 16, 2016).

the northwest. The Siliguri corridor (a narrow stretch of land), also referred to as the "chicken neck," is the only stretch of land that connects the Northeast with "mainland" India. Socioculturally, this is one of the most diverse regions in the world. NEI is home to over 220 ethnic groups and as many languages belonging to multiple language families. Agriculture is the primary source of livelihood in the region. In the hills, shifting cultivation referred to as *jhum* is practiced, while settled cultivation is the norm in the plains. The hills are home to a large number of tribes spread over different states while the plains are overwhelmingly non-tribal.

Method of cultivation has been a marker of civilizational norms where settled cultivation has been the yardstick of transformation from the "state of nature" to the "state of civilization." Those areas and people beyond the purview of settled cultivation were in the "state of nature" and neither their proprietary rights nor their method of cultivation were considered to be legal by the colonial state. So until land is privately owned and is under settled cultivation, it fails to produce value and thereby is not regarded as property (Harvey 2011). In NEI, those communities who practice shifting cultivation on community land face similar categorization in the eyes of the state. This chapter discusses the role of the state in dealing with community land, and the ramifications in hill areas of the region. As the state failed to take into account the usufruct practices of the communities (tribal) in the hills, institutional dispossession of the communities from their land became a consequence of the development agenda of the state. The chapter elaborates how such institutional dispossession has been manifested in different states in NEI. For example, a process of selective interpretation of customary rights governing community land by the emerging elites among the Khasi tribe (a matrilineal tribe) in Meghalaya paved the way for increasing land concentration in their hands; whereas the absence of a cadastral survey and record of rights over community land leads to the non-acceptance of proprietary rights of the communities while appropriating these lands for development projects in Tripura. Likewise, the vacillating understanding by the state of "custodian" and "owner" of community land created conflicting claims over lands among several hill communities in Manipur, which continues to fuel instability in the state. To provide a contrasting view of the hill areas, three states are selected for our analysis. While Meghalaya and the hill areas of Tripura (under the Tripura Tribal Autonomous District Council) are governed by the Sixth Schedule of the Indian Constitution,[1] which provides autonomy to the communities to govern themselves, Manipur is not governed by the schedule. By comparing the scenarios among the states, the chapter provides critical insights into the modes of dispossession of the communities from their community land and highlights that, under the development agenda pursued by the state, constitutional protection falls short in checking usurpation of community resources in NEI.

In Section 13.2, we deal with the epistemology of the gap in the perceptions between the community and the state in terms of land as a resource, which will help in theorizing land in the hill areas of the region. In Section 13.3, the intuitive nature of the communities to continue with the practice of shifting cultivation and the counterintuitive effort by the state through policy initiatives in discouraging such cultivation is discussed. The state's agenda of comparing shifting cultivation with petty production and simultaneously encouraging the adoption of policies to replace it with settled cultivation and/or plantation agriculture is an age-old phenomenon. The consequences of such an approach on the communities are also discussed in this section. The constitutional provision in the form of the Sixth Schedule for the protection of tribal rights, providing them autonomy to govern themselves, for example, through the formation of the district councils and the acceptance of customary laws, has been an important milestone in Indian administration. But the gaps between the provisions in the Schedule and the practices in administering the same, particularly related to community resources (land) are discussed in Section 13.4. Selected judicial cases are referred to here to highlight the contradictions in the schedule areas. Section 13.5 deals with the modes of dispossession in the three selected states of the region. While selective interpretation of the customary laws under the Sixth Schedule helps emerging elites in the Khasi society to usurp community land in Meghalaya, thereby aggravating the problems of landlessness and indebtedness in Meghalaya, the absence of a cadastral survey and therefore title deeds on community land has resulted in communities being dispossessed from these lands in Tripura. In Manipur, the vacillating nature of the state in recognizing the role of the chiefs as well as equating their rights as "owners" instead of "custodians" has led to conflicting claims of various communities over community lands. While the discussion in the chapter focuses on community land and the dispossession of the communities from them due to various institutional measures, in Section 13.6, we relate this understanding to the emerging hydropower sector in Arunachal Pradesh, which is becoming the latest casualty in terms of separation of the communities from their resources and thereby dispossession. This hill state (a non-scheduled state under Panchayati Raj, which is a decentralized form of governance by a committee at the village level) in NEI bordering China is presently undergoing a spree of dam building for hydropower generation, showcased as a development initiative through state–market collusion to foster growth and prosperity in the region. What has been the experience of the communities in the state in this process? Is it leading to a new form of dispossession in the contemporary era? In the concluding section we sum up the experiences of institutional dispossession of the communities in the hills of NEI and foresee that the collusion of state and capital in the contemporary era of globalization may lead to further dispossession of the remaining community resources in the region.

13.2 Theorizing Land in the Hills

In addition to the geophysical and climatic basis of economic differentiation between the hills and the plains in NEI, the land systems are also markedly different. In the hills, shifting cultivation is practiced, resulting in a subsistence-based social economy, compared to surplus-yielding settled cultivation in the plains. Our theoretical understanding of the land system in the hills of the Northeast gets obfuscated if we see the same through the lens of market principles. The difference in the understanding of land between the two economies and its people is thereby essentially historico-epistemological. First, land in the subsistence economy is held under the stewardship of a chief (tribal leader) and the community members enjoy only customary usufruct rights, not ownership right, over such land. The real value of land can thereby never be expressed only in terms of transaction value. In a market-based economy, on the other hand, land is primarily a market good having transaction value. The ownership of land is based on individual rights defined by positive law. This constitutes a fundamental difference between the two economic systems (Herskovits 1964). Second, in a subsistence economy, land has a dual connotation. One, devolution of property is from the community to the individual and the devolution itself is subject to the control of the community; and two, in ontological terms, individual rights are subsumed within the community rights since no person or group can have property in anything except as it is acknowledged by the relevant community and therefore property is never private (Taylor 1966). Community property being an organic part of the social economy, any change in it also brings about change in production relations (Commons 1957). Introduction of positive law delinks the social economy of land from the community, which had been established historically and sanctified by customary laws. In other words, positive laws become a source of dispossession of the communities from their community resources.

The individual rights regime in the hills of the Northeast was first established by colonial jurisprudence. The application of the jurisprudential principles of the colonizers over the customary laws of the communities is therefore critical for a larger theoretical understanding of the community land system. The colonial administrators, following Weber's empirico-positivistic notion, legitimized force to capture the land and land-based resources (Roy Burman 1987). They conveniently used the jurisprudential principles of *res nullius* (meaning, that which is not assigned by the sovereign belongs to the sovereign) and *terra nullius* (meaning no man's land) (Roy Burman 1987) and found nothing unusual in prefixing "waste" to any uninhabited and/or unused land (or any land that did not accrue revenue to the Crown) and appropriated it for revenue generation (Chakraborty 2012: 8). For the tribal communities, on the other hand, social behavior rather than force was the basis of customary rights over land. The right to use land resources was embedded in the customary law, which was steeped in the ethnohistory of the tribes. These customary laws, according to Roy Burman (1987), were based on the jurisprudential principle of *lex loci rei sitae* (meaning law of the people where the asset exists). The jurisprudential principles of *res nullius* and *terra nullius* of the colonial state on the one hand and *lex loci rei sitae* of the people on the other, marked the fundamental difference between the two concepts of rights.[2] While the colonials individuated the rights on land, the tribal concept of right was communal. Due to the usufruct nature of land rights, community members did not treat any unused land as "waste" since this was subject to re-utilization at a later time (see Table 13.1). Therefore, despite the formal variations in the customary land ownership patterns among the different tribes in the region, there were marked commonalities in the essence of land ownership systems in terms of use of tools and techniques of production, non-accumulation of effective surplus, absence of rent-revenue, communal labor, labor reciprocity, and other activities on community land (Ganguli 2008).

This epistemological difference in perception and practice between the colonial state and the communities in the hill areas of the region laid the foundation for appropriation of community resources, particularly community land,

Table 13.1. Perception gap on land: Customary vs. positive law

Customary	Positive
Right is customarily acquired	Rights are legally acquired
As such, non-alienable	Liable to alienation and appropriation
Largely community ownership	Owned by the state
Ownership subsumed in community	Individual right is the basis of economic growth
Place value of land	Space value of land
Provides identity to the ethnic group	Is cosmopolitan, melting pot of identity
Community is the person	Community is not legal person
Land is a cultural place	Land is a commercial space
Land for subsistence product	Land for surplus product
Land is non-revenue yielding	Land is revenue yielding
Product yielding	Product and revenue yielding
Oral/customary grant of land	Legal tenure
Symbol of community solidarity	Symbol of power, wealth, and status
Community labor	Hired or tenant labor
For subsistence	For earning rent, interest, and profit
Natural regeneration	Value addition, technology, and capital intensity
Jural fragmentation of land use	Uni-jural structural model of land use
Nonwage-based land use	Wage-based land use

which facilitated the process of dispossession through various modes. In the subsequent sections, we discuss the modes of dispossession. It is interesting to note that the process of detachment of the community from the community land went hand in hand with the process of discouraging the traditional methods of production associated with such lands. So first we highlight how the process of discouraging shifting cultivation (*jhum* cultivation) in the name of petty production added to dispossession and then discuss the process of institutional dispossession in the hill areas of the region.

13.3 Differing Perspectives on Shifting Cultivation

The petty production mode of shifting cultivation did not receive official patronage of the colonial state due to its nonsurplus nature, market imperfection, and the absence of economy of scale. The petty production of shifting cultivation was discouraged for the sake of generating revenue from the agricultural sector. It provided the common alibi to the colonials to discourage shifting cultivation and introduce cash crop cultivation (Dutta 1983; Scott 2009).[3]

The postcolonial Indian state followed the colonial policy of discouraging shifting cultivation in postcolonial economic reconstruction. The Indian Council of Agricultural Research (ICAR), the state-supported organization for agricultural development and extension services, discarded shifting cultivation as environmentally degrading and promoted the three-tier model of land use in the hills (see Table 13.2) (Ashokvardhan 2004). This ICAR model, however, was contested for its labeling of shifting cultivation as an environmentally degrading activity. It was critiqued by Ramakrishnan (1993), who showed that the ICAR model was framed without looking into the village structures, settlements, and livelihood systems of different hill dwellers. According to Ramkrishnan, "the activities of industrial man," and not the tribal people, was responsible for destroying the forests in the hill areas of the region. The "big science" approach followed by the policy makers could not succeed in replacing *jhum* cultivation, as the hill dwellers did not show

Table 13.2. The three-tier model of land use and conservation of ICAR

Slope	Approximate percentage of total land area	Land use	Conservation measures
Top	33	Forestry	—
Middle	33.5	Horticulture	Half-moon terracing for horticultural plants
Lower	33.5	Agriculture	Bench terracing

Note: ICAR = Indian Council of Agricultural Research.

Source: Ashokvardhan (2004)

interest in those measures. He reiterated that *jhum* was a way of life and not an evil practice (Ramakrishnan 1993).

In spite of this, from the report of the Commissioner of Scheduled Castes and Scheduled Tribes in 1955 to the Draft Perspective Plan for Development of the Forestry Sector 2007 (Eleventh Five-Year Plan, 2007–12), shifting cultivation faced the wrath of the state as environmentally degrading and an inefficient economic practice. The approach of the state towards land use in the hills has been influenced by capitalist relations that were tilted toward sedentary cultivation, cash cropping, horticulture, and industrial use of land. The state formulated unfriendly policies toward shifting cultivation, de-recognized the community as a person, and rationalized land tenure through individuation and land deeds. The Draft National Tribal Policy (2004–6) promoted individual ownership that defeated the fundamentals of the economy of shifting cultivation, its usufruct mode, and the incipient collective consciousness associated with this practice. The National Afforestation Program of the Ministry of Environment and Forest (MoEF) during the Ninth (1998–2002) and Tenth (2002–7) Five-Year Plans and the National Mission on Horticulture (2005–6) also became instrumental in discouraging *jhum* lands for horticulture and commercial timber.

The policy measures adopted by the states to discourage *jhum* in the post-colonial era were critiqued by various studies.[4] Karna found that

> the new technology and strategy having been geared to goals of production, with secondary regard to social imperative have brought about a situation in which elements of disparity, instability and unrest are becoming conspicuous with the possibility of increase in tension. (Karna 1987: 5)

In an interview with Roy Burman, Datta (1994) found that wet rice cultivation and dairy farming propagated as an alternative to *jhum* cultivation had far-reaching consequences in the hills. Roy Burman remarked:

> most of the states in the north-east have adopted programmes of extending wet rice cultivation and dairy farming in the hills. But in most of the hill areas where these programmes have been introduced, wet rice cultivation is generally done by migrant Hindu and Muslim peasants hailing at one time or the other from Bangladesh, similarly dairy farming in the hills is frequently dependent on Nepali labour. This is happening not because the hill men are lazy or unenterprising. But these operations are based on experiences passed down from generation to generation on crop planning and activities concerning crops in different stages by relating to soil type and water management, ethno-meteorology; animal behaviour in dairy farming, indigenous system of treatment of animal diseases and so on. Gender related social organisation of labour and time budgeting for various activities are involved.... (Datta 1994: 61)

Lack of proper understanding and appreciation of the dynamics of shifting cultivation on the part of the state caused increased dispossession of tribal land and its transfer to non-tribal people. This at times made the region politically volatile. Although the nontribals were legally not entitled to own land in the tribal areas of the region, due to their traditional skills in wet rice cultivation, they often became the de facto owners of land through informal leasing-in land of the tribals and cultivating the same land for years together. Over the years these de facto owners slowly expanded their area of cultivation and craftily became the *de jure* owners through state patronage and/or with the blessings of the tribal power brokers. This has become a common phenomenon in the hills. This shift in policy toward settled cultivation gave a boost to dispossession of community land and the tribal people suffered from a perceived identity threat. Undeniably this has been one of the major factors in the tribal outcry for protection of an ethnic "homeland" and which also gave fillip to insurgency in the region (Baruah 2005).[5]

13.4 Institutional Dispossession

Earlier we discussed how epistemological differences in perception and practice between the community and the state have led to different patterns of understanding and utilization of community resources, particularly community land in the hill areas. In this section we will show how the institutions that were created to deal with the state–community contradictions performed during the postcolonial era. Could these institutions provide safeguards to avoid usurpation of community resources? Did these institutions protect the interest of tribals at large from dispossession and deprivation? Or, did they create newer contradictions of authority and agency obfuscating the very purpose of their creation?

The process of introducing new laws governing land in the hills of NEI started during the colonial era.[6] They were enacted to introduce positive laws with little safeguard for community land and communitarian practices compared to the customary laws. The policy shift toward sedentary cultivation gave a boost to private rights in land that subordinated the customary and community land ownership system, and invited the market forces in land. This disturbed the social economy of the tribal areas. The colonial overseers brought the chiefs and headmen of the tribes under colonial control through "structured subordination" and initiated the process of transforming the chiefs from custodians of community land to *proto* owners. The process, which started during the colonial era, became more acute in the postcolonial state and let loose perennial state–community conflict over land rights in the hills of Northeast India. Here, we discuss the role of the traditional institutions such as the district councils and their dealings with community land in select hill areas of the region to highlight the gap in the avowed principle of autonomy and self-governance of the institutions under the Sixth Schedule and the actual process of administration in these areas. A brief discussion on the legal pronouncements helps us to understand these differences related to governing community land under the councils.

The Bordoloi Subcommittee (the precursor to the Sixth Schedule) registered anxiety of the hill people over the protection of their land from being alienated. The report stated that:

> there was an emphatic unanimity of opinion among the hill people that there should be control of immigration and allocation of land to outsiders, and that such control should be vested in the hands of the hill people themselves... We recommend that the Hill Districts should have powers of legislation over occupation or use of land other than land comprising Reserved Forest under the Assam Forest Regulation of 1891 or other law applicable. (Hansaria 2005: 230–1)

On the forest issue, the report stated:

> we would strongly emphasize that in questions of actual management, including the appointment of forest staff and the granting of contracts and leases, the susceptibilities and the legitimate desires and needs of the hill people should be taken into account, and we recommend that the Provincial Government should accept this principle as a part of its policy. (Hansaria 2005: 230–1)

On the question of *jhum* cultivation, the report stated that:

> the tribes should have the right of deciding for themselves whether to permit *jhum* cultivation or not... While therefore we feel strongly that *jhuming* should be discouraged and stopped whenever possible, no general legislative bar can be imposed without taking local circumstances into account. (Hansaria 2005: 230–1)

The land implications of the Sixth Schedule are laid down in Section 3 (1) (a) of the schedule, which establishes that the district council can make law with respect to allotment, occupation, or use, to promote the interest of the inhabitants. But these provisions under the Sixth Schedule had multiple interpretations and each judicial pronouncement gave conflicting signals regarding the right of these traditional institutions to enact laws for the protection of community land under their jurisdiction.[7] This puts a question mark on the extent of legislative competence of the institution of the district council. This became a contentious issue and a series of claims and counterclaims got underway between these traditional institutions and the state governments (Hansaria 2005). This did little to help in preserving the community land for the usufruct use of the community. A disjointed cocktail of loopholes in the customary law and the attempt to simultaneously unleash the "animal spirit" of the positive law by the state resulted in dispossession.

This process of dispossession has been manifested in different ways in the hills of NEI. For example, Meghalaya exhibits a case of dispossession due to selective interpretation of customary laws by the elites, while in Tripura dispossession is associated with the absence of record of rights on community land; Manipur shows how competing claims on community land by different tribes lead to dispossession. These select state scenarios are discussed in the next section.

13.5 Community Land, Customary Laws, and Dispossession

In Meghalaya there has been a progressive transformation of the *Ri-raid* land to *Ri-kynti* land. The root of this problem was in the customary law itself. In Khasi customary laws, ownership of land is common while occupancy right is individual. So if an occupant leaves the land uncultivated for a sufficiently long period, he loses the right over it but if one makes permanent improvement and continues to occupy it, in the course of time that land is recognized as the occupant's property. In postcolonial times, when the opportunities to earn cash from nonagricultural sources increased, this provision of the customary law favored the rich. The resource-poor land users failed to bring about improvements in land, and were unable to claim private ownership over community land. Similarly, the rich who introduced cash crops under the tutelage of the state could employ large numbers of workers and inputs and made permanent improvements on community land, transforming them into private lands (Nongbri 1995). The *Bakhrawbatris* (traditional nobleman) in the Khasi Hills, being the custodian of community lands, were the first to establish *Ri-kynti* right (individual rights on land) followed by others in the *Raid Durbar* (traditional assembly of the Khasi tribal community). This in effect created cleavage of ownership between those who owned and occupied community land and those who did not, that is, the tribal masses at large (Datta 1992).

The process of land dispossession was speeded up over time due to population increase, improvements in transportation, and growing demand for land for cash crop cultivation and extraction of industrial raw materials such as coal, limestone, uranium, and timber. The emerging tribal "haves" created big enclosures in *Raid* land and facilitated the outsiders (meaning tribals outside the *Raid*) to grab community land. It is reported that those who lost on land were the people from the *Raid* whereas those who gained were from the towns (Sen 1987). This phenomenon in the Khasi Hills was not much different in the Garo villages (another major tribe in Meghalaya, where there emerged a class of absentee landed gentry, following the same path) (Majumdar 1983).

To arrest the trend of land dispossession, the Land Reforms Commission instituted in Meghalaya (1974) recommended codification of customary rights, cadastral mapping, and a record of rights of land to protect the interest of the weak from unscrupulous land grabbers. The recommendation remained largely on paper as the influential landholding elites foiled the initiative of the Commission, on apprehension that the land reform would lead to state appropriation. Bezbaruah (2007) found that waning community ownership and absence of institutional reforms for a formal property rights regime has resulted in growing differentiation in landholding and has led to the emergence of landlessness among the tribal population groups in the region.

Protection of the rights of the tribals from nontribals remained the cornerstone of politics in the hill areas, mainly under the provisions of the Sixth Schedule of the Constitution. However, with the passage of time in the name of development of tribal areas the transfer of land to nontribals became an increasing phenomenon. The development initiatives through schemes such as plantations on one hand, and nonrecognition of the community rights over resources on the other, created conditions that not only enhanced socioeconomic differentiation within the tribes (and among them as well) but also created enough space for direct or indirect interventions by the nontribals in the hill areas in NEI. The emerging tribal elites who benefited from such development initiatives vacillated between two ends—tribalism and pan-Indianism—choosing options that suited their narrow interests. Recently, the Meghalaya Transfer of Land (Regulation) Amendment Act 2012 allowed the transfer of land through sale, gift, exchange, mortgage, lease, and so on to the nontribals. The Khasi Hills Autonomous District Council issued No Objection Certificates (NOC) for mortgage of land to financial institutions and issued land *patta* (deeds) to private individuals. Educational institutions came under this exemption with the Meghalaya Private Universities (Regulated & Establishment & Maintenance of Standard) Act 2012. This provoked civil society activism by the umbrella organization called the Social Organizations of Meghalaya Against Land Alienation (SOMALA), which demanded revocation of the same (Jyrwa 1988).

Community land as a social safety net is fast disappearing among the Khasis. With no capital access and declining returns from overcultivated land, many rural Khasis gave up cultivation and became wage laborers. Absentee landlordism increased as the wealthier and the "urban" tribals started up plantations and agribusinesses with institutional credit. They maintained a nexus with the office bearers in the traditional institutions, for example *durbars*, where the business interests made the custodians of these institutions act as land brokers. The official attempts to conduct a land survey therefore faced resistance from the *durbars* and the emerging elites within the Khasi society (Sharma 1984). Thus, the Sixth Schedule, which was designed to protect the community resources in the tribal areas, has fallen short of its desired objectives. Our analysis reveals that the institutions of *durbar* and the district councils, which should have acted as a bulwark against usurpation of community land, have failed in their objectives. Moreover, the threat of land alienation, which was earlier believed to be from nontribals, has become more ominous with the emergence of elites within the tribal societies. Under such a scenario, privatization of community resources, particularly land, has become the order of the day in these areas.

While the dispossession in Meghalaya shows the lack of ability of the traditional institutions to protect the interest of the community by preserving their community land, the scenario in Tripura points to another dimension of dispossession of the community from their land, which occurs due to the state's non-recognition of the community's proprietary rights on community land. The community land has never been under cadastral survey and there are no land titles for these lands, which in the eyes of the state remain the only source of proprietorship over such land. This thereby became a major source of dispossession of the tribes from their ancestral land. Both the history of migration of nontribals to the state as well as the enactment of legislation aggravated the process of dispossession in the hill areas of Tripura state. The process of discouraging *jhum* cultivation through introduction of rubber plantations as well as settler-based plow cultivation reversed the land system of the tribal people and their communal organization of labor. This privileged the non-tribal migrant farmers, reducing the tribals to agricultural laborers.

It is interesting to note that Section 187 of the Tripura Land Revenue and Land Reforms Act 1960 imposed restrictions on transfer of land from tribals to nontribals without prior permission of the collector. But the caveat was that it only recognized the individually owned and registered lands. As most of the tribal lands were community owned and thereby beyond the purview of registration and records, it became the source for alienation. This made it easy for the state to take over community land as *khas* land and to settle the same with nontribals in the name of refugee rehabilitation. The refugees, who were otherwise traditionally more used to settled cultivation, were encouraged to buy land and were provided with incentives for that purpose. In the process, the tribals lost large swathes of land to the Bengali refugee settlers. Some among them even set up "land cooperatives," such as the Swasti Samity in North Tripura, which violated the Tribal Reserves regulations and took over large areas of land in connivance with the bureaucrats (Bhaumik 1996). Dispossessed from their ancestral land, the tribal youth gravitated toward insurgency and to the "*Sengkrak*" (clenched fist) movement (Bhaumik 1996). In some parts of south Tripura, as much as 60 percent of the tribal lands were alienated to nontribals (Debbarma 2008), which further added fuel to the process of dispossession and dislocation in these societies.

The same fallacy was exhibited when the state acquired tribal land for dams in the state.[8] In the early 1970s, the Dumbur dam submerged over 23,530 acres of land, most of which belonged to the tribal communities. They were not compensated as their lands were not registered. Although the project files mentioned only 2,558 individual landowning displaced families (13,000 persons), studies found that 8,000–9,000 families (40,000–50,000 persons) were displaced without rehabilitation or compensation. Similarly, the Gumti hydroelectric project submerged large stretches of tribal arable lands in the Raima Valley in South Tripura. About 26,000 acres of land, including 24,000 acres of cultivable land, were submerged and 5,000 tribal families were displaced by the project (Fernandes 1995).

This is thereby another case where the introduction of positive laws on land in a milieu of communal ownership led to dispossession and deprivation among the tribals in the state. The more the state tried to bring about change through development in these areas the greater was the delinking of the community from the social economy of community land.

When the institution of district council was created under the Sixth Schedule of the Indian Constitution, the major focus was the protection of culture and tradition of the tribal people based on their customary practices and usages. It was similarly thought of in case of protection of tribal landholdings. But contradictions emerged once they were operationalized. On the one hand, the tribal chiefs deemed this to be a threat to their authority and status within their clan through state intervention; and on the other, a lack of understanding of the state of the difference between "custodian" and "owner" of community land/resources created the opportunities for usurpation of community resources from within the tribes (Karlsson 2011). The scenario in the hills of Manipur shows these operational contradictions, which is discussed in this section.

With the introduction of the Manipur Hill Peoples' (Administration) Regulation 1947 and the Manipur (Village Authorities in the Hill Areas) Act 1956, the conflict between customary law and positive law escalated. These Acts were opposed by the Kuki chiefs (Kuki tribes are one of the important inhabitants in the Manipur Hills, the other being the Nagas) as they viewed this as a move by the state to erode their traditional rights and privileges, to reduce their "control" over community land. They feared that over time these lands would be transferred to nontribals from the Manipur plains. To allay their apprehension, the Manipur (Village Authorities) in Hill Areas Act was subsequently amended in 1983, stipulating that no transfer or allotment of land could be made to nontribals without the previous permission in writing of the deputy commissioner of the concerned district. The deputy commissioner had to secure consent of the village council as per the rules of the district council. However, in operational terms, neither the safeguards in the acts nor the district council could arrest land alienation and dispossession of the ordinary Kuki tribesmen who enjoyed the usufruct rights in these lands. The

Kangpokpi–Kanglatongbi range in Senapati district of Manipur is one such living example where the habitat of the Kukis was settled by the more "entrepreneurial" nontribals by managing to acquire "*patta*" lands in these hill areas.

Again, there were complications when the state tried to win over the chiefs through palliatives such as ownership over community land. The "custodians" of the traditional lands sought to be treated as "owners" by bringing in the issue of compensation for seceding their rights over these lands. The enactment of the Manipur Hill Areas (Acquisition of Chief's Rights) Act, 1967 has been a step in this direction. This act stipulated compensation based on (a) the amount of land under the chiefs; (b) total number of households within the chiefdom; and (c) whether compensation was to be given in installments or as a lump-sum. Despite these allurements, the act was opposed by the Kuki chiefs, who took a defensive stand in favor of retaining the institution of chieftainship. In the process, neither the act was implemented nor was the compensation paid. In the stalemate, the distinction between the "custodian" and "owner" of community lands that the state wanted to enact was utilized by the chiefs and they posed themselves as the owners who were now de facto licensed to usurp these lands for personal favors (Ray and Kamkhenthang 1991). This also led to dispossession of ordinary tribesmen from their community land.

In a nutshell, what appears in the hills of Manipur is a fragile order. On one hand, there is a concerted effort by the emerging tribal elites to interpret the customary laws in their favor and on the other, the state is bent on applying the yardsticks of the plains to deal with the community land in the hills in order to entrench its reach in these areas. This has given rise to competing claims of ownership of hill land, namely among the chiefs (in the Kuki-Chin Group); the community and village (in case of the Naga group); the state government (through Manipur Land Revenue and Land Reforms Act), and the district council (for which a movement is ongoing in the hills to bring these areas under the Sixth Schedule of the Indian Constitution). Under such a situation of competing claims the overlapping issues related to community land remains unreconciled, leading to land-related conflicts and instability in the hill areas of the state.

These examples from the hill regions of the Northeast suggest that once there is an attempt by the state to introduce positive laws over community lands, it leads to a change in production relations in these areas. Community land being an organic part of the tribal social economy, the introduction of the positive laws delinks the social economy of land and results in dispossession. In this regard, what also appears from our analysis in this chapter suggests that constitutional provisions and safeguards (like the Sixth Schedule) to protect the traditional way of life and landholdings of the tribes under the *hegemony of capital* have little success in preventing dispossession of the tribes from their historically established communal resources sanctified by customary laws.

In the next section, we deal with the ongoing issue of dispossession of the tribal masses from their communal resources being enacted by the collusion of state and market in the name of generating eco-friendly hydropower in Arunachal Pradesh.

13.6 Emerging Issues with Hydropower

In our section on theoretical understanding we analyzed how the application of different jurisprudential principles of *res nullius* and *terra nullius* by the colonial and the postcolonial state and *lex loci rei sitae* by the tribals resulted in the fundamental difference between the individuated and communal concepts of rights over land. A similar scenario is unfolding in the hills of Arunachal Pradesh in the name of generating hydropower through the accumulation of water rights and resources and subsequent dispossession of the ordinary tribesmen from the *milieu* in which they are embedded through ages. It seems that the state is bent on defining property rights in the name of "development" (read for hydropower generation) in such a way that "the owners of hydropower plants acquire the basic fuel," that is, water in this case, "almost free of cost," which then minimizes their operational costs vis-à-vis other sources of power generation (Baruah 2012). Again, although it is true that the tribes in the hills of Arunachal Pradesh may belong to the land through which the river Brahmaputra flows (under different names), if the state chooses, as it also appears to do to use the logic of private property, the same land being communally owned may not belong to the inhabitants at all. So these people can then be "readily cast as uninhabitants" who as tribals can be shown to have "residual presence from a pre-capitalist era" (Nixon as quoted by Baruah 2012: 49). In such a case, the issue of compensation can easily be denied or fixed, if at all, at insignificant levels. Thereby both accumulation and dispossession can be dealt with at minimum cost. Sensing the opportunity, both the state and the markets are in collusion and the spree of activities related to hydropower generation in Arunachal Pradesh seems to be an indication in this direction.

The Central Electricity Authority (CEA) assessed the hydroelectric potential of the northeastern region (NER, including Sikkim) to be 63,257 megawatts (MW). Out of this estimate, only 1,011 MW, or 1.6 percent of the total, has been developed so far (see Table 13.3). In terms of basin-wise hydroelectric potential, the Brahmaputra basin has the highest potential and so 168 projects were planned along its course (Table 13.3). Once the potential was identified the search turned to a *modus operandi* to realize the potential (Menon and

Table 13.3. Hydropower potential* assessed and developed in NER (in MW)

State	Potential Assessed	Potential Developed
Arunachal	50,328	281
Assam	674	250
Manipur	1,784	105
Meghalaya	2,394	185
Mizoram	2,196	0
Nagaland	1,574	91
Tripura	21	15
Sikkim	4,286	84
Total NER	63,257	1,011

Note: * Excluding power units below 3 MW.
Source: Menon and Kohli (2005: 22)

Kohli 2005). The Northeast Business Summit in Mumbai during 2002, the 50,000 MW Hydro Initiative of the Ministry of Power in 2003, and the Pasighat Proclamation on Power in 2007 were some of the initiatives used. By 2010, the government of Arunachal Pradesh had already allotted 132 projects to companies in the private (120) and public sectors with a total installed capacity of 40,140.5 MW. Ironically, in almost every agreement, huge monetary advances were paid upfront "which greatly compromised the manner in which subsequent clearances took place as such projects were considered a fait accompli by both the developers and the state government" (Vagholikar and Das 2010).

Generating hydropower involves huge social, cultural, economic, and environmental costs. The impact assessment of hydropower projects either ignored or showed little willingness to consider these costs. If we highlight this gap with the example of shifting cultivation, a common practice among the tribes in the state, it becomes illustrative. Project documents related to impact assessments of hydropower often marked the submergence area within their catchments as "degraded." That those areas, which they termed "degraded," would either come under cultivation at the end of the *jhum* cycle or have been left fallow after a cycle was beyond their comprehension since they had little understanding of shifting cultivation. Their idea of what is "big" and what is "small" was similarly vexed. The project assessments often stated that only a "small" number of people would be displaced in Arunachal Pradesh, but what is small for mainland India may be very "big" for the affected tribes in the state. In the words of Dr. Mite Lingi, chairman of the Idu Indigenous Peoples Forum:

> the "small displacement" argument to sell these projects as being benign needs to be confronted. The entire population of the Idu Mishmi tribe is around 9,500 and at least 17 large hydel projects have been planned in the Dibang Valley in Arunachal Pradesh. As per this faulty argument (of small number of affected

> people), little social impact will be indicated even if our entire population is supposedly displaced. (Vagholikar and Das 2010)

Ultimately the land, forest, and water resources that are utilized for commercial purpose are nothing but community resources in the hills of the NEI. This raises several questions, such as, who owns the resources? Who decides their utilization? And who benefits from their use? In the words of Akhil Gogoi of Krishak Mukti Sangram Samity:

> it is a matter of our rights over our natural resources. These resources are being handed over to power companies and our rivers transformed dramatically by political decisions taken in New Delhi and within state governments of the region. This requires a political response from people of the region and that will be our focus in the coming days. (Vagholikar and Das 2010)

One who is conversant with the NER will recollect that during the birth of various insurgencies in the region the questions raised were not much different.

Thus this state–market collusion in the name of hydropower generation is becoming hegemonic, ignoring the enormous social and environmental costs both at the site of the projects as well as the downstream areas. In this conflict, dispossession of tribal people from their communal resource base in their ancestral land is a foregone conclusion.

13.7 Conclusion

Following the logic of wastelands (as discussed in Section 13.2),[9] the colonial officials considered community property as the sign of a poor institution, one that hindered investment and growth. Community dispossession in NEI therefore also became conditional to accumulation. This has been the "similarity" of the region with the history of global capitalism. But the historico-epistemological gap between positive law applied by the colonial and postcolonial state and the customary laws followed by the tribal populations over community land creates social contradictions and contestation with no real scope for resolution. Various modes of dispossession through the introduction of alien institutions in the region throw light on the "uniqueness" of accumulation in the region. This chapter has highlighted that state policies of land use and cropping pattern, purchase, lease, and mortgage have already become common. The lease market has sprawled in the hills and both the "*crafty* locals" (Datta 1992) and "outsiders" have begun to take maximum advantage of this situation. Moreover, our analysis also shows that dispossession of community lands has become acute either because of non-recognition of the customary laws or their selective interpretation aided by the lack of record of rights on these lands. So a process of trivializing the economic and social function of community land has simultaneously enhanced the threat to food and livelihood security as well as dislocation of the tribes from their sociocultural base. One of the resultant effects of such dislocations in the region has been the ever-increasing number of political or ethno-nationalist movements, both constitutional and extra-constitutional, in the name of the right to protect resources through the demand for separate "homelands" for the different tribal communities. Today when land and resources are viewed collusively by the state and business as a *spatial fix* for corporate investment to further capital accumulation, increasing dispossession of whatever is left of communal resources in the Northeast seems inevitable.

Notes

1. A subcommittee under the chairmanship of Sri Gopinath Bordoloi, the then premier of Assam was constituted in February 1947 to report on the North East Frontier (Assam) Tribal and Excluded Areas. This subcommittee extensively toured the region to gain experience from the people in the region on many issues particularly related to land, natural resources, forests, and so on and their control over them. The threat of these resources being exploited by non-tribal communities was also ascertained. The committee submitted its recommendations during July 1949 to the chairman of the Advisory of Fundamental Right and gave a number of recommendations related to the continuation of shifting cultivation, forestry, control of immigration, legislative measures for protection of the tribals, and so on. The report and recommendations became the Sixth Schedule of the Indian Constitution.
2. Eugen Ehrlich was one of the earliest jurisprudential analysts to interpret the dichotomy between the "state law" and "societal law" and in the process contributed significantly toward the development of legal sociology and legal pluralism theory. According to Ehrlich, the monopoly of the state in creation of legal norms appeared only in modern times, as a precept of political theory and natural law doctrines in the wake of the absolutist state. Reviewing Ehrlich's *Living Law* Michel Coutu opines that societal law is based upon cooperation and voluntary adherence, whereas state law is grounded in "domination" and "constraint." For a detailed analysis see Coutu (2009). Sociological approaches to law highlight the non-uniqueness of law, which is reflected in the words of Michael Freeman, "a vision of law as but one method of social control" (2006: 1). He quotes Brian Z. Tamanaha (which reflects Ehrlich's notion of societal law) to validate this point as a rebuttal to standard conceptual jurisprudence, "what law is and what law does cannot be captured in any single scientific concept. The project to devise a scientific concept of law was based upon a misguided belief that law comprises a fundamental category. To the contrary law is thoroughly a cultural construct, lacking any universal essential nature. Law is whatever we attach the label law to" (2006: 5). For further details see Freeman (2006: 1–5).

3. James Scott provides an insight regarding the role of the state towards the people practicing swidden (shifting) cultivation in the South East Asian uplands. According to Scott, "swiddening has been anathema to all state makers, traditional or modern" (Scott 2009: 77). He states that "the overriding reason behind such policies, it appears, has been the states need to use such land for permanent settlement, to realize for itself the revenue from the extraction of natural resources, and to bring such non-state people finally to heel" (p. 78). Here it is interesting to note that states with diverse ideological beliefs have practiced in similar ways to encourage sedentary cultivation vis-à-vis swidden cultivation. Thailand, Vietnam, and Burma have a poor record of torturing and dispossessing population groups who practiced swidden cultivation. For a detailed discussion see Scott (2009).

   Regarding the introduction of cash crops and plantations in the hills of the region, it is interesting to note that David Scott, the British administrator, introduced potato cultivation in the Khasi Hills in the early 1850s. In 1853, the total potato cultivation was 30,000 *maunds* which increased to 185,000 *maunds* in 1872 or more than six times since 1853. For details see Dutta (1983).
4. A number of studies have found cases of land concentration and landlessness among the tribals of North East India, particularly the Khasi-Jayantias of Meghalaya. Prominent among these studies are Vincent (1979), Mathew and Nair (1983), Dutta and Datta (1986), and Datta (1984).
5. The Fact Finding Report of Manab Adhikar Sangram Samity (MASS, a human rights activist group based in Assam) concerning Karbi Anglong district of Assam highlights the linkages between the change in land use pattern and the perceived threats to "homeland," "identity," and "security" among the tribal population groups in the district. As the nontribals possess the occupational skills for wet rice cultivation and as plantation crops come more under cultivation (with the connivance of the tribal elites), they become the *de jure* owners of such land. Now, in the event of any schism between the tribals and the nontribals the situation boils down to the issue of "identity-threat" for the tribals at the hand of the nontribals and therefore securing the land (read "homeland") from the clutches of the *de jure* owners becomes their "sacred" duty. This process led to a situation where the Karbi youths are influenced by the Karbi insurgent groups with the aim of driving out the nontribals from their homeland and on the other, the nontribal cultivators (mainly Hindi speaking people) informally looked to the Indian security personnel stationed in those areas as their protectors, as they shared common ethnic ties. This resulted in spiraling of insurgency and violence in the region. For a detailed discussion see Baruah (2005).
6. The Bengal Eastern Frontier Regulation 1873, the Assam Forest Regulation, 1891, Land Acquisition Act 1894, the Assam Land and Revenue Regulation, 1886, the Chin Regulation 1896 and a host of other regulations.
7. In Sunil Dev *vs.* State of Tripura 2002 (1) GLT 538 it was decided that under clause (a) of paragraph 3 (a), the District Council can make law controlling the occupation or use of any land within the tribal area, it can make laws fixing minimum ceiling limit of allotment, prohibiting conversion of agricultural land into nonagricultural and preventing diversion of land. It can compel the landholders to keep some portion

for grazing purposes and for promotion of social forestry. It can set up its own infrastructure to supervise the proper, fruitful, and beneficial occupation and use of land. It can also make law compelling the landholder to keep some place vacant within its boundaries and constitute a community farm for the benefit of poorer sections. In Tarima Kanta Das *vs.* Karbi Anglong District Council, 1989(1) GLR 147 and Udaldas Panika Prahlad Chandra Das *vs.* Karbi Anglong District Council, 1990(1) GLR 78 it was stated that land comprised in autonomous districts, however, does not belong to the District Council and is owned by the state government (Hansaria 2005).

These conflicting judgments regarding the rights of the District Council to frame rules concerning land often created an impression in the mind of the elected Councilors that they can legislate like state legislators. Justice Hansaria, commenting on the legislative power of a district council and referring to the case the District Council of United Khasi and Jaintia Hills *vs.* Sitimon Sawin (1971) 3 SCC 708: AIR 1972 SC 787: (1972) 1 SCR 398, that had come up for consideration before the Constitution Bench of the Supreme Court, stated that the power of the District Councils to make law were extremely limited unlike the Parliament or the State Legislatures (Hansaria 2005).

8. There are similar incidents about loss of land, particularly community land of the tribes, due to dam construction. In fact, the Chakmas who are attributed with conflicting nationalities and statelessness in NER and South East Asia also have a similar history of dispossession (other than the political causes) due to the construction of the Kaptai dam in the river Karnaphuli located in Chittagong Hill Tracts (CHT) in 1964. For details see Bhaumik (1996), Chaudhury (2011), Prasad (2012).
9. For example, Locke's famous statements, such as "what is produced from one acre of enclosed land is ten times more than which are yielded by an acre of land, of equal richness, lying waste in common." Similarly, "he that encloses and has a greater plenty of the conveniences of life from 10 acres than he could have from 100 acres left to nature. . . ." For details see Chakraborty (2012: 5).

References

Ashokvardhan, C. (ed.). 2004. *Socio-economic Profile of Rural India, Vol. II: NE India,* New Delhi: Concept Publishing House.

Baruah, S. 2005. *Durable Disorder: Understanding the Politics of Northeast India,* New Delhi: Oxford University Press.

Baruah, S. 2012. "Whose River Is It Anyway? Political Economy of Hydropower in the Eastern Himalayas." *Economic and Political Weekly,* 47(29): 41–52.

Bezbaruah, M. P. 2007. "Land Tenure System in North East India: A Constraint for Bank Financing?" *Dialogue,* 8(3). Available at: <http://www.asthabharati.org/Dia_Jan%2007/Bez.htm> (accessed May 21, 2014).

Bhaumik, S. 1996. *Insurgent Crossfire: North East India.* New Delhi: Lancer.

Chakraborty, G. 2012. "Roots and Ramifications of a Colonial Construct: The Wastelands in Assam." Occasional Paper 39, Institute of Development Studies Kolkata, Kolkata.

Chaudhury, A. 2011. "Unheard Voices: The Stateless Chakmas in Arunachal Pradesh." Available at: <http://refugeewatchonline.blogspot.in/2011/02/unheard-voices-stateless-chakmas-in.html> (accessed December 24, 2013).

Commons, J. R. 1957. *Legal Foundations of Capitalism*. Madison, WI: University of Wisconsin Press.

Coutu, M. 2009. "Review of Hertogh Marc (ed.) *Living Law: Reconsidering Eugen Ehrlich*." *Osgoode Hall Law Journal*, 47(3): 587–93.

Datta, P. S. 1984. "Emerging Differentiation in a Traditional Tribal Economy." *Social Research*, 4(3–4): 1–7.

Datta, P. S. 1992. *India's North-East: A Study in Transition*. New Delhi: Har-Anand Publications.

Datta, P. S. 1994. *North East As I See It*. New Delhi: Omsons Publications.

Debbarma, S. 2008. "Refugee Rehabilitation and Land Alienation in Tripura." In *Land, People and Politics: Contest over Tribal Land in Northeast India*, edited by W. Fernandes and S. Barbora, Guwahati: North Eastern Social Research Centre and International Workgroup for Indigenous Affairs, pp. 113–27.

Dutta, P. N. 1983. "The Emergence and Role of the Middle Class in the Khasi-Jaintia Hills 1824–1950." In *The Emergence and Role of Middle Class in North East India*, edited by B. Datta Ray, Delhi: Uppal Publishing House, pp. 107–49.

Dutta B. B. and Datta, P. S. 1986. "Land Holding Pattern among the Khasi-Jayantias: The Tradition and Deviation." In *Alienation of Tribal Land and Indebtedness*, edited by B. N. Bordoloi, Gauhati: Tribal Research Institute, pp. 149–56.

Fernandes, W. 1995. "Indian Tribals and Search for an Indigenous Identity." In *Tribals in India, Development, Displacement and Discontent*, edited by A. K. Singh and M. K. Jabbi, New Delhi: Council for Social Development, pp. 62–79.

Freeman, M. (ed) 2006. *Law and Sociology*, vol. 8. New York: Oxford University Press.

Ganguli, J. B. 2008. "Impact of North East Indian Tribal Values and Culture on the Hill Region's Socio-Economic Development in the Pre-Plan Period." In *Society, Politics and Development in North East India. Essays in Memory of Dr. Basudeb Datta Ray*, edited by A. Ray and S. B. Chakraborti, New Delhi: Concept Publishing House, pp. 115–21.

Hansaria, V. 2005. *Sixth Schedule to the Constitution*. New Delhi: Universal Law Publishing Company.

Harvey, D. 2011. "The Future of the Commons." *Radical History Review*, 109: 101–7.

Herskovits, M. 1964. *Economic Anthropology: The Economic Life of Primitive Peoples*. New Delhi: Eurasia Publishing House.

Jyrwa, E. 1988. "Critical Assessment of the Working of the Khasi Hills Autonomous District Council in Meghalaya." In *Power to the People of Meghalaya*, edited by M. N. Karna, L. S. Gassah, and C. J. Thomas, New Delhi: Regency Publication, pp. 89–96.

Karlsson, B. G. 2011. *Unruly Hills: Nature and Nations in India's Northeast*. New Delhi: Orient Blackswan.

Karna, M. N. 1987. "Agrarian Studies in North-East India." In *Land Relations in North East India*, edited by B. B. Dutta and M. N. Karna. New Delhi: Peoples Publishing House, pp. 5–13.

Majumdar, D. N. 1983. "The Emerging Middle Class Among the Garos." In *The Emergence and Role of Middle Class in North East India*, edited by B. Dutta Ray, New Delhi: Uppal Publishing House, pp. 181–91.

Mathew, T. and Nair, M. K. S. 1983. "Tribal Mode of Production (in Meghalaya) in Transition." Department of Economics NEHU, Shillong (mimeo).

Menon, M. and Kohli, K. 2005. *Large Dams for Hydropower in Northeast India*. New Delhi: Kalpavriksh-SANDRP.

Nongbri, T. 1995. "Report on Meghalaya." Paper presented at the workshop on "Land Reforms—Agenda for the North East," held at Gauhati, organized by Land Reforms Unit, Lal Bahadur Shastri National Academy of Administration, Mussoorie, April 3–5.

Prasad, C. 2012. "Between Nation and Nationality: Chakma Refugees in Arunachal Pradesh." Available at: <http://www.asthabharati.org/Dia_Jan012/chun.htm> (accessed November 8, 2013).

Ramakrishnan, P. S. 1993. *Shifting Agriculture and Sustainable Development: An Interdisciplinary Study from North-Eastern India*. UNESCO-Man and Biosphere Series, vol. 10. New Delhi: Oxford University Press.

Ray, A. K. and Kamkhenthang, H. 1991. "Land Reforms and Organizational Problems: The Chin-Kukis of Manipur," Paper presented at the seminar "Organising the Rural Poor for Land Reforms," March 14–16, Hyderabad, National Institute of Rural Development.

Roy Burman, B. K. 1987. "Issues in Land Reforms in N-E Tribal Area." *Mainstream*, August 1 and 8.

Scott, J. 2009. *The Art of Not Being Governed: An Anarchist History of Upland Southeast Asia*. New Haven, CT: Yale University Press.

Sen, S. 1987. "Land as Property: Its Significance in the Traditional Society and Polity in Khasi and Jaintia Hills." In *Land Relations in North East India*, edited by B. B. Dutta and M. N. Karna, New Delhi: Peoples Publishing House, pp. 92–7.

Sharma, B. D. 1984. "Shifting Cultivators and Their Development." *Journal for Social Science & Humanities*, 2(2): 1–36.

Taylor, J. F. A. 1966. *The Masks of Society*. New York: Meredith.

Vagholikar, N. and Das, P. J. 2010. "Damning the Northeast: Juggernaut of Hydropower Projects Threaten Social and Environmental Security of Region." Available at: <http://www.kalpavriksh.org/images/EnvironmentandDevelopment/NorthEastDams/Damming%20Northeast%20India_Single%20page%20format.pdf> (accessed November 5, 2013).

Vincent, K. 1979. *Socio-Economic Study of Bhilymbong*. Chennai: Christian Literature Center.

Postscript

Land, Livelihoods, and Late Capitalist Development

Anthony P. D'Costa

The place of land in society has taken on various shades of significance. Land in noncapitalist or precapitalist societies was a source of livelihood that produced a subsistent lifestyle all over the world. Agricultural productivity was low while wars, tributes, and natural calamities ensured subsistence living. That changed fundamentally as a new form of economic organization emerged, namely profit-seeking, market-driven production in trading towns and cities and the rise of agrarian capitalists in the countryside. Capitalism as an economic system became embedded in a wide variety of social formations but the pursuit of capital accumulation became the name of the game. Land was no longer simply a source of livelihood but an asset that would yield returns over time. Increasing agricultural productivity and surplus labor expelled from the countryside became the basis for modern economic growth and structural transformation through industrial development. Primitive accumulation has been the *modus operandi* for reducing surplus labor, by separating the peasants from their land, and a source of agrarian dynamism, leading to the rise of a wage-dependent proletariat.

Not every society experienced this kind of primitive accumulation. Western Europe was fortunate to have the new world in North and South America and Oceania as almost a systemic repository for surplus labor, sent out as emigrants. The colonies such as India, on the other hand, did not have this option of sending away their surplus labor. Indentured labor in the nineteenth century could not exhaust the labor supply, while migrant labor to the Gulf countries in the twentieth and twenty-first centuries are only temporary movements. Peasants remained and held on to their land, eking out a livelihood as taxes and appropriations rearranged land tenure systems and served the industrial, economic, and state interests of the colonizers. Primitive accumulation was incomplete and lagging even as the march of capitalist industrialization ushered in a new postcolonial India. Land was needed for large-scale industrialization and infrastructure building and the state had the final say about the use of land for such purposes. But by and large, land remained valued for providing livelihoods to hundreds of millions of small farmers and landless workers. The selective rise of industry, rising agricultural productivity, and the growth of cities and towns in India has contributed to rural–urban migration, with land increasingly taking on commodity status. However, the level and pace of Indian urbanization remains lower than that in Latin America and many East and Southeast Asian countries. India remains an agrarian society without an agrarian revolution and a transition without the extensive and intensive industrialization China has accomplished, even as India positions itself as one of the fastest-growing and largest economies in the world.

The place of land in India in this new economic organization has also changed. But the terms of the debate are no longer between pre-/noncapitalism and capitalism or capitalism and socialism but rather within the orbit of global, predominantly neoliberal capitalism. This does not mean that vestiges of noncapitalist relations of production do not exist or that substantial parts of Indian petty commodity production are not outside the wage–profit nexus. It also suggests the possibility of antisystemic movements that could reshape neoliberal capitalism into something more benign. However, under lagging primitive accumulation, land is no longer playing out its historic role as the originator of capital and a source of accumulation although the dispossession and displacement of peasants continue unabated.

Rather than a headlong rush to capitalize on land as a source of high productivity-led agricultural income (which is occurring in small pockets under contemporary contract farming and cash crops), land has become a burden to many. Farmer indebtedness due to increasing reliance on industrial inputs, fragmentation of land to uneconomic sizes, drought and other unfavorable weather conditions, water salinity, and so on have all contributed to reduce the viability of land as a source of income in many parts of India. Yet people hold on to land since the alternatives to agriculture are few. While there has been diversification of the rural economy through nonagricultural activities and the demand for construction labor in urban areas has acted as a vent for some of the surplus labor, Indian agriculture remains underdeveloped. The unviability of land in a context of commodification of land and class differentiation has also meant the emergence of noncultivating households who secure land for rental income. This of course has some interesting implications for agrarian dynamism in India should such ownership be or become widespread. But that development itself would be consistent with the increasing commodification of land, which, without the inherent agrarian dynamism that is so necessary for India, is unlikely to be able to siphon off the rural surplus labor to the cities, which at the current juncture do not have the capacity to absorb them gainfully.

The value of land has a new twist in the context of the rise of a consuming middle class and rapid growth of Indian cities, fueled by domestic capitalist maturity, local and transnational finance capital, and a state that is more capital friendly. Far from the moral economy of the tribal communities for whom land is both sacred and a source of livelihood, minerals and forests are seen as resources to be profited from. Domestic and foreign mining and metallurgical industries see natural resources to be exploited for firm expansion, while the Indian state sees growth as necessary for its own legitimacy, by anticipating job creation and economic development. For small farmers, land is the source of subsistence while dispossession and displacement are not desired. But as land exchange becomes routine more for its nonagricultural use, its economic value also rises. Not all farmers cling to their land or want to be tied to the countryside. They are willing to sell so long as the price is right and an alternative livelihood is available. In this vortex of land exchange, agrarian crisis, and urban-based growth, many farmers willingly sell, while others are coerced to vacate with small compensation. The recent contestation between landowners, cultivators, and the state has been over prices just as it has been over just compensation and rehabilitation. Those who leave, voluntarily or otherwise, often end up joining the ranks of the rural and urban petty commodity producer sector that remains a persistent feature of the Indian economy despite high rates of growth.

The persistence of the petty commodity sector is not surprising. The absence of a rural dynamic and large-scale labor-absorbing industrialization limits employment possibilities outside the petty commodity sector. Employment-wise, the formal sector in India is relatively small. One principal reason for this is India's late entry into capitalist industrialization. With rapid technological change underway, late entry compels both strategic and compulsive leapfrogging. This was already evident by state-led industrialization in post-independence India where import substitution industrialization jump-started an agrarian economy but failed to either transform agriculture or create formal industrial jobs on a large scale. Capital intensity in heavy industries was already high relative to India's labor-abundant agrarian economy. With the diffusion of digital technologies and globalization, Indian firms have been pushed to adopt recent vintages of technologies and adopt automation like their global counterparts. Not surprisingly, the result has been further increases in capital intensity among Indian industries in a milieu of surplus labor. Politically this outcome can be read as capital on the offensive. With technology and automation, formal labor is put on the defensive through increasing contract, part-time, and temporary work. While the virtues of leapfrogging are observable on economic growth and business profits, employment creation remains daunting. Even the services sector, which makes up for India's manufacturing slack, comprises the vast low-value labor-absorbing petty commodity sector. The dynamic technology, education, and skill-intensive services sector for which India is globally visible does not generate large-scale employment.

The Indian state is not a passive actor in this unfolding of late capitalism—far from it. It is both a taker and giver of land and livelihoods. For all practical purposes the Indian state is a democratic state and is answerable to its citizens—rich or poor. It compensates the dispossessed and displaced while it gives land or facilitates land taking for development. Like all states its autonomy is relative even as it enjoys the power of eminent domain when it comes to the use of land for public purpose. However, the ability to legally wrest control over land from the public is not without contestation. Through legal acts the state has made it possible to give the dispossessed and displaced some compensation but its quest for development in a capitalist system remains intact, albeit checked by the limits of its own capacity. In fact, the legal provisions facilitate land taking by setting up the precedent for a more bureaucratic approach to transferring land. Hence, capitalist development and dispossession are expected to continue but not in the same historic way as in early capitalism. The Indian state must now politically manage the fallout of dispossession and displacement even as it facilitates land taking for national capital accumulation.

Late capitalism in India as it has unfolded is not the capitalism experienced earlier by today's rich economies. It is a curious mix of lagging primitive accumulation, leapfrogging industries, and a vast persistent petty commodity sector. Elsewhere I have labeled such uneven and combined development as the workings of compressed capitalism (D'Costa 2014, 2016) that accounts for the ongoing but tentative process of dispossession with few livelihood alternatives. It also implies the necessity of political management of the dispossessed in a democratic India even as the resources devoted to it have declined.

The structural dilemma, however, remains. Late capitalist development in a deregulated global economic environment today has made transforming surplus labor into relative scarcity a virtual impossibility. Technological change and rising productivity, while contributing to growth, also contribute to labor substitution through automation. Late developers are not immune from this development as businesses are compelled to adopt similar competitive strategies. In the end, the classical transition anticipated from successful primitive accumulation to capitalist industrial expansion appears to have come to a dead end. The sobering fact that India adds roughly a million new workers every month poses an employment and livelihood challenge that no state can meet, nor can a dynamic private sector (domestic or foreign) invest enough to employ such an expansion in the workforce. How India will absorb and accommodate them is anyone's guess, given the persistence of the petty commodity sector and limited formal employment. State intervention and high economic growth thus far have not made a dent in what Sanyal (2007: 249) calls a "dark space."

Whether the state will politically manage this "dark space," or not manage it, are two daunting possibilities that are hard to imagine. Will the state become repressive, tearing asunder the very democratic fabric that it has sewn together since the Republic's founding? Or will the frustration bubbling beneath the veneer of aspirations result in revolutionary fervor, or something else? As the more prosperous state of Gujarat shows, the economic and social divide between the landless and landowners has become not just pronounced but predatory, a perfect cocktail for fascism (Breman 2010: 335). Should the many cities in India and elsewhere that indicate the spread of the *lumpenproletariat*—often linked to the ranks of the dispossessed (Saunders 2011)—be seen as an anomaly or a permanent feature of the world economy? As jobs disappear and self-employment and self-exploitation become the only avenues for desperate livelihoods, politics could take on a whole new meaning, whose boundaries and practices at the moment remain undefined and untested but unquestionably tense. In the end, under late capitalism, land and livelihoods are contested, conflicts bloody, and dispossession inevitable, but a classic agrarian transition and capitalist development of an earlier era is beyond the realm of our immediate imagination.

References

Breman, J. 2010. "The Political Economy of Agrarian Change in India." In *Routledge Handbook of South Asian Politics: India, Pakistan, Bangladesh, Sri Lanka, and Nepal*, edited by P. R. Brass, Oxon: Routledge, pp. 321–36.

D'Costa, A. P. 2014. "Compressed Capitalism and Development: Primitive Accumulation, Petty Commodity Production, and Capitalist Maturity in India and China." *Critical Asian Studies*, 6(2): 317–44.

D'Costa, A. P. 2016. "Jobless Growth Even in Labor-Abundant India: Capitalist Transition and the Challenges of Employment." Paper Presented for the Panel on "Global Capitalism, National Experiences, and Changing Labor Markets in Asia" at the Association of Asian Studies Annual Conference, Washington State Convention Center, Seattle, March 31–April 3, 2016.

Sanyal, K. 2007. *Rethinking Capitalist Development: Primitive Accumulation, Governmentality and Post-Colonial Capitalism*. New Delhi: Routledge India.

Saunders, D. 2011. *Arrival City: The Final Migration and Our Next World*. Toronto: Vintage Canada.

Index

Printed and bound by CPI Group (UK) Ltd, Croydon, CR0 4YY